recent advances in phytochemistry

volume 18

Phytochemical Adaptations to Stress

RECENT ADVANCES IN PHYTOCHEMISTRY

Proceedings of the Phytochemical Society of North America
General Editor: Frank A. Loewus *Washington State University, Pullman, Washington*

Recent Volumes in the Series

Volume 11 The Structure, Biosynthesis, and Degradation of Wood
Proceedings of the Sixteenth Annual Meeting of the Phytochemical Society of North America, Vancouver, Canada, August, 1976

Volume 12 Biochemistry of Plant Phenolics
Proceedings of the Joint Symposium of the Phytochemical Society of Europe and the Phytochemical Society of North America, Ghent, Belgium, August, 1977

Volume 13 Topics in the Biochemistry of Natural Products
Proceedings of the First Joint Meeting of the American Society of Pharmacognosy and the Phytochemical Society of North America, Stillwater, Oklahoma, August, 1978

Volume 14 The Resource Potential in Phytochemistry
Proceedings of the Nineteenth Annual Meeting of the Phytochemical Society of North America, Dekalb, Illinois, August, 1979

Volume 15 The Phytochemistry of Cell Recognition and Cell Surface Interactions
Proceedings of the First Joint Meeting of the Phytochemical Society of North America and the American Society of Plant Physiologists, Pullman, Washington, August, 1980

Volume 16 Cellular and Subcellular Localization in Plant Metabolism
Proceedings of the Twenty-first Annual Meeting of the Phytochemical Society of North America, Ithaca, New York, August, 1981

Volume 17 Mobilization of Reserves in Germination
Proceedings of the Twenty-second Annual Meeting of the Phytochemical Society of North America, Ottawa, Ontario, Canada, August, 1982

Volume 18 Phytochemical Adaptations to Stress
Proceedings of the Twenty-third Annual Meeting of the Phytochemical Society of North America, Tucson, Arizona, July, 1983

A Continuation Order Plan is available for this series. A continuation order will bring delivery of each new volume immediately upon publication. Volumes are billed only upon actual shipment. For further information please contact the publisher.

recent advances in phytochemistry

volume 18

Phytochemical Adaptations to Stress

Edited by

Barbara N. Timmermann

The University of Arizona
Tucson, Arizona

Cornelius Steelink

The University of Arizona
Tucson, Arizona

and

Frank A. Loewus

Washington State University
Pullman, Washington

PLENUM PRESS • NEW YORK AND LONDON

Library of Congress Cataloging in Publication Data

Phytochemical Society of North America. Meeting. (23rd: 1983: University of Arizona)
Phytochemical adaptations to stress.

(Recent advances in phytochemistry; v. 18)
"Proceedings of the Annual Meeting of the Phytochemical Society of North America . . . held July 5–8, 1983, at the University of Arizona, Tucson, Arizona"—T.p. verso.
Includes bibliographical references and indexes.
1. Plants—Metabolism—Congresses. 2. Plants, Effect of stress on—Congresses. 3. Plants—Adaptation—Congresses. 4. Botanical chemistry—Congresses. I. Timmermann, Barbara N. II. Steelink, Cornelius. III. Loewus, Frank Abel, 1919– . IV. Title. V. Series.
QK861.R38 vol. 18 581.19′2 s 84-9842
[QK887] [582′.01]
ISBN 0-306-41720-0

Proceedings of the Annual Symposium of the Phytochemical Society of North America on Phytochemical Adaptations to Stress, held July 5–8, 1983, at the University of Arizona, Tucson, Arizona

A Division of Plenum Publishing Corporation
233 Spring Street, New York, N.Y. 10013

Printed in the United States of America

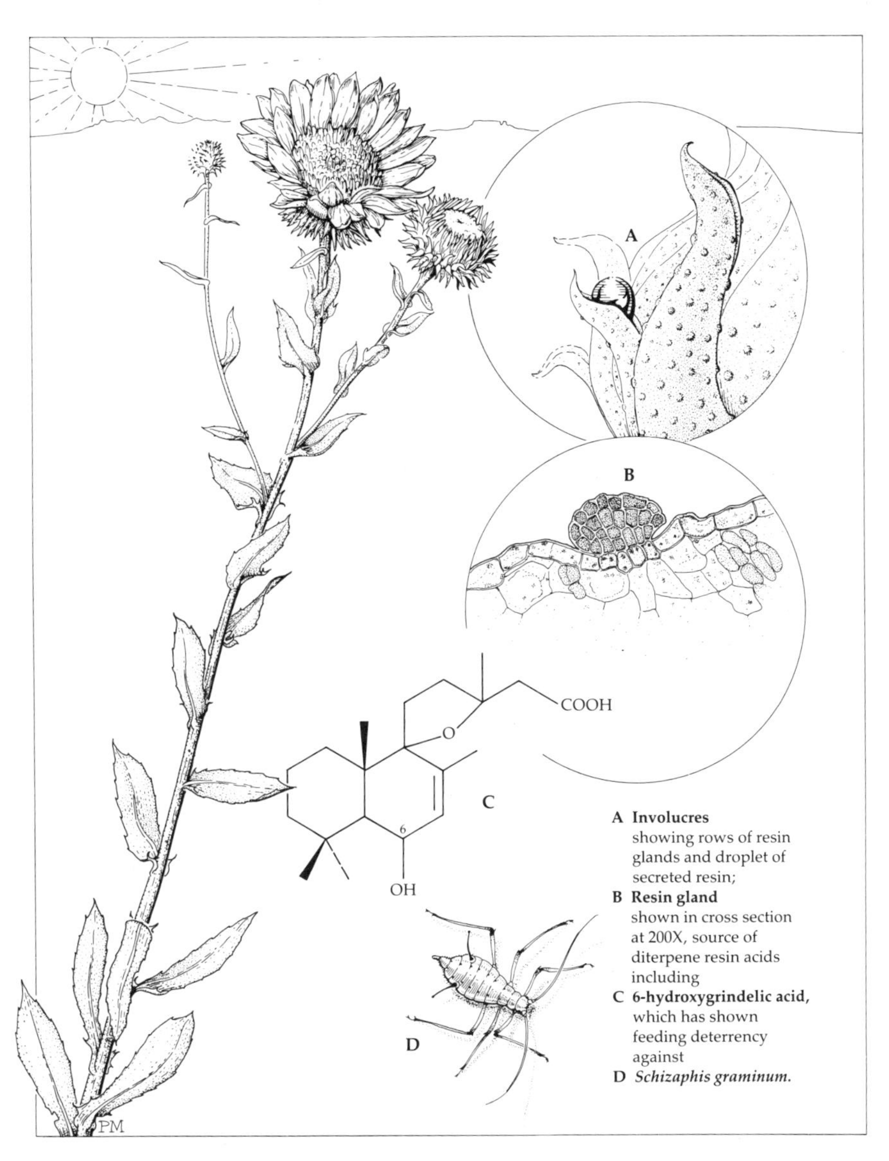

Grindelia camporum Greene

Resins produced by the arid-adapted species *G. camporum* appear to have ecological significance as well as potential economic value. (Drawing by Paul Mirocha, Office of Arid Lands Studies, University of Arizona, Tucson, Arizona.)

PREFACE

This volume is based on the proceedings of the Phytochemical Society of North America's 23rd Annual Meeting on "Phytochemical Adaptations to Stress" which was held at the University of Arizona, Tucson, July 5-8, 1983. It contains a series of articles which focus on our current knowledge on the production of secondary (natural) metabolites by higher plants in response to biological and physiological stresses.

The editors of this volume are deeply indebted to a number of people and organizations for their support and contributions which were critical to the success of this scientific meeting.

Generous grant support was provided by the Agricultural Research Service of the United States Department of Agriculture. Additional financial support came from the Phytochemical Society of North America. Indispensable services and personnel were donated by the Departments of Chemistry and Pharmaceutical Sciences, the College of Agriculture and the Office of Arid Lands Studies of the University of Arizona. Special recognition is due to Paul Mirocha of the Office of Arid Lands Studies for his drawing of the frontispiece and the superb photograph on the jacket. The Division of Conferences and Short Courses of the University of Arizona deserves credit for its pivotal role in maintaining a well-run and pleasant conference. Many other volunteers gave their time and energy to make the Symposium a success; we wish to mention two from the Department of Pharmaceutical Sciences, Brian Weck and Catherine L. Buckner.

Moreover, there are three members of the Organizing Committee who deserve special thanks. They are Ms. Suzanne Weck, Department of Chemistry; Dr. Robin F. Bernath, Department of Entomology, and Dr. R. Phillip Upchurch, College of Agriculture. Without their foresight and perseverance, this Symposium would never have materialized.

Typing of draft copy, correction of proof, assembly and preparation of camera ready material was done by Ms. Diana Thornton, Institute of Biological Chemistry, Washington State University. We are most grateful for her dedicated efforts.

Finally, we are very grateful to all of the authors who contributed to this book for their cooperation and comprehensive articles where they review the latest development of research in their respective fields.

December, 1983

B. N. Timmermann
C. Steelink
F. A. Loewus

CONTENTS

Chapter One

INTRODUCTORY CHAPTER

BARBARA N. TIMMERMANN[a] AND CORNELIUS STEELINK[b]

Department of Pharmaceutical Sciences
and Office of Arid Lands Studies[a]
Department of Chemistry[b]
University of Arizona
Tucson, Arizona 85721

Higher plants possess a myriad of adaptive mechanisms for surviving a variety of abiotic and biotic stresses. No single volume or symposium can ever attempt to review this extensive subject in detail. By focusing on the phytochemical responses - specifically the effects of stress on plant secondary metabolites - this symposium has delineated a specific aspect of this important subject. A broad review of the effects of certain types of stresses on the production of secondary metabolites in whole plants and tissue cultures is discussed in two chapters of this volume. These reviews are supplemented by other chapters describing specific types of stresses on the fate of selected metabolites.

Although these studies represent widely divergent disciplines, there are many common themes in all of them. Most of these studies derive their inspiration from a desire to improve economic crops necessary to human welfare and survival[1] and to increase the yields of useful secondary metabolites especially in flavoring and pharmaceutical applications.[2] For this goal, scientists have employed controlled conditions to subject plants to cold hardening, mechanical perturbation, mineral deficiencies, synthetic bioregulators, genetic manipulation and other stresses that occur in nature. They have carefully monitored the effects of stresses mainly on metabolites that are necessary to healthy plant growth or human usage. The reader may be interested to find in Chapters two, four, seven, nine and ten that certain plant defenses against specific stresses also defend those plants against other stresses.

The effects of abiotic and biotic stresses on the production of plant secondary metabolites are separately discussed in this volume, although the line of demarcation between the two types is sometimes not easy to define. Mechanical perturbation, extreme temperature effects, water deficiencies, and light are easily identified as abiotic. Predator and pathogen damage and genetic manipulation are considered biotic. But nitrogen and potassium deficiencies, bioregulators, osmolarity and ethylene effects are not clearly defined in either category. They may be causes of plant stress from internal or external sources. But to whatever category the stress is assigned, the effect of stress is dramatically reflected in the plants by the production of secondary metabolites. Extensive reviews in Chapter five (DiCosmo and Towers) and Chapter ten (Gershenzon), illustrate the complexity of these effects.

In their review, DiCosmo and Towers note the classic definition of stress in plants as "the external constraints which limit the rate of dry matter production of all or parts of the vegetation". While this may be of special interest to agronomists, it is not very useful to phytochemists. Therefore, these authors prefer the following definition: "stresses are those external constraints that limit the normal production of secondary metabolites in plants". This is a definition which is followed by the authors in this volume.

DiCosmo and Towers focus on cultured plant cells as a model system for studying the effects of stress on the regulation of synthesis of secondary (natural) products. Cultured cells can be strictly monitored and therefore have advantages over intact plants. On the other hand, many cultured plant cells fail to show the normal pattern of secondary metabolite characteristics of intact plants.[3] With this warning, the authors survey the multitude of stress effects induced by nutrients, light, temperature, pH, aeration, antibiotics,fungal elicitors, etc.

On the other hand, Gershenzon reviews the effects of stress on intact plants, with special emphasis on water deficits and nutrient deficiencies upon the production of sulfur containing compounds, cyanogenic glycosides, alkaloids, phenolics, and terpenoids. The cost of the production of secondary metabolites is discussed in relation to the synthesis and storage of these compounds in plants, as well as the adaptive value of such compounds to higher and lower plants and effect on different organs of the plant.

Despite the complexity of this subject, our understanding of the different chemical and biochemical mechanisms as well as the functional roles of plant secondary metabolites has increased over the past few years. In many instances, this understanding has also led to practical applications. Specific examples of stress and phytochemical responses are described by several authors.

In separate chapters Smith and Wyn Jones examine the effects of salt stress on plant metabolites from different points of view. Certain metabolites are known to accumulate in response to salt stress.[4,5] Smith focuses on how potassium deficiency triggers enzyme activation in the biochemical pathway leading to putrescine formation in many plants whereas Wyn Jones looks at the compartmentalization of certain cations (K^+ and Na^+) and organic solutes (cytosolutes) between the cytoplasm and vacuoles in higher plants growing in hypersaline environments.

Both authors discuss the adaptive role of organic solutes in the plant cell in maintenance of cell turgor, stabilization of pH, and protection of protein and cell wall membranes. It is interesting to note that production of the cytosolute putrescine, as well as sorbitol and betaines, is also mediated by water and temperature stresses.

Hardening of plant tissue is discussed in Chapters four (Jaffe and Telewski) and six (M. Christiansen). The hardening is caused by mechanical perturbation (i.e. wind damage) leading to thickening or thigmomorphogenesis[6] in one case and by cold temperatures[7] in the other. Jaffe and Telewski have observed that mechanical stress affect the growth and development of plants by producing two major metabolites: callose (a β-glucan) and the endogenous phytohormone ethylene. In fact, the external application of ethylene can cause thickening also. Christiansen notes that plants exposed to cold temperatures respond by modifying their cell wall lipids. Specifically, the content of unsaturated fatty acids is increased upon chilling. This is the basis for the commercial cold hardening process for seedlings, which makes the plant more resistant to cold temperatures.

In both chapters, it is noted that hardening has an adaptive advantage in increasing the ability of a plant to withstand other stresses. Thus, hardening due to mechanical perturbation makes the plant more resistant to freezing and drought. This also makes them less susceptible to injury. Christiansen notes that oxygen is essential to the production of unsaturated lipids and poses an interesting question: does air pollution affect this process and does it affect membrane permeability? The biochemical sequence of events leading to ethylene and callose production as well as the critical enzyme precursors to linolenic acid are examined in detail in both chapters.

Predators, pests and parasites are responsible for immense crop damage and economic losses. These organisms can certainly be regarded as an important natural stress to which many higher plants have adapted by the production of defensive secondary products in specific tissues and glands.[8,9]

Chapter seven (A.A. Bell) is concerned with the defensive roles of volatile terpenes in addition to flavonoid glycosides, cyclopropenoid fatty acids, terpenoid aldehydes, and tannins in different species of wild and cultivated cotton (Gossypium). Also treated in this chapter are the physiological and genetic regulations of these compounds including the identification and localization of major genes on specific chromosomes controlling these important plant defenses.

Xerophytic plants have evolved numerous adaptive strategies which enable them to survive the harsh environmental conditions including the production of diverse secondary metabolites, some of which show potential for commercialization as renewable resources of phytochemicals.

Yokoyama, et al. (Chapter eight) describe the effects of several synthetic bioregulators on the induction of synthesis of polyisoprenoids in guayule (Parthenium argentatum) and show among others, how dichlorophenoxy-triethylamine (DCPTA) enhances rubber production in this plant that has been identified as a good candidate for commercial rubber production in the near future. Natural rubber from guayule appears to be an excellent substitute for *Hevea* rubber[10] and its resin by-product has been characterized and found to have potential uses in the coating industry.[11]

The induction of rubber production by the use of bioregulators is explained in this chapter to occur as a result of mechanisms that involve gene derepression. Moreover, similar mechanisms are described in detail for different types of bioregulators also affecting the biosynthesis of tetraterpenoids in grapefruit and other citrus fruits.

Chapter nine describes some of the phytochemical adaptations of arid-adapted plant species in which special emphasis is placed upon the production, localization, and possible ecological roles that diterpene acids have in resinous members of the Astereae. Chemical analyses of the crude resin of *Grindelia* have resulted in the isolation of labdane-derived diterpene acids in very high concentrations. The discovery of this chemical group in arid-adapted plant species seems to be of considerable economic significance since these compounds have chemical and physical properties similar to the acid resins used in the naval stores industry.

The economic development of arid-adapted plants which produce chemicals with practical value, could involve the establishment of biomass sources in marginal lands where the growth of traditional food and fiber crops is not economical. These plants could provide useful chemicals as substitutes for non-renewable petroleum-derived sources.[12]

Additional study of the different physical and biological types of stress affecting the production of plant secondary products is warranted in order to understand the mechanisms of synthesis and the roles played by these important metabolites in plants.

REFERENCES

1. PALEG LG, D ASPINALL, eds 1981 Physiology and Biochemistry of Drought Resistance in Plants.

Academic Press, New York 492 pp
2. BARZ W, E REINHARD, MH ZENK, eds 1977 Plant Tissue Culture and its Bio-Technological Application. Springer Verlag, Berlin, New York 419 pp
3. TABATA M 1977 Recent advances in the production of medicinal substances by plant cell cultures. In W Barz, E Reinhard and MH Zenk, eds. Plant Tissue Culture and its Bio-Technological Applications. Springer Verlag, Berlin, New York pp 3-16
4. FLORES HE AND AW GALSTON 1982 Polyamines and plant stress: activation of putrescine biosynthesis by osmotic shock. Science 217: 1259-1261
5. STEWART GR AND F LARHER 1980 Accumulation of amino acids and related compounds in relation to environmental stress. In BJ Miflin, ed, Amino Acids and Derivatives, The Biochemistry of Plants, Vol 5 Academic Press, New York pp 609-635
6. JAFFE MJ 1980 Morphogenetic responses of plants to mechanical stimuli or stress. BioScience 30:239-243
7. LYONS JM, D GRAHAM, JK RAISON, eds. 1979 Low Temperature Stress in Crop Plants. Academic Press, New York 565 pp
8. MABRY TJ, JE GILL 1979 Sesquiterpene lactones and other toxic terpenoids. In GA Rosenthal and DH Janzen, eds, Herbivores: Their Interaction with Secondary Plant Metabolites, Academic Press, New York pp 501-537
9. HEDIN PA, ed, 1983 Plant Resistance to Insects. American Chemical Society Symposium Series 208. 375 pp
10. FOSTER KE, WG MCGINNIES, JG TAYLOR, JL MILLS, RR WILKINSON, FC HOPKINS, EW LAWLESS, J MALONEY AND RC WYATT 1980 A Technology Assessment of Guayule Rubber Commercialization, Final Report to Grant No. PRA 78-11632, National Science Foundation, Washington, DC
11. BELMARES H, LL JIMENEZ AND M ORTEGA 1980 New rubber peptizers and coatings derived from guayule resin (Parthenium argentatum Gray) Ind Eng Chem Prod Res Dev 19: 107-111
12. HOFFMANN JJ 1983 Arid lands plants as feedstocks for fuels and chemicals. CRC Crit Rev Plant Sci 1: 95-116

Chapter Two

PUTRESCINE AND INORGANIC IONS

TERENCE A. SMITH

Long Ashton Research Station
University of Bristol
Long Ashton, Bristol, BS18 9AF, U.K.

INTRODUCTION

The diamine putrescine and the polyamines spermidine and spermine are now known to be ubiquitous, and studies with micro-organisms, animals, and plants have shown the importance of these amines in metabolism and growth.[1-4] Their ability to interact with the anionic phosphate residues in nucleic acids and membranes is well established. However in some biological systems, especially in higher plants, they also appear to be concerned in ionic relations, and it is this aspect which is the subject of the present review. The structures of the di- and polyamines are given in Figure 1. The function and metabolism of the polyamines in plants have been surveyed recently.[1,4-9]

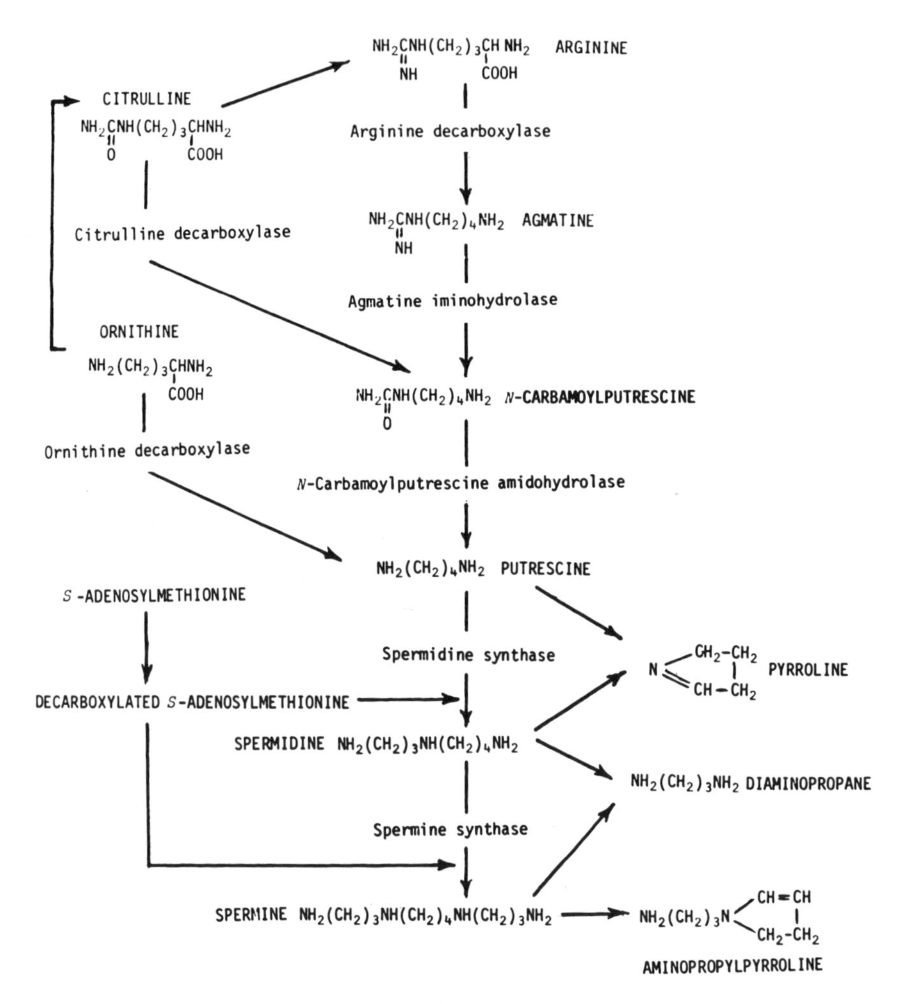

Fig. 1 Biosynthetic pathways of putrescine and the polyamines in higher plants.

PUTRESCINE AND POTASSIUM NUTRITION

The function of K in plants is still incompletely understood, despite the fact that this element is required in relatively large amounts for normal growth.[10] During the course of an intensive study of the effects of K deficiency in higher plants, F.J. Richards and his co-workers investigated the changes of the free ninhydrin-positive compounds separated by paper chromatography using barley as a model system. During this work Richards and Coleman[11] discovered that K deficiency in barley causes a dramatic accumulation of the diamine, putrescine. In extracts of plants grown under extreme deficiency this amine was the most intense ninhydrin-positive spot on their paper chromatograms. In moribund tissue the concentration was estimated to be at least 23 μmol/g dry weight and later work[12] has shown that the concentration in barley can rise up to 140 μmol/g dry weight (1.2% of the dry matter) in extreme K deficiency (see Table 9). Putrescine then accounts for at least 20% of the total nitrogen. In a subsequent paper Coleman and Richards[13] observed that putrescine also accumulates in K-deficient wheat and red clover leaves. Table 1 illustrates the universal nature of this response to K deficiency amongst higher plants.[14] Coleman and Richards[13] showed that maximum putrescine accumulation coincided with the appearance of the severe symptoms characteristic of K deficiency in barley. Putrescine was found at an earlier stage in plants supplied with nitrate and phosphate as the ammonium salts than in those to which the corresponding Ca salts were given. These workers also found that the putrescine accumulated mainly in the tops rather than the roots and none was found in the protein fraction. In K deficient plants putrescine concentration was shown to be reduced on supplying K, Rb, or Na in decreasing order of effectiveness. Moreover, on

Table 1. Putrescine, spermidine and spermine content of the leaves of plants grown under normal (+K) and potassium-deficient (-K) conditions (expressed as μmol/g fresh weight). Adapted from reference 14.

	Age	Putrescine		Spermidine		Spermine	
	days	+K	-K	+K	-K	+K	-K
Barley	29	0.13	1.75	0.19	0.34	0.05	0.04
Maize	31	0.44	3.38	0.12	0.31	0.03	0.06
Radish	31	0.08	0.98	0.31	0.47	0.04	0.05
Pea	47	0.01	0.63	0.40	0.54	0.12	0.11
Blackcurrant	200	0.05	1.09	0.10	0.20	0.04	0.03
Tobacco	58	0.08	0.48	0.07	0.16	0.04	0.04

feeding putrescine to the partially cut leaves of high K (normal) barley plants, symptoms similar to those found in K-deficient plants could be reproduced. This provided presumptive evidence for the possibility that some symptoms of K deficiency could be attributed to the accumulated putrescine.[13] The early work on this subject has been reviewed in references.[15-17]

Biosynthesis of Putrescine

Initial experiments on the biosynthetic pathways of putrescine in barley implicated ornithine as a precursor. However radioactivity from ornithine was also readily incorporated into arginine and citrulline, and the possibility that the latter amino acids might be precursors could not therefore be eliminated.[18]

Further investigation of extracts of K-deficient barley plants demonstrated the presence of agmatine, the decarboxylation product of arginine.[19] Accumulation of agmatine was greatly stimulated by K deficiency and this led to the suggestion that arginine is the precursor of putrescine with the intermediate formation of agmatine. The discovery of agmatine accumulation in K-deficient red clover leaves suggested that this biosynthetic route is common throughout the plant kingdom. Feeding arginine to cut barley shoots yielded agmatine, and agmatine yielded putrescine, confirming the close metabolic relationship of these compounds. Agmatine feeding also caused the appearance of *N*-carbamoylputrescine, the decarboxylation analogue of citrulline, and synthetic *N*-carbamoylputrescine fed to cut barley shoots was readily hydrolyzed to putrescine. *N*-Carbamoylputrescine was therefore formed as an intermediate between agmatine and putrescine, though this amine is not usually detectable in barley, even when grown under extreme K deficiency.[20] However, *N*-carbamoyl-putrescine is accumulated by K-deficient *Phaseolus*[21] and *Sesamum*[22] in the absence of added precursor (see later). Some of the metabolic pathways involving putrescine which are found in higher plants are shown in Figure 1.

Effect of Potassium Deficiency on Enzymes Involved in Putrescine Formation

Arginine decarboxylase, the enzyme forming agmatine from arginine, was detected in extracts of barley leaves, and activity of this enzyme was shown to be enhanced by K deficiency. When putrescine was increased 30 to 60 fold by K deficiency on a dry weight basis after 9 to 12 weeks of growth, agmatine increased about 6-fold and the arginine decarboxylase activity was about 2.2-fold greater in the K deficient leaves on a fresh weight basis.[23] Leaf extracts of oats grown in diurnal illumination for only 3 to 4 weeks in K-deficient nutrient showed arginine decarboxylase activity which was 3- to 4-fold greater than corresponding extracts of the normal plants. By contrast with these normal plants grown in the light, activity of the arginine decarboxylase in K-deficient oat seedlings grown in continuous dark was quite small.[24]

In a range of cereals grown in a normal K nutrient, arginine decarboxylase activity in leaf extracts was low in rice, maize, wheat, and rye (less than 30 pkat/g fresh weight). Activity in barley ranged from 190 (for normal K nutrition) to 420 (for K-deficient nutrition) pkat/g fresh weight. In oats, activity was relatively greater, ranging up to 2,700 pkat/g fresh weight for K-deficient plants 3 to 4 weeks old. However, after 26 days of growth the K-deficient barley leaves possessed 4.5 μmol/g fresh weight of putrescine, while the K-deficient oat leaves had only 2.6 μmol/g fresh weight of putrescine. Indeed, the K-deficient medium reduced the growth of the barley by 50% but growth of the oats was reduced by only 25% compared to the growth in the standard medium.[24] It appears that oats are more resistant to K deficiency than barley, and that the resistance is not dependent on the accumulation of putrescine in the oats.

Activity of N-carbamoylputrescine amidohydrolase, the enzyme hydrolyzing N-carbamoylputrescine to putrescine, was also found to be increased by K deficiency. After 11 weeks of growth, extracts of K-deficient barley leaves had 48 pkat/g fresh weight, while extracts of leaves from the standard medium had 18 pkat/g fresh weight.[25] Agmatine iminohydrolase, the enzyme forming N-carbamoylputrescine, could not be detected in barley extracts, though in maize extracts where it could be demonstrated, K-deficiency showed no apparent effect on its activity. However putrescine was increased on average by K-deficiency only by 6-fold in these plants, suggesting that the enzyme system leading to putrescine formation was not greatly stimulated by K-deficiency in this experiment.[26]

Since feeding arginine causes agmatine accumulation,[19] an increase in free arginine, initiated for instance by protein hydrolysis in conditions of stress, could also cause agmatine accumulation since the arginine decarboxylase is operating below its Km, at least in barley. This in turn could cause formation of N-carbamoylputrescine and putrescine by mass action. In the absence of information on arginine concentration, any increase in agmatine or putrescine could therefore be attributed to protein hydrolysis, without necessarily being induced by an increase in the activities of the enzymes concerned in putrescine formation. The ratio for the increase in agmatine (1:2.4) was less than that for arginine (1:2.8) in K deficient 16-day-old barley plants although the ratio of the increase for putrescine was 1:4.1.[27] However, for six-week-old barley grown in both Mg and K deficiencies the ratio of increase with deficiency became progressively greater in the metabolic sequence arginine-agmatine-putrescine suggesting a real increase in enzyme activities within this biosynthetic pathway[28] (Table 2). The relationship found in the analysis of the amino acid and amine concentrations in K-deficient *Sesamum* leaves[29] (see later) similarly indicated an increase in the activity of the enzymes of the postulated biosynthetic pathway (Figure 2).

The Enzymes of Diamine Formation and Breakdown (Fig. 1)

Arginine decarboxylase (ADC; L-arginine carboxy-lyase; EC 4.1.1.19). This enzyme occurs widely in bacteria and plants. In bacteria both constitutive and adaptive forms

Table 2. Dry weights, arginine, agmatine and putrescine content of the leaves of normal, magnesium and potassium deficient barley shoots, aged 6 weeks. Adapted from reference 28.

Treatment	Control	-Magnesium	-Potassium
Dry weight (%)	8.5	10.0	18.7
ratio	1	1.2	2.2
Arginine (μmol/g fr. wt.)	0.13	0.19	0.26
ratio	1	1.4	2.0
Agmatine (μmol/g fr. wt.)	0.23	0.52	1.08
ratio	1	2.5	4.8
Putrescine (μmol/g fr.wt.)	0.20	1.26	2.45
ratio	1	6.5	12.6

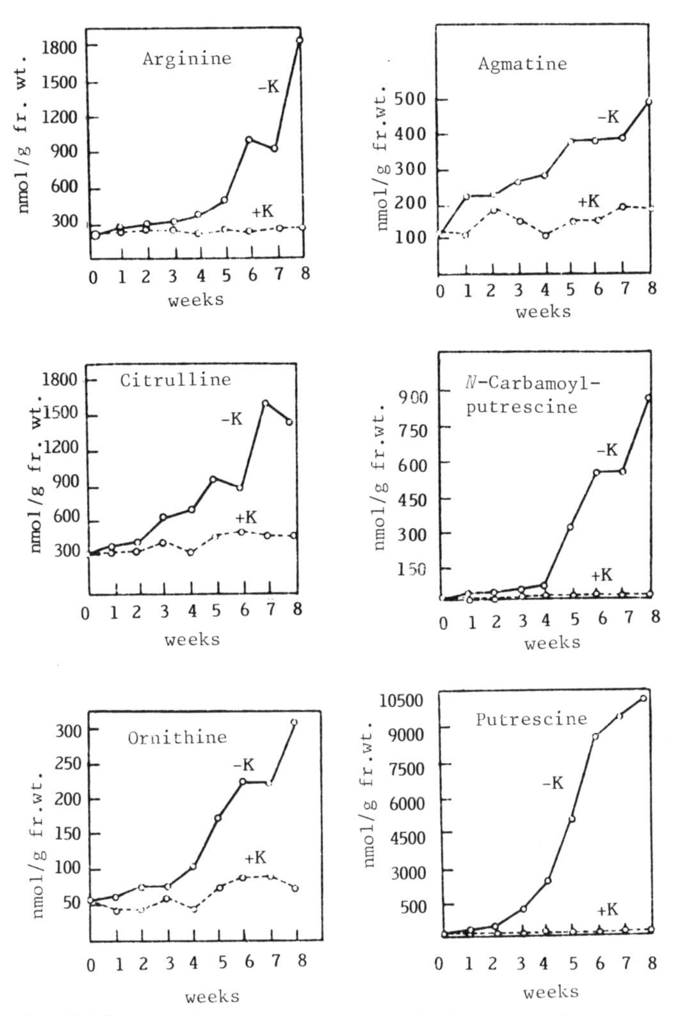

Fig. 2 Effect of potassium nutrition on the concentration of putrescine and of its potential precursors in Sesamum leaves. Reference 29.

are found.[30] The enzyme has been purified and characterized from barley,[23] oats,[24] rice,[31] and Lathyrus.[32]

After partial purification, ADC from oats was separated by gel chromatography into two fractions, with molecular weights 195,000 (A) and 118,000 (B). Fraction A was twice as active as fraction B in both normal and K-deficient plants despite the greater activity (x5) of the latter source. The properties of the two fractions were similar. The pH optimum was 7 to 7.5 and the Km was 3×10^{-5}M. The enzyme showed absolute specificity for L-arginine; activity was at least 20,000-fold greater with arginine than with ornithine. The enzyme was strongly inhibited by D-arginine, canavanine and NSD 1055 (an inhibitor specific for pyridoxal phosphate enzymes), and pyridoxal phosphate stimulated activity by about 30%. Ca and Mg ions inhibited the enzyme by 50% at 20 mM. Putrescine and the polyamines showed only moderate inhibition at 10mM, but agmatine reduced activity to 30%.[24]

Although ADC from oats was purified to apparent homogeneity, the barley enzyme was only partially purified.[23] The pH optimum for the barley enzyme was very broad, ranging from 6.5 to 9 for a purified extract and from 5.5 to 8 for a crude preparation. After exhaustive dialysis against a K-free buffer, activity on adding K was unchanged. The increase in activity with K deficiency cannot therefore be attributed to a direct inhibition due to the K ions in the normal plants. Again the enzyme showed absolute specificity for L-arginine, and D-arginine was an inhibitor.[23] Unlike the adaptive bacterial enzyme,[30] the barley enzyme was apparently not substrate-induced.[33]

Like the oat enzyme, ADC of rice embryos could also be separated by gel chromatography into two fractions with molecular weights 174,000 and 88,000, and this is almost precisely correct for dimer and monomer. The dimeric fraction showed the greatest activity. It is possible that oat ADC is also similarly composed of two sub-units; if so, the enzymes from both oats and rice exist mainly as the dimer. The optimal pH for the rice arginine decarboxylase was 8.0 and the Km was 3×10^{-4}M. Agmatine (10 mM) reduced activity to 43%. The greatest inhibition was shown by 1mM spermine, which reduced activity to less than 50%. Abscisic acid (1mM) inhibited the activity in rice seedlings to 8% of the control.[31]

ADC from Lathyrus appears to be a hexamer with identical sub-units, each with a molecular weight of 36,000. The optimum pH was 8.0 and the Km was 1.7×10^{-3}M. The activity was inhibited competitively by various amines including agmatine, but the greatest inhibition was found with spermine and arcain. Like the rice and oat enzymes, ADC from *Lathyrus* showed a requirement for a sulfydryl agent.[32] The effect of K deficiency on the rice or the *Lathyrus* enzymes has not yet been established.

Agmatine deiminase (agmatine iminohydrolase, EC 3.5.3.12). This enzyme, which hydrolyzes agmatine to *N*-carbamoylputrescine, was first detected in maize[26] and later purified and characterized from that plant.[34] The enzyme purified 7,300-fold was homogeneous, with a molecular weight of 85,000 and was formed from two sub-units, each with a molecular weight of about 43,000. The optimum pH was 6.5 and no activity was found with arginine, creatine, creatinine or glycocyamine. Agmatine iminohydrolase has also been found in seedlings of *Helianthus*, oats,[26] *Glycine max* (soybean)[35] and *Arachis hypogaea*[36] and in cabbage leaves.[26]

N-Carbamoylputrescine amidohydrolase (N-carbamoyl-putrescine N^5-carbamoyldihydrolase, EC 3.5.1.-). This enzyme, which is responsible for the hydrolysis of *N*-carbamoylputrescine to putrescine, has been detected in leaves of wheat, rye, oats, maize, pea, radish, sunflower, and barley.[25] In barley, the enzyme has a pH optimum of 7 to 8 and it is inactive with putrescine, allantoin, citrulline, and *N*, *N*'-dicarbamoylputrescine.[25] The enzyme has now been purified about 70-fold from maize shoots. The molecular weight was 125,000 and like the enzyme from barley, it showed a high substrate specificity.[37]

The reason for the occurrence of *N*-carbamoylputrescine as an intermediate in putrescine biosynthesis during the hydrolysis of agmatine by a two-step process is not easily explained, although a similar system is also found in some bacteria (reviewed in reference 20).

Putrescine synthase. A multi-functional enzyme system in *Lathyrus sativus* effects the following overall reaction:

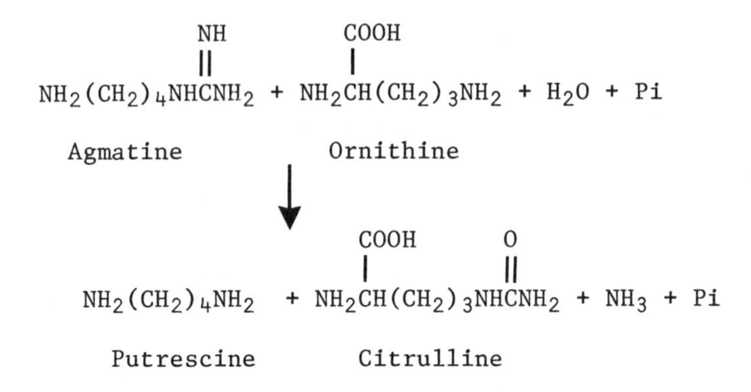

This enzyme is dependent on the presence of Pi[38] and it is therefore distinct from N-carbamoylputrescine amidohydrolase, which does not appear to require phosphate ions.[25,37] The putrescine synthase pathway is probably energetically advantageous to the plant since the carbamoyl group is conserved and the transfer occurs via carbamoyl phosphate.

Decarboxylases for ornithine (EC 4.1.1.17), lysine (EC 4.1.1.18) and citrulline (EC 4.1.1.-). In animals, putrescine is formed primarily by decarboxylation of ornithine, and arginine decarboxylation is very rare.[24,39] By contrast, in plants, putrescine appears to be formed mainly via agmatine and N-carbamoylputrescine. However in some plants including barley (where the activity is in the nucleus),[40,41] oats,[42] mung bean[43] and tobacco,[44] tomato and potato,[44-46] recent work has demonstrated the formation of putrescine by the direct decarboxylation of ornithine. The effect of K deficiency on ornithine decarboxylase is not yet established, although in tobacco roots its activity appears to be relatively unaffected by K deficiency (Table 3.)[47]

In experiments by Young and Galston, arginine decarboxylase activity was found to be increased more than 5-fold in the leaves of K-deficient oat seedlings. However, although particulate ornithine decarboxylase was increased 2-fold, soluble ornithine decarboxylase was decreased 2-fold.[48]

Lysine decarboxylase which forms the diamine cadaverine is widespread in higher plants.[49] Cadaverine occurs especially in legumes, though in peas, cadaverine concentration, unlike putrescine, was not increased by K deficiency.[50]

Table 3. Effect of nutrient deficiency on the activities of enzymes concerned in nicotine biosynthesis in tobacco roots. The activities are expressed as pkat/mg protein. Adapted from reference 47.

Enzyme	Ornithine decarboxylase	Putrescine *N*-methyl transferase	*N*-Methylputrescine oxidase
Complete	1.3	12.0	113
- Nitrogen	19.8	1.7	31
- Phosphorus	3.6	1.6	9.7
- Sulfur	1.2	3.5	24
- Calcium	2.0	3.7	169
- Magnesium	2.0	4.8	64
- Boron	2.0	6.4	64
- Potassium	1.0	4.5	63

Citrulline decarboxylation was demonstrated in extracts of K deficient *Sesamum indicum* leaves[22,29] and it may occur in sugar cane.[51] Citrulline decarboxylase has also been demonstrated in *Escherichia coli*.[52]

Amine oxidases. Di- and polyamines may be destroyed by amine oxidases, amongst which two main classes have been distinguished. Diamine oxidase occurs widely in higher plants, but principally in the legumes. It is a copper-containing enzyme and it attacks the primary amino groups with a very broad specificity, forming pyrroline from putrescine and aminopropylpyrroline from spermidine.[53,54] Polyamine oxidase in higher plants has so far been found only in the leaves of members of the family Gramineae which includes the cereals. It is specific for spermidine and spermine, oxidizing these polyamines at the secondary amino group to give diaminopropane, together with respectively pyrroline and aminopropylpyrroline.[55] This enzyme is probably associated with the cell wall.[56] In barley the activity of the polyamine oxidase depends on the nitrogen source. In seedlings grown in the dark, greatest activity was found with a nitrate medium while in the light greatest activity was shown with an ammonium medium.[57] Oxidation of spermine, though not of spermidine, by polyamine oxidase from oat leaves is promoted by high salt concentration *in vitro*. With 1M sodium phosphate buffer, activity is double that with only 0.1M sodium phosphate buffer, and similar stimulation was obtained on adding 1M sodium chloride to 0.1M sodium phosphate buffer.[58] Aminopropylpyrroline and diaminopropane are apparently at a greater concentration in Ca-deficient plants and in plants grown under saline conditions than in control plants grown in standard media.[28] This is compatible with the *in vivo* inhibition of the enzyme by Ca ions and its promotion by salts which has been demonstrated *in vitro*.[58]

The Accumulation of Putrescine in Mineral Deficiency

Barley. Some of the earlier experiments on barley have already been reported in the introductory section of this review. Typical K deficiency symptoms were found only when the phosphate concentration of the nutrient medium was in excess of the K concentration. Hackett et al.[12] showed that for large amounts of putrescine to be accumulated a similar relationship was necessary. The greatest concentration of agmatine occurred when K was low, but agmatine also accumulated when P was deficient in a plus-K medium when no significant increase in putrescine was found. Arginine decarboxylase was increased by both phosphate and K deficiency while *N*-carbamoylputrescine amidohydrolase was increased significantly only by K deficiency.[59]

After the sixth or seventh leaf stage, fluctuations in the level of putrescine were negatively correlated with changes in temperature, while fluctuations of arginine and agmatine were positively correlated. The older leaves had a greater amine content than the younger leaves. Within a leaf, amines and amino acids increased in concentration from base to tip, while K and phosphate showed an opposite gradient. Mildew infection caused a significant increase in putrescine (x2) and agmatine (x1.5) in leaves of K-deficient barley.[60]

The concentration of agmatine was increased by deficiencies of K (x10), P (x3), Ca (x3), S (x3), Mg (x2), and Mn (x2), but of these treatments only K deficiency increased the putrescine concentration significantly.[59] However, in testing the effect of a series of nutritional conditions in later work Mg deficiency was found to cause a considerable increase in putrescine concentration[28] (Table 4).

In subsequent work, Mg deficiency increased putrescine concentration in barley and in some other plants, but K deficiency was by far the more important factor in determining putrescine accumulation[27,28] (Tables 2

Table 4. Putrescine (expressed as μmol/g fresh weight) in the leaves of barley plants grown in various nutritional conditions. Adapted from reference 28.

Treatment	Age	
	4 weeks	6 weeks
- Nitrogen (N)	0.071	0.038
N as nitrate	0.062	0.025
N as ammonium	0.157	0.292
" " " + potassium chloride	0.367	0.321
" " " - sulfur	0.077	0.132
" " " - phosphorus	0.146	0.104
" " " - calcium	0.409	0.349
" " " - magnesium	1.001	0.701
" " " - potassium	3.104	2.850

and 5). Agmatine was also increased by K and Mg deficiencies (x5 and x2.5 respectively).[28]

Since Mg is required in smaller amounts than K, it is more difficult to produce visual deficiency symptoms comparable with those found with K deficiency. It is therefore not easy to equate relative deficiency, and under equal relative deficits plants deficient in Mg might even be expected to produce putrescine at a concentration comparable with that found in K deficiency.

The antifungal hordatines found in barley are derived from coumaroylagmatine by dimerization. The concentration of hordatines A and B (aglycones) was increased more than 6-fold by K deficiency while the concentration of hordatine M (glycosides of hordatines A and B) was doubled. In this experiment agmatine was increased about 13-fold by the K deficiency.[61]

Other members of the Gramineae. In the leaves of maize plants grown in a nutrient solution from which K was omitted, putrescine concentration was over 50 times greater than in those grown in the standard medium. However, addition of only 10 mg K per liter of medium (0.4 meq/l) caused the loss of putrescine almost to the control value.[62] Both agmatine and putrescine were increased in Italian rye grass (*Lolium multiflorum*) grown in a K-deficient soil, by comparison with a soil to which KCl was added at up to 240

Table 5. Effect of magnesium and potassium deficiencies on the putrescine concentration (expressed as µmol/g fresh weight) of radish, pea and spinach plants. Adapted from reference 27.

Plant	Age	Treatment	Putrescine
Radish	40 days	control	0.21
		- magnesium	0.51
		- potassium	1.66
Pea	34 days	control	0.27
		- magnesium	1.23
		- potassium	2.03
Spinach	64 days	control	0.27
		- magnesium	0.24
		- potassium	0.86

mg of K per kg.[63] Moreover, K deficiency increased putrescine in rice seedlings.[64,65]

Sesame. In addition to agmatine and putrescine, K-deficient *Sesamum indicum* also accumulates *N*-carbamoylputrescine. Radioactive *N*-carbamoylputrescine is formed directly by decarboxylation of carbamoyl-[^{14}C]citrulline in extracts of K-deficient *Sesamum* leaves. No activity could be found in extracts of leaves from plants grown in standard nutrient solution.[22] The increase in arginine, citrulline, and ornithine (Fig.2) with K deficiency which is probably due to protein hydrolysis also contributes to the accumulation of the amines. However, the increase in *N*-carbamoylputrescine and putrescine is considerably greater than the increase in agmatine. Arginine feeding caused an increase in agmatine but not in *N*-carbamoylputrescine or putrescine. In fact, in this plant citrulline was the best precursor of putrescine.[29]

Fruit. Forshey and McKee[66] found putrescine in leaves of K deficient apple, pear, prune and grape, and putrescine feeding reproduced the symptoms characteristic of K deficiency. Injection of agmatine and cadaverine failed to reproduce these symptoms. Gur and Shulman[67] similarly found putrescine accumulation in the leaves of K-deficient apple when the roots were at 25 to 30°C (optimum for growth) though at 35°C putrescine accumulation no longer occurred. An enhanced putrescine concentration was also found in K-deficient orange fruit peel,[68] and in K-deficient banana leaves.[69]

Tobacco. Putrescine accumulation occurred in tobacco plants in response to K deficiency in all leaves and shoots, but P deficiency was also found to cause putrescine accumulation, though only in the lower leaves. Ca, Mg, Fe, Mn, B, and S deficiencies also caused smaller increases of the putrescine concentration. Roots of the tobacco plants grown in the standard medium contained much putrescine and this concentration was not increased by K or P deficiency.[70]

Increased accumulation of putrescine in K- or S-deficient tobacco plants found by Yoshida was attributed in part to reduced degradation of putrescine in the deficient plants. However the pool of putrescine accumulated in the deficient plants is greater, leading to greater isotope dilution and apparently a smaller turnover. Detached leaves and roots formed putrescine from [2-^{14}C]ornithine and [U-^{14}C]arginine with equal facility, though since the pool size of arginine in leaves is much greater than the pool size of ornithine it is likely that the putrescine is formed mainly by decarboxylation of arginine.[71,72] Delétang found that the concentration of total nicotine alkaloids was reduced by 50% in tobacco plants deficient in K.[73] This may be related to a reduction of activity of certain key enzymes (putrescine N-methyltransferase and N-methylputrescine oxidase) involved in nicotine biosynthesis (Table 3).[47] Putrescine is incorporated into nicotine at a slower rate in K-deficient tobacco plants, although free putrescine and the various putrescine and spermidine hydroxycinnamic acid amides increase in concentration in these conditions (Table 6).[73] Deficiencies of K, Ca, Mg, or P caused accumulation of the amides caffeoylputrescine, caffeoylspermidine, and di-caffeoylspermidine in tobacco leaves.[74] The occurrence and biosynthesis of cinnamic acid amides in tobacco and other plants has been reviewed by Smith et al.[75]

Peas. Although putrescine was increased 6-fold in 27-day-old K-deficient pea seedlings, there was no effect on the concentration of cadaverine, a diamine which is also found in this plant and in other legumes.[50] Potassium deficiency in the shoots of peas induced the activity of diamine oxidase about three-fold,[76] and this increased diamine oxidase activity is probably substrate-induced.[76,77] Minář has suggested that resistance to K deficiency found in legumes is associated with the occurrence of the diamine oxidase (see next section).[78]

Table 6. Content of total and free amines (putrescine + spermidine) (in terms of dry weight, as parts per thousand), in tobacco leaves collected at the end of the vegetative stage from plants grown with two levels of potassium. The bound amines are calculated by difference (total - free amines). Adapted from reference 73.

Location of leaves	Potassium ion level, meq/liter					
	0.25	7	0.25	7	0.25	7
	Amine content, ppt dry wt.					
	Total		Free		Bound	
Lowest	4.3	0.3	3.7	0.3	0.6	0
Median	6.2	0.4	5.4	0.2	0.8	0.2
Highest	7.1	0.4	5.4	0.2	1.7	0.2
Cymal	4.3	0.4	4.1	0.1	0.2	0.3

<u>Tissue culture</u>. Smith et al.[79] sampled K-deficient suspension cultures of Paul's Scarlet rose tissue 8 and 14 days after inoculation (Table 7). Putrescine concentrations were enhanced on both occasions in the K-deficient media. A smaller increase was found in spermidine. However spermine, which was present in trace amounts was enhanced only on day 8. No growth could be obtained from the deficient cells on subculture. Investigation of the effects of K deficiency in a wide range of plants suggested that the concentrations of polyamines spermidine and spermine are relatively unaffected, while putrescine is increased sharply (see also Table 1), despite the fact that putrescine is the precursor of the polyamines (Fig. 1).[1]

Cinnamoylputrescine concentration in tobacco cell suspension cultures was inversely correlated with the phosphate concentration of the medium although there was a direct linear correlation of growth with phosphate.[80,81]

Cucumber cotyledons in organ culture showed a considerable reduction of putrescine (50% loss) on incubation for 72 h with 50 mM K by comparison with a K-free control in which putrescine increased 2-fold during this period.[82] The potassium chloride treatment also decreased the activity of the arginine decarboxylase in the cotyledons.[83] Spermidine and spermine were both increased by the K treatment (1.4-fold and 2.1-fold respectively).[82] Potassium chloride

Table 7. Effect of potassium deficiency on the amine content of suspension cultures of Paul's Scarlet rose cells. Cultures were grown at three potassium concentrations, and the amines estimated after 8 days and 14 days. Adapted from reference 79.

Age days	Potassium concentration in medium (meq/l.)	Fresh weight (g/flask)	Dry weight (g/flask)	Putrescine	Spermidine	Spermine
				(μmol/g, fresh wt.)		
8	10	8.3	0.69	0.176	0.059	0.004
	1	7.9	0.68	0.292	0.079	0.008
	0	7.9	0.67	0.407	0.111	0.013
14	1	8.9	0.47	0.085	0.055	0.003
	1	8.2	0.47	0.292	0.059	0.003
	0	7.8	0.45	0.327	0.093	0.003

also reduced the putrescine concentration of the cotyledons of germinating lettuce seeds.[84]

Use of Putrescine as a Marker of Potassium Deficiency

From work with many species of higher plants it is clear that the elevation of putrescine in K-deficient plants can be quite considerable, and the proportional increase in putrescine concentration is much greater than the fall in K concentration resulting from deficiency. Moreover amongst the nutrient deficiencies there is a distinct specificity for K, only Mg deficiency causing a similar though apparently smaller increase. For these reasons the possibility of using putrescine as a marker for K deficiency has been considered by several workers. On the basis of an experiment with apple seedlings, Hoffman and Samish[85] suggested using this technique for establishing the optimum concentration of K. They noted that excess of K, like deficiency, also causes putrescine accumulation, and they proposed that the K concentration in the growth medium should be adjusted to provide a minimum concentration of putrescine in the leaves (Fig. 3). (The effect of excess salt in causing putrescine accumulation is considered in a later section).

A similar study in lucerne (Fig. 4) showed that putrescine was a sensitive indicator of the K requirements. The lowest putrescine concentration was found only in plants producing more than 90% of the maximum yield in terms of dry matter. Magnesium deficiency also caused putrescine

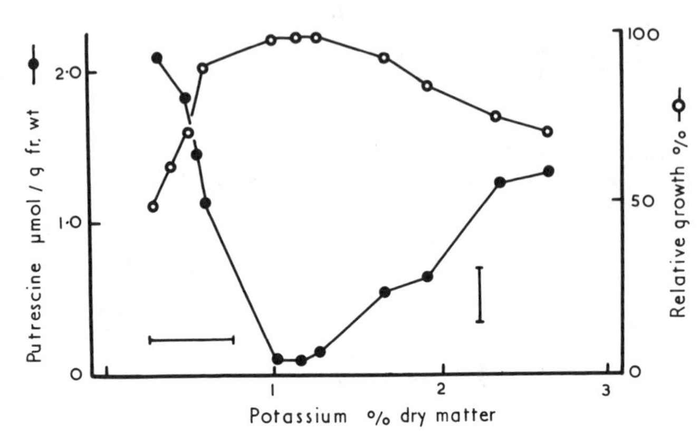

Fig. 3 Variation of putrescine and relative growth (maximum = 100%) with potassium concentration in leaves of apple seedlings. Adapted from reference 85.

accumulation but K deficiency was by far the most important factor regulating putrescine concentration. It was established that minimum values for putrescine occurred with 3% K in the dry matter of the leaves. Visual symptoms of K deficiency became apparent at less than 1.5% K in the dry matter, but putrescine accumulation was significant at about 2.5% leaf K and below. Increasing the K up to 5.5% of the dry matter did not provide further growth, nor did it change the putrescine concentration.[86] Moreover, although pea and bean (Vicia) plants grown in Mg- and K-deficient nutrients showed no visual symptoms, the putrescine concentration was raised significantly in both deficiencies. This suggests again that putrescine is a better guide to K deficiency than visual symptoms in legumes.[27] Even so, for some non-leguminous plants, e.g. blackcurrant, rye grass, visible symptoms may be a better indicator of K deficiency than putrescine analysis.[63,87]

Certainly the possibility of increased putrescine concentration both with a deficiency and an excess of K necessitates care in the interpretation of the results of putrescine analysis. Moreover, the effect of the ammonium ion on putrescine concentration (see later) could also cause complications.

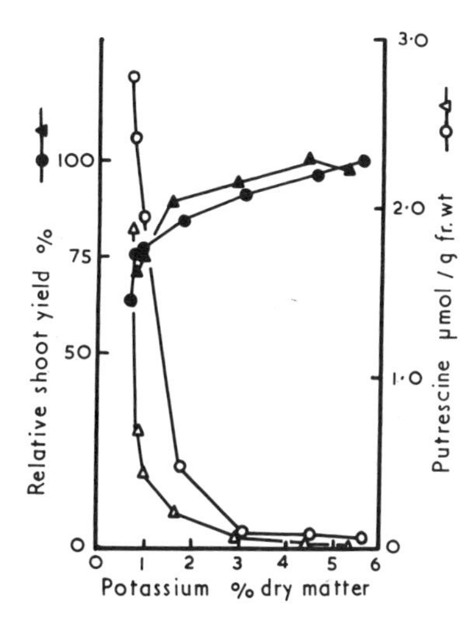

Fig. 4 Variation of putrescine (●, ▲) and relative shoot growth (○, △) (maximum = 100%) with potassium concentration in the leaves of lucerne (*Medicago sativa*). Plants were magnesium-deficient (○, ●) or were given adequate magnesium (△, ▲). Adapted from reference 86.

Ionic Balance in Potassium Deficiency

In a study of the effect of K deficiency on putrescine and agmatine in blackcurrant leaves the minimum K leaf content necessary to suppress symptoms and to provide maximum growth was given 1 meq/l of K (Table 8).[87] The difference in leaf K between that found at this concentration and that found in the lowest K supply was 35 µmol/g

Table 8. Putrescine and potassium content of blackcurrant leaves (expressed as µmol/g fresh wt). Each value is the mean of 16 independent analyses on samples taken after 11 and 16 weeks' growth. Adapted from reference 87.

Concentration of potassium in medium (meq/l)	Putrescine	Potassium	Leaf fr wt
0	5.39	29.1	1.59
0.125	2.17	32.2	1.78
0.25	0.60	37.6	1.89
0.5	0.03	44.5	2.11
1	<0.03	64.1	2.40
2	<0.03	117.4	2.59

fresh weight. The amount of putrescine at the lowest K concentration was 5.4 μmol/g fresh weight, while the putrescine formed at 1 meq/1 K was negligible. Since the putrescine has two cationic sites, the lack of K in the deficient plants was therefore compensated to the extent of about 30% by putrescine formation in terms of ionic balance. However using the index of 0.5 meq/1 K, the deficit is compensated by 70%. Similarly, at 0.25 and 0.125 meq/1, the compensation is by 112% and 170% respectively. The latter values in excess of 100% suggest that in substituting for K, putrescine is not as efficient as the inorganic ion. Excess putrescine may also be required to balance the organic acid which can accumulate in deficiency (see later).

A similar relationship was found on recalculating the data of Hackett et al. [12] (Table 9) using the sum of putrescine and agmatine. Compensation for the three lower K concentrations in the nutrient media were respectively 45%, 105%, and 140%. However without knowing the full ionic balance of cations and anions in these K deficient systems it may be unwise to pursue this interpretation. The difficulties inherent in this mode of explanation may be seen in Figure 7 (see below) where acidification with sulfur dioxide caused a considerable increase in organic acids.

Moreover, the effect of K deficiency on organic acid content in higher plants varies according to species. Potassium deficiency causes loss of organic acid in lettuce, no change in tomato or rye grass, and a gain by broad bean, cabbage, beet, and potato.[88]

Table 9. Putrescine, agmatine and potassium (expressed as μmol/g fresh weight) in barley leaves, recalculated from the data given by Hackett et al. (1965), as means of samples taken on three occasions, from plants grown with two levels of phosphate in the media. Adapted from reference 12.

Concentration of potassium in medium (meq/1)	Putrescine	Agmatine	Potassium
0.01	16.4	9.9	24
0.04	11.3	6.7	36
0.15	3.1	1.7	65
0.59	0.5	1.2	133

EFFECT OF NITROGEN SOURCE

The work of Le Rudulier and Goas has shown clearly that greater putrescine accumulation occurs with ammonium rather than nitrate as nitrogen source. Working with soybean seedlings grown in the light and dark they demonstrated an increase of about 100-fold in the putrescine concentration in ammonium by comparison with nitrate nutrient (Table 10).[89] However analysis showed that the K concentration is twice as great in the nitrate- as in the ammonium-fed plants, and this led to the supposition that the putrescine accumulation could depend on a K deficiency in the ammonium-fed plants. However experimental evidence showed that with an identical K concentration in the tissues, ammonium-based nutrients induced a much greater putrescine concentration than nitrate-based nutrients.[90] A similar accumulation of putrescine was found with ammonium in barley (Table 4), maize, wheat, tomato, tobacco, pea, pepper, petunia, and egg plant.[28,91-94]

The effects of K and ammonium nutrition on putrescine and cadaverine levels in the leaves of tobacco and tomato growing in a greenhouse soil mixture was studied by Hohlt et al.[91] In the absence of additional K, putrescine and cadaverine concentrations increased with increasing ammonium ion. On adding 10mM K, putrescine became undetectable in tobacco, but the cadaverine concentration was doubled. Addition of 10mM K to the tomato plants caused no significant change in putrescine but the cadaverine was considerably increased with the greatest ammonium concentration (80 mM ammonium). Under these conditions the cadaverine concentration reached the surprisingly high value of 140 μmol/g fresh weight. This increase of cadaverine on addition of low concentrations of K (10 mM) seems to have no analogies in other work.

[5-^{14}C]Ornithine was converted more rapidly to putrescine in ammonium- than in the nitrate-fed soybean seedlings, though ornithine is rapidly incorporated into arginine in plants grown in either medium.[95] Arginine decarboxylase could be found in extracts of the soybean seedlings[96] but ornithine and citrulline decarboxylase could not be detected.[35,96] Agmatine was transformed into putrescine at least in part with the intermediate formation of N-carbamoylputrescine, and the speed of agmatine metabolism by this pathway was accelerated in the ammonium

Table 10. Amines in soybean seedlings (expressed as μmol/g dry weight) grown for 15 days in dark or light with nitrate (NO_3^-) or ammonium (NH_4^+) as nitrogen source (tr = trace). Adapted from reference 89.

		Roots		Hypocotyls		Epicotyls	
Amine		NO_3^-	NH_4^+	NO_3^-	NH_4^+	NO_3^-	NH_4^+
Dark	Putrescine	2.3	58.9	0.2	26.1	1.0	48.1
	Cadaverine	0.3	9.0	0.6	6.1	3.4	1.9
	Spermidine	2.7	5.1	0.8	3.8	4.1	3.2
	Spermine	tr	tr	tr	2.0	tr	tr
Light	Putrescine	0.6	53.6	0.4	40.1	0.8	42.3
	Cadaverine	1.3	7.1	0.6	3.9	tr	tr
	Spermidine	2.7	6.1	0.9	2.8	2.8	5.8
	Spermine	tr	tr	tr	tr	tr	tr

medium.[97] In extracts, the activity of agmatine iminohydrolase was shown to be about 20% faster in the ammonium-grown plants than in the nitrate-grown plants.[35] Putrescine and agmatine were oxidized more slowly in plants grown on an ammonium-based medium than in plants grown on a nitrate-based medium.[98,99] The feeding of ammonium nutrients leads to acidification of the medium due to the rapid uptake of the ammonium ion. Clearly this results in complications in the study of the effect of the ammonium ion (see following section). However, in aseptic culture in which a constant pH of 6.5 to 7 was maintained using solid buffers of Ca_3PO_4 and $CaCO_3$, it was established that any reduction of ammonium level reduces the putrescine concentration. In darkness any K increase caused putrescine reduction but this effect was less clear in the light. Le Rudulier and Goas conclude from these experiments that the effect of the ammonium ion cannot be due to an overall reduction of the K concentration, though a localized depletion within the cell was not eliminated. However they favour the possibility that the ammonium ion has a direct effect on putrescine formation.[90] It now seems likely that this ammonium effect is another example of the role of putrescine as a metabolic buffer (see later).

EFFECT OF ACID FEEDING

Several amino acid decarboxylases including arginine decarboxylase are formed when bacteria are grown in an acid medium. It is likely that the amines produced by these decarboxylases are advantageous to the bacteria by raising the pH to values which are more conducive to growth. It was therefore anticipated that putrescine may be formed in higher plants in response to reduced pH of the medium. To investigate this possibility barley seedlings were grown in water culture, and after the seventh day from germination the water of the medium was exchanged for 25 mM hydrochloric acid (Fig. 5). Enzyme activity of these seedlings was compared with activity from other seedings grown in water as a control. After three days acid feeding, activity of both arginine decarboxylase and N-carbamoylputrescine amidohydrolase had increased by a factor of two. In a similar feeding experiment using sulfuric acid, arginine decarboxylase activity was also almost doubled. The effect was apparently not due merely to injury caused by the toxic acids since heating the seedlings for 16 h at 45°C which induces roughly the same degree of injury as the acid, caused less than 10% increase in activity of the arginine decarboxylase.[33] Experiments in which 10 mM HCl was fed to Cucumis sativus cotyledons showed a similar doubling of arginine decarboxylase activity.[83] In other experiments, peeled oat leaf segments incubated for 8 h on nutrient media of pH 4 contained six times the amount of putrescine of a control at pH 6. Levels of the polyamines spermidine and spermines were

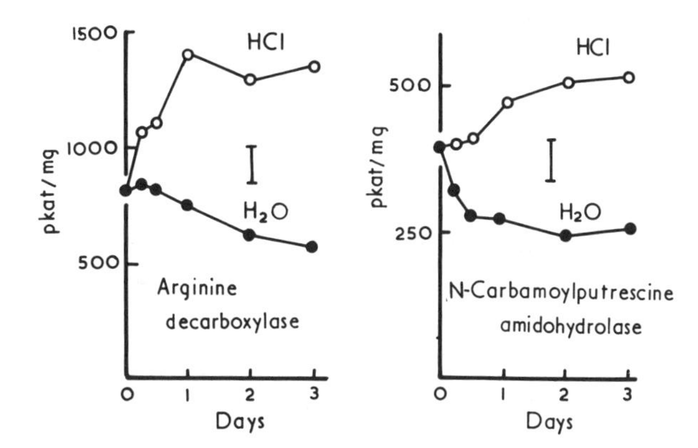

Fig. 5 Arginine decarboxylase and N-carbamoylputrescine amidohydrolase activity in the first leaves of barley seedlings fed 25 mM hydrochloric acid (O) and in the distilled water controls (●). Samples were taken at intervals after the seventh day, when the acid feeding commenced. The vertical lines represent the least significant difference at 5 per cent level of probability. Reference 33.

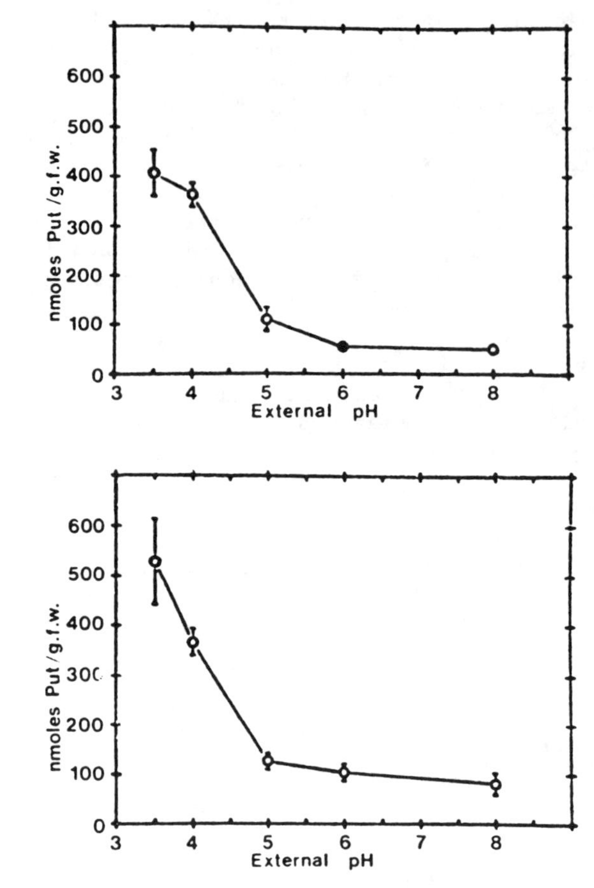

Fig. 6 Response of endogenous putrescine concentration to external pH in excised oat leaf segments (upper) and pea leaf discs (lower) incubated for 8 h. Error bars represent standard deviation. Reference 100.

unchanged. On feeding [^{14}C]arginine to the peeled oat leaf segments at pH 5 and transfer to either pH 4 or pH 6 buffer showed that the metabolism of arginine to putrescine is much faster at pH 4 [100,101] (Fig. 6). Moreover, diurnal variations in putrescine in leaves of *Kalanchoe blossfeldiana* show a strong correlation with changes in malic acid.[102]

In pea seedlings, both putrescine and spermidine increased in response to SO_2 fumigation, though a much greater response was obtained with plants which had been grown in a medium in which nitrogen was supplied as ammonium rather than as nitrate. With 0.3 ppm SO_2 in ammonium-grown plants the putrescine was increased 13-fold and the spermidine 4-fold on a dry weight basis.[94,103]

In continuous SO_2 pollution at 0.1 ppm, a concentration which did not induce necrosis, leaves of Phaseolus vulgaris showed an increase in the putrescine concentration on the 12th day. This had declined considerably by the 16th day. This putrescine peak coincided with a peak also shown by arginine. Spermidine levels were virtually unchanged, but SO_2 pollution caused a considerable loss of spermine.[104] Increases in polyamines and basic amino acids were positively correlated with increases in organic acids (Fig. 7).

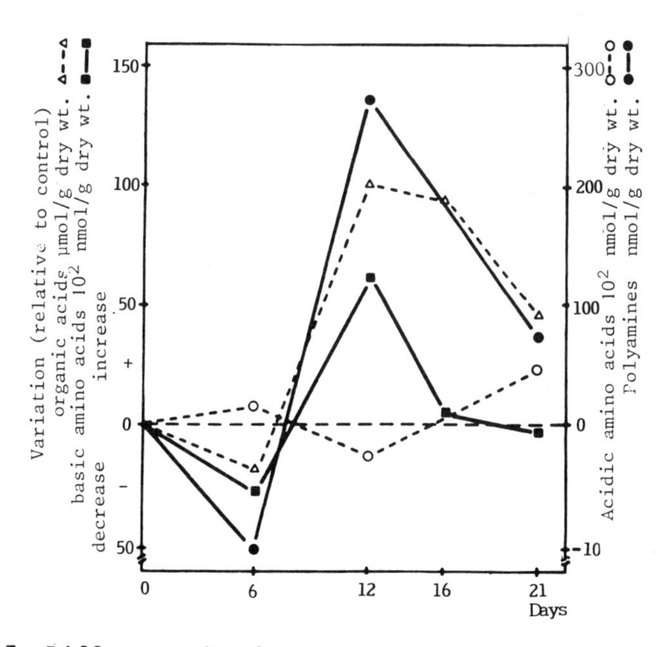

Fig. 7 Difference in the content of organic acids, basic and acidic amino acids, and polyamines between the leaves of Phaseolus seedlings exposed to 0.1 ppm sulphur dioxide and control leaves from untreated plants as a function of the number of days of pollution. Reference 104.

EFFECT OF SALT

The accumulation of diamines in saline environments has been studied intensively by Russian workers over many years. In experiments described by Strogonov in which cut shoots of ten-day-old *Vicia faba* plants were fed with NaCl at 0.3M, a considerable accumulation of putrescine was found.[105] Moreover, when putrescine was fed to these leaves and shoots, lesions similar to those found with salt toxicity were induced. On suddenly raising the salinity of the intact plants with 0.3M NaCl, large amounts of putrescine were formed in the bean plants, although sunflower and barley showed little response even with 0.4M NaCl. Amines other than putrescine were also detected in *Vicia* plants grown in saline conditions. Ashour and Thalouth[106] found that when *Vicia faba* plants which had been grown with a high salt concentration in the medium were sprayed with 100 ppm putrescine dihydrochloride, the growth of the plants was improved (see also ref. 9).

Prikhod'ko and Klyshev[107,108] showed that cadaverine was greatly enhanced in five-day old pea seedlings on growing in a saline medium. Simultaneously, various cadaverine-derived alkaloids could also be detected. These included aphylline, lupinine, and anabasine, some of which are found characteristically in the Chenopodiaceae, a family in which halophytes are common. The cadaverine concentration declined in 10-day-old pea seedlings. However, Anderson and Martin[109] found that pea seedlings grown for 13 days in saline nutrient contained 55 nmol/g fresh weight of cadaverine, while in those grown in tap water, the cadaverine concentration was even slightly greater (70 nmol/g fresh weight). Moreover, these authors were unable to detect the alkaloids found by Prikhod'ko and Klyshev[107], although these differences may be explained by the greater age of their seedlings.

In cotton, arginine concentration declined with Na_2SO_4 salinity while the putrescine concentration increased (Table 11). However in saline conditions the arginine was converted more slowly to putrescine, and putrescine was further converted to spermidine more slowly. This implies that putrescine arises from an alternative precursor (possibly ornithine) in cotton in these conditions.[110] Although labeled putrescine was metabolized more quickly in cotton plants grown in sulfate salinity, this was not effected by

Table 11. Effect of Na_2SO_4 salinity on the composition and metabolism of leaves from 36-day-old cotton seedlings (mean of 2 replicates). Adapted from reference 110.

Unit of assay	Component	Control	Na_2SO_4 concentration	
			42 mM	84 mM
% of dry wt.	Na^+	0.1	2.1	5.1
	K^+	2.5	1.9	1.7
	Ca^{2+}	2.0	2.0	1.2
μmol/g, dry wt.	Arginine	10.2	5.7	0
	Putrescine	8.1	11.1	31.0
[^{14}C]arginine feeding				
% of total activity	Putrescine + agmatine	5.0	3.8	5.9
	unknowns	0	1.9	47.7
[^{14}C]putrescine feeding				
% of total activity	Spermidine	0.3	0	0.1
	Putrescine residue	17.6	12.9	0.6

conversion of the putrescine to spermidine (Table 11). Moreover this increased putrescine metabolism was apparently not attributable to an increase in amine oxidase, since activity of this enzyme declined with salinity.[105] Similarly, activity of diamine oxidase in *Vicia faba* was decreased by severe NaCl salinity.[111]

In *Oenothera biennis*, salt concentrations high enough to cause leaf necrosis induced putrescine (6-10 fold), spermidine (3-5 fold) and spermine (5-10 fold) by comparison with the controls.[112] Apple leaves containing K ions in excess of 1% showed an increased putrescine content (Fig. 3).[85]

In mung bean seedlings, salt stress caused putrescine accumulation in the shoots, but the activities of arginine and ornithine decarboxylases were inhibited in the leaves.

However a 2- to 8-fold increase in ornithine decarboxylase was detected in the roots.[113]

The accumulation of putrescine in saline environments by higher plants is apparently not always easily demonstrated, and conditions for the formation of this diamine may be subject to factors which are at present unknown and uncontrolled. Jäger and Priebe[114] studied putrescine formation on growing pea and maize seedlings in nutrients containing various salts. The amount of putrescine per shoot was found to be diminished with NaCl treatment both for pea and maize, though an increase of 10 to 30% was found when expressed in terms of the dry weight. However, in later work no putrescine, spermidine, or spermine accumulation could be demonstrated on a dry weight basis on increasing the salinity of the nutrient medium up to 750 mM NaCl, in *Vicia faba*, *Atriplex* spp., or *Salicornia europaea*.[115]

The mean putrescine concentration in the leaves of *Limonium vulgare* plants grown on the seashore was not significantly different from the putrescine concentration of the garden-grown plants. However the putrescine content of the roots and stems of garden-grown *Limonium* during the flowering stage was 10-fold greater on a dry weight basis than the putrescine content of the plants grown on the sea-shore.[116] Unlabeled agmatine reduced the labeling of putrescine on feeding [^{14}C]arginine, suggesting that putrescine biosynthesis occurs from arginine via agmatine in *Limonium*.[117] However, only relatively small changes in arginine and agmatine metabolism could be found on adding NaCl to the feeding solutions.[118,119]

Strogonov et al.[111] and Shevyakova[120] suggest that with moderate salinization, putrescine formation is of adaptive significance. However, they propose that high salt concentrations liberate adsorbed amines and thereby contribute to the breakdown of normal function. When introduced into the plant at 10 to 100 μM, putrescine may act as a growth stimulant, but at a concentration in excess of 1 mM (1 μmol/g fresh weight) it appears to act as a toxin.[120] Even so, in ammonium nutrition putrescine may accumulate to quite high concentrations without the manifestation of toxicity symptoms.[93] Putrescine may be accumulated in sub-cellular compartments (e.g. vacuoles) in physiological stress. Putrescine which is fed to a plant however may penetrate parts of the cell which are normally inaccessible to it with consequent toxicity or even growth stimulation, depending on the concentration.

It is possible that Na ions when in excess displace K ions from a specific site, or prevent uptake of K ions by the plant. Estimation of K in the shoots of plants grown in saline conditions should establish if the latter possibility is true. In cotton leaves fed with Na_2SO_4 or NaCl, Shevyakova et al. found that the K level declined by 32% and 40% respectively (Table 11).[110] However in this experiment putrescine increased only under conditions of Na_2SO_4 salinization and it was concluded that K deficiency is not necessarily a cause of putrescine accumulation in saline conditions. The fact that the endogenous level of putrescine can also rise with a K surplus makes a specific role for K in putrescine accumulation unlikely.[120]

Although some workers find considerable accumulation of putrescine in saline conditions, others find no accumulation or even a loss of putrescine on salinization. There is an absence of consensus of the effect of salinity, even for the same plant species. Applying a high salt stress to an unadapted plant apparently causes putrescine accumulation in some instances though it is difficult to generalize. Moreover, it is even possible that the accumulation of putrescine by glycophytes in saline conditions when it occurs has no beneficial function and merely represents an abnormal response by the plants to an environment to which they have no ability to adapt successfully.

GENERAL DISCUSSION

Putrescine accumulation has now been demonstrated in K and Mg deficiencies, with ammonium excess and with acidification, of which SO_2 fumigation is probably one example, and sometimes with high osmolarity. Coleman and Richards[13] realized that putrescine, being an organic cation, may be formed by the plant to replace the inorganic cation K, when this is deficient. On the basis of experiments in which barley seedlings showed increased arginine decarboxylase and N-carbamoylputrescine amidohydrolase activities on being fed with acid, Smith[16] suggested that putrescine may be formed in K deficiency in response to a reduction in pH. On this hypothesis putrescine formation was therefore a system for the maintenance of a constant cytoplasmic pH. This homeostatic mechanism also explains the greater accumulation of putrescine with the ammonium ion as nitrogen source, by comparison with nitrate, as described earlier. The assimilation of the ammonium ion produces at least one excess proton per ammonium ion, and at least one excess of OH^- per nitrate ion.[121-123] Cytoplasmic pH is therefore reduced by ammonium assimilation and increased by nitrate assimilation, and these tendencies may be compensated by putrescine synthesis or elimination, with additional adjustment provided by changes in other ionic components (e.g. organic acids). Salinity could therefore cause putrescine accumulation by differential accumulation of anions and cations, with the overall trend causing acidity of the cytoplasm. This is therefore an extension of the mechanism causing the formation of amines by bacteria on growing in an acid medium.[30] The apparent absence of a significant shift in pH of the cell sap with K deficiency[33] demonstrates that the homeostatic system, whatever its mechanism, is very efficient. It is of interest that Escherichia coli, when grown in a medium in which Na ions substitute for K ions, large amounts of putrescine were

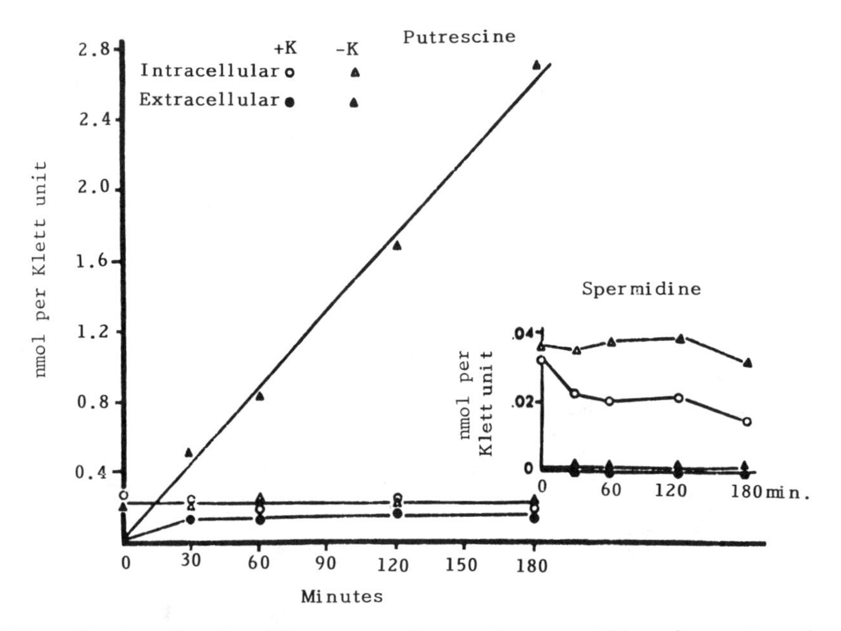

Fig. 8 Synthesis of putrescine and spermidine by potassium (+K) and sodium (-K) cultures of Escherichia coli. Reference 124.

secreted into the medium, though the intracellular putrescine and the spermidine and spermine concentrations remained unchanged (Fig. 8).[124] Conversely, high pH was found to cause a loss of putrescine in E. coli.[125] Moreover, Chlorella emersonii showed a significant rise in putrescine concentration on K starvation (Kordy and Maiss, unpublished).

Cations also inhibit putrescine biosynthesis in animal cells. Na, K, or Mg ions at 10 to 20 mM inhibited the induction of ornithine decarboxylase normally found on inoculation into a fresh medium in L 1210 (mouse leukaemia) cells.[126]

The accumulation of putrescine in higher plants caused by Mg deficiency implies that putrescine may also substitute for Mg in some of its functions. Moreover, Hurwitz (cited in ref. 127) reported that Mg deficiency caused spermidine accumulation in Escherichia coli, and in Aspergillus nidulans an inverse relationship between Mg and spermidine was found.[128] Indeed, Mg and di- and poly-amines are activators of processes involving nucleic acid function, notably protein synthesis.[129]

Recent work by Flores, Galston and Young[101,130-132] has shown that the putrescine content of oat leaf cells and protoplasts increases up to 60-fold on exposure to 0.4 to 0.6 M sorbitol for 6 h. Leaves of other cereals (barley, maize, wheat, wild oats) showed the same effect. The putrescine is synthesized by a pathway involving increased arginine decarboxylase, while ornithine decarboxylase remains unchanged (Fig. 9). Mannitol, proline and betaine have a similar effect. Addition of the arginine decarboxylase inhibitor difluoromethylarginine at 1mM prevented the increase in putrescine, while difluoromethylornithine had little effect on the system (Fig. 9). Although arginine decarboxylase activity increased 2- to 3-fold in stressed leaves in both light and dark, the increase in putrescine in the dark was only 1/3 of that in the light. This appears to be due to arginine limitation in the darkened leaves. The increase in putrescine found shortly after treatment with sorbitol and the similar increase in putrescine bound by Strogonov[105] shortly after treatment with the NaCl may be both related to the same mechanism, since both could cause an increase in intracellular salt concentrations.

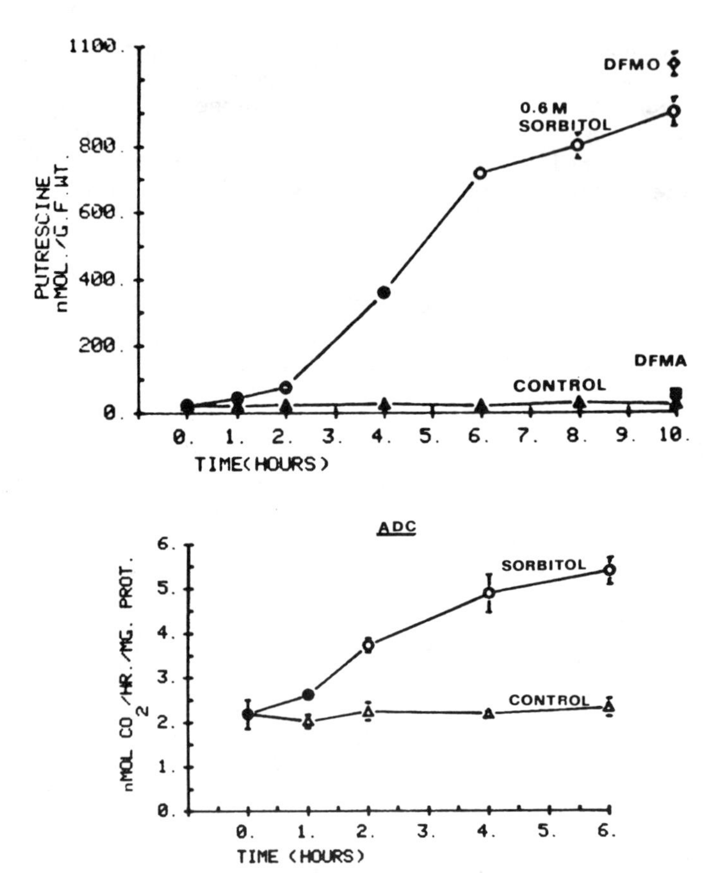

Fig. 9 Upper - Time course of changes in putrescine in illuminated oat leaf segments in the presence and absence (control) of sorbitol (0.6 M). The putrescine concentration in sorbitol containing 1 mM difluoromethylornithine (DFMO) and 1 mM difluoromethylarginine (DFMA) on the 10th day are also shown. Bars represent ±SEM. Lower - Time course of changes in arginine decarboxylase activity in illuminated oat leaf segments in the presence and absence (control) of 0.4 M sorbitol (Flores and Galston, unpublished).

Bacterial and animal cells and slime molds seem to respond to increased osmolarity by reducing the putrescine concentration, in contrast with plants. In *Escherichia coli* grown in a medium with increased osmolarity produced by

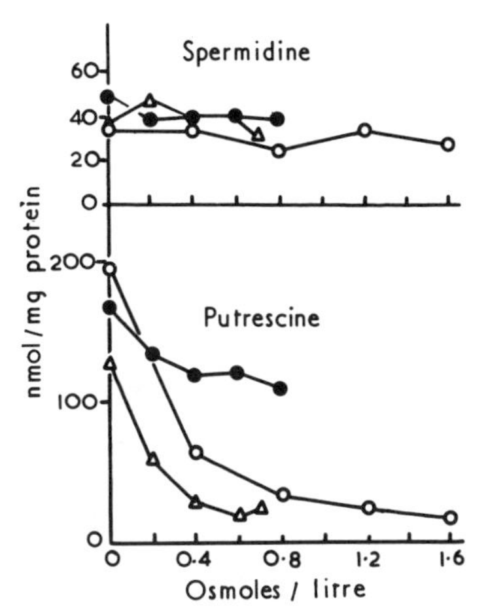

Fig. 10 Effect of osmolarity of the medium on polyamine content of Escherichia coli. An inoculum was diluted 1:100 in nutrient broth containing NaCl(O) sucrose (Δ) or glycerol (●) and the cultures grown to an optical density of 0.3 to 0.4. Adapted from reference 127.

NaCl, KCl, $MgCl_2$ or sucrose, intracellular putrescine concentration declined, while the concentration of cellular spermidine showed little change (Fig. 10). A sudden increase in osmolarity caused a rapid excretion of putrescine which was dependent on the presence of K. Potassium could not be replaced by Na, ammonium, Rb, or Mg ions. Putrescine excretion appears to be coupled with K uptake but K uptake is not dependent on putrescine excretion. It is unlikely that the putrescine excretion is due to membrane damage as spermidine is not lost to the medium. Since the effect is rapid it is likely that the excretion is caused by a change in turgor pressure which may cause a temporary cessation of nucleic acid and protein synthesis. Glycerol, which does not cause putrescine excretion penetrates the cell rapidly and would not be expected to reduce the turgor pressure. It is suggested that putrescine excretion plays a role in balancing the internal ionic strength during K uptake. The K uptake appears to be part of a mechanism concerned with increasing cell turgor, to balance the external osmoticum.[127]

The role of putrescine in osmoregulation in E. coli has been further investigated by the use of difluoromethylornithine (DFMO), a specific inhibitor of ornithine decarboxylase. Salt inhibits ornithine decarboxylase in *E. coli* and this enzyme is more important than arginine decarboxylase at low ionic strength. DFMO therefore inhibits putrescine formation to a greater extent in diluted media.[124] However DFMO had no significant effect on the intracellular concentration of Na^+ or K^+. Therefore, although increased ionic strength reduces putrescine content, decreased putrescine content does not cause an intracellular increase in inorganic ions in *E. coli*.[133]

Like the *E. coli* system, putrescine declined on increasing the osmolarity of the medium in animal cells grown in culture.[134-136] Similarly, increased osmotic concentrations induced a transient reduction of ornithine decarboxylase activity by conversion to a less active form of the slime molds *Dictyostelium* and *Physarum*.[137-8] Other aspects of the relation of polyamines with cations in animals and micro-organisms are reviewed by Canellakis et al.[139] It has been suggested[140] that proline and other compounds like the betaines, which accumulate in response to cold, salt, drought[141-143] could protect labile cell constituents (e.g. enzymes), and this hypothesis has been supported by experimental evidence.[144,145] The possibility that putrescine protects polyanionic groups (nucleic acids or membranes) under the circumstances of K stress is therefore not implausible. It has been proposed that proline, glycine betaine and possible also sugars, which accumulate with increased salinity and drought are stored as non-toxic metabolites to restore the osmotic balance, and under these circumstances this may also be a reason for putrescine accumulation.[146] Salinity and desiccation may induce both putrescine and proline accumulation in higher plants. However, although proline has been found to accumulate to high concentrations in K deficiency[147] this is not a general response.[148-150]

Many enzymes, especially those concerned in protein synthesis, are activated by K which is required at a relatively high concentration for greatest effect.[146,151] The stimulation of protein synthesis by polyamines is well established, but it would be of interest to determine whether putrescine could substitute for K^+ in other enzyme systems. It is relevant in this connection that particulate starch synthase from sweet corn, a K-requiring enzyme, is also stimulated by spermine.[152] Another possibility is that putrescine accumulates as a relatively non-toxic form of nitrogen, perhaps sequestered in the vacuole, when protein synthesis cannot be sustained. This is also one possible explanation for proline accumulation in drought[142] although it is likely that there are several ways in which proline and putrescine are beneficial to the plant in conditions of stress. The intracellular location of putrescine is still unknown, though it would be of interest in this context. Techniques for the isolation of intact vacuoles may help to answer this problem.

The activation of the membrane-bound F_1 ATPase from Vigna mitochondria by putrescine and polyamines at physiological concentrations and its dependency on the concentration of inorganic cations suggests a regulatory mechanism for polyamines in the accumulation of inorganic ions. The ATPase showed this interaction only in association with the membrane, and the binding sites for the Mg^{2+} and the polyamines appeared to be localized on the membrane.[153] ATPase obtained from rabbit kidney microsomes is similarly activated by spermine.[154]

Di- and polyamines prevent solute leakage from plant cells, probably through their interaction with cell membranes.[155,156] Putrescine and the polyamines also allow improved growth in conditions of stress[9,157-159] and in certain biological systems putrescine and other di- and polyamines will stimulate growth even in the apparent absence of stress.[6,9] These amines are also known to stabilize nucleic acids, stimulating protein synthesis and increasing the accuracy of transcription and translation.[160]

Putrescine concentration rises rapidly in response to increased growth in animals[2] and in plants,[1] probably due to its requirement in polyamine synthesis. Putrescine biosynthesis is therefore also initiated by factors unrelated to environmental stress. In higher plants, unambiguous evidence for such accumulations has hitherto been obtained mainly in tissue culture experiments, though the control of arginine decarboxylase by the phytochrome system in pea seedlings provides strong support for this theory.[161,162]

Further progress in understanding the function and mechanism of putrescine biosynthesis in relation to mineral nutrition might be made by the use of amino acid decarboxylase inhibitors, difluoromethylornithine and difluoromethylarginine. These have already been applied widely to studies of mammalian polyamine metabolism[2] and have recently been used in the investigation of higher plant systems.[130] The intracellular location of putrescine and of the inorganic ions, and the possible protective effect of putrescine on proteins, nucleic acids and membranes would also be of considerable interest. These questions justify thorough investigation in view of the worldwide nature of the problems of crop production in K deficiency and high salinity.

SUMMARY

The diamine putrescine has been shown to accumulate in a wide range of higher plants when grown in K-deficient nutrient, especially when nitrogen is supplied as the ammonium ion. The putrescine is formed mainly from arginine via agmatine and N-carbamoylputrescine, and the enzymes in this pathway are activated in K deficiency. Putrescine concentration in the leaves may be useful as an indicator of K deficiency, especially in legumes where the putrescine accumulates before deficiency symptoms are visible. Putrescine is also accumulated in Mg deficiency though to a smaller extent. The enzymes of putrescine formation are also activated by feeding mineral acid suggesting a function of putrescine in ionic balance and pH control. Induction of putrescine biosynthesis in some higher plants by increased osmolarity is in contrast to the response found in animal and bacterial cells where high osmolarity caused a decline in putrescine.

ACKNOWLEDGEMENTS

The author is very grateful to Dr. E. J. Hewitt F.R.S. for his interest and encouragement.

REFERENCES

1. WAREING PF (ed) 1982 Polyamines. In Proc 11th Int Conf Plant Growth Substances pp 449-494
2. PEGG AE, PP McCANN 1982 Polyamine metabolism function. Am J Physiol 243: C212-C221
3. BACHRACH U, A KAYE, R CHAYEN (eds) 1983 Advances in Polyamine Research. Raven Press, New York Volume 4
4. GALSTON AW 1983 Polyamines as modulators of plant development. BioScience 33:382-388

5. SMITH TA 1977 Recent advances in the biochemistry of plant amines Prog Phytochem 4:27-81
6. SMITH TA , N BAGNI, D SERAFINI FRACASSINI 1979 The formation of amines and their derivatives in plants In: EJ Hewitt, CV Cutting eds, Nitrogen assimilation of plants. Academic Press, London New York pp 557-570
7. SMITH TA 1980 Plant amines. In EA Bell, BV Charlwood, eds Encyclopedia of Plant Physiology, New Series Springer-Verlag, Berlin 8:433-460
8. SMITH TA 1981 Amines. In EE Conn ed, Biochemistry of Plants. Academic Press, New York 7:249-268
9. SMITH TA 1982 Polyamines as plant growth regulators. Br Plant Growth Regulator Group News Bull 5:1-10
10. CLARKSON DT, JB HANSON 1980 The mineral nutrition of higher plants. Annu Rev Plant Physiol 31:239-298
11. RICHARDS FJ, RG COLEMAN 1952 Occurrence of putrescine in potassium-deficient barley. Nature (London) 170:460
12. HACKETT C, C SINCLAIR, FJ RICHARDS 1965 Balance between potassium and phosphorus in the nutrition of barley. 1 The influence on amine content. Ann Bot (London) 29:331-345
13. COLEMAN RG, FJ RICHARDS 1956 Physiological studies in plant nutrition. XVIII Some aspects of nitrogen metabolism in barley and other plants in relation to potassium deficiency. Ann Bot (London) 20:393-409
14. SMITH TA 1970 Putrescine, spermidine and spermine in higher plants. Phytochemistry 9:1479-1486
15. SMITH TA 1968 The biosynthesis of putrescine in higher plants and its relation to potassium nutrition. In EJ Hewitt, CV Cutting eds, Recent Aspects of Nitrogen Metabolism in Plants. Academic Press, London, New York pp 139-146
16. SMITH TA 1970 The biosynthesis and metabolism of putrescine in higher plants. Ann NY Acad Sci 171:988-1001
17. SMITH TA 1971 The occurrence, metabolism and functions of amines in plants. Biol Rev Cambridge Phil Soc 46: 201-241
18. COLEMAN RG, MP HEGARTY 1957 Metabolism of DL-ornithine-2-^{14}C in normal and potassium-deficient barley. Nature (London) 179: 376-377
19. SMITH TA, FJ RICHARDS 1962 The biosynthesis of putrescine in higher plants and its relation to potassium nutrition. Biochem J 84: 292-294

20. SMITH TA, JL GARRAWAY 1964 N-Carbamoylputrescine - an intermediate in the formation of putrescine in barley. Phytochemistry 3: 23-26
21. CROCOMO OJ, MA CATTINI, EA ZAGO 1974 Accumulation of amines and amino acids in relation to the potassium level in leaves of beans (Phaseolus vulgaris). Arq Biol Tecnol 17: 93-102 (Cited in Chem Abst 83: 203810)
22. CROCOMO OJ, LC BASSO, OG BRASIL 1970 Formation of N-carbamylputrescine from citrulline in Sesamum. Phytochemistry 9: 1487-1489
23. SMITH TA 1963 L-Arginine carboxy-lyase of higher plants and its relation to potassium nutrition. Phytochemistry 2: 241-252
24. SMITH TA 1979 Arginine decarboxylase of oat seedlings. Phytochemistry 18: 1447-1452
25. SMITH TA 1965 N-Carbamylputrescine amidohydrolase of higher plants and its relation to potassium nutrition. Phytochemistry 4: 599-607
26. SMITH TA 1969 Agmatine iminohydrolase in maize. Phytochemistry 8: 2111-2117
27. BASSO LC, TA SMITH 1974 Effect of mineral deficiency on amine formation in higher plants. Phytochemistry 13: 875-883
28. SMITH TA 1973 Amine levels in mineral deficient Hordeum vulgare leaves. Phytochemistry 12: 2093-2100
29. CROCOMO OJ, LC BASSO 1974 Accumulation of putrescine and related amino acids in potassium-deficient Sesamum. Phytochemistry 13: 2659-2665
30. ABDELAL AT 1979 Arginine catabolism by microorganisms. Ann Rev Microbiol 33: 139-168
31. CHOUDHURI MM, B GHOSH 1982 Purification and partial characterization of arginine decarboxylase from rice embryos (Oryza sativa L.). Agr Biol Chem 46: 739-743
32. RAMAKRISHNA S, PR ADIGA 1975 Arginine decarboxylase from Lathyrus sativus seedlings. Eur J Biochem 59: 377-386
33. SMITH TA, C SINCLAIR 1967 The effect of acid feeding on amine formation in barley. Ann Bot (London) 31: 103-111
34. YANAGISAWA H, Y SUZUKI 1981 Corn agmatine iminohydrolase. Plant Physiol 67:697-700
35. LE RUDULIER D, G GOAS 1980 Biogenèse de la N-carbamylputrescine et de la putrescine dans les plantules de Glycine max (L.) Merr. Physiol Veg 18: 609-616

36. SINDHU RK, HV DESAI, BI NAIK, SK SRIVASTAVA 1980 Regulation of agmatine iminohydrolase activity in germinating ground nut seeds. Phytochemistry 19: 19-21
37. YANAGISAWA H, Y SUZUKI 1982 Purification and properties of N-carbamylputrescine amidohydrolase from maize shoots. Phytochemistry 21: 2201-2203
38. SRIVENUGOPAL KS, PR ADIGA 1981 Enzymic conversion of agmatine to putrescine in Lathyrus sativus seedlings. J Biol Chem 256: 9532-9541
39. SMITH TA 1981 Biosynthesis and metabolism of polyamines in plants. In DR Morris, LJ Marton eds, Polyamines in Biology and Medicine, Raven Press New York, pp 77-82
40. PANAGIOTIDIS CA, JG GEORGATSOS, DA KYRIAKIDIS 1982 Superinduction of cytosolic and chromatin bound ornithine decarboxylase activities of germinating barley seeds by actinomycin D. FEBS Lett 146: 193-196
41. KYRIAKIDIS DA 1983 Effect of plant growth hormones and polyamines on ornithine decarboxylase activity during the germination of barley seeds. Physiol Plant 57: 499-504
42. KAUR-SAWHNEY R, LM SHIH, HE FLORES, AW GALSTON 1982 Relation of polyamine synthesis and titer to aging and senescence in oat leaves. Plant Physiol 69: 405-410
43. ALTMAN A, R FRIEDMAN, N LEVIN 1982 Arginine and ornithine decarboxylases, the polyamine biosynthetic enzymes of mung bean seedlings. Plant Physiol 69: 876-879
44. HEIMER YM, Y MIZRAHI 1982 Characterization of ornithine decarboxylase of tobacco cells and tomato ovaries. Biochem J 201: 373-376
45. COHEN E, YM HEIMER, Y MIZRAHI 1982 Ornithine decarboxylase and arginine decarboxylase activities in meristematic tissues of tomato and potato plants. Plant Physiol 70:544-546
46. KAUR-SAWHNEY R, LM SHIH, AW GALSTON 1982 Relation of polyamine biosynthesis to the initiation of sprouting in potato tubers. Plant Physiol 69: 411-415
47. YOSHIDA D 1973 Effects of nutrient deficiencies on the activities of enzymes catalyzing nicotine synthesis. Bull Hatano Tobacco Exp Sta 73: 239-243
48. YOUNG, ND, AW GALSTON 1983 Expression and biosynthesis of arginine decarboxylase in

K^+-deficient oat seedlings. Plant Physiol (suppl.) 72: 137 abst 786
49. SCHOOFS G, S. TEICHMANN, T HARTMANN, M WINK 1983 Lysine decarboxylase in plants and its integration in quinolizidine alkaloid biosynthesis. Phytochemistry 22: 65-69
50. SMITH TA, G WILSHIRE 1975 Distribution of cadaverine and other amines in higher plants. Phytochemistry 14: 2341-2346
51. MARETZSKI A, M THOM, LG NICKELL 1969 Products of arginine catabolism in growing cells of sugar cane. Phytochemistry 8: 811-818
52. AKAMATSU N, M OGUCHI, Y YAJIMA, M OHNO 1978 Formation of _N_-carbamylputrescine from citrulline in _Escherichia coli_. J Bacteriol 133: 409-410
53. YANAGISAWA H, E HIRASAWA, Y SUZUKI 1981 Purification and properties of diamine oxidase from pea epicotyls. Phytochemistry 20: 2105-2108
54. RINALDI A, G FLORIS, A FINAZZI-AGRO 1982 Purification and properties of diamine oxidase from _Euphorbia_ latex. Eur J Biochem 127: 417-422
55. SUZUKI Y, H YANAGISAWA 1980 Purification and properties of maize polyamine oxidase: a flavoprotein. Plant Cell Physiol 21: 1085-1094
56. KAUR-SAWHNEY R, HE FLORES, AW GALSTON 1981 Polyamine oxidase in oat leaves: a cell-wall localized enzyme. Plant Physiol 68: 494-498
57. SMITH TA 1976 Polyamine oxidase from barley and oats. Phytochemistry 15: 633-636
58. SMITH TA 1977 Further properties of the polyamine oxidase from oat seedlings. Phytochemistry 16: 1647-1649
59. SINCLAIR C 1967 Relation between mineral deficiency and amine synthesis in barley. Nature (London) 213: 214-215
60. SINCLAIR C 1969 The level and distribution of amines in barley as affected by potassium nutrition, arginine level, temperature fluctuation and mildew infection. Plant Soil 30: 423-438
61. SMITH TA, GR BEST 1978 Distribution of the hordatines in barley. Phytochemistry 17: 1093-1098
62. VASCONCELLOS CA, JM FORTES, J FERNANDES, ZT SANTOS, LC BASSO, E MALAVOLTA 1977 Occurrence of putrescine in leaves of potassium deficient maize var. Piranao. Rev Ceres 24: 88-93

63. NOWAKOWSKI TZ, M BYERS 1972 Effects of nitrogen and potassium fertilizers on contents of carbohydrates and free amino acids in Italian rye grass. II Changes in the composition of the non-protein nitrogen fraction and the distribution of individual amino acids. J Sci Food Agr 23: 1313-1333
64. MITSUI S, K KUMAZAWA 1961 Dynamic studies on the nutrient uptake by crop plants. XXXIII Amino acids and other nitrogenous compounds in rice roots under different nutritional conditions. Nippon Dojo-Hiryogaku Zasshi 32: 57-59 (Chem. Abst. 57: 11570)
65. MATSUZAKA Y, T SHIRAI, K IMAZUMI 1962 Potassium deficiency of paddy rice plant. I Nutrient uptake and metabolism. Nippon Dojo-Hiryogaku Zasshi 33: 125-128 (Chem. Abst. 59: 12113)
66. FORSHEY CG, MW McKEE 1970 Effect of potassium deficiency on nitrogen metabolism of fruit plants. J Am Soc Hort Sci 95: 727-729
67. GUR A, Y SHULMAN 1971 The influence of high root temperature on the potassium nutrition and on certain organic constituents of apple plants. Recent Adv Plant Nutr 2: 643-656
68. BAR-AKIVA A 1975 Effect of potassium nutrition on fruit splitting in Valencia orange. J Hort Sci 50: 85-89
69. STEWARD FC, AC HULME, SR FREIBERG, MP HEGARTY, JK POLLARD, R RABSON, RA BARR 1960 Physiological investigation on the banana plant. Ann Bot N S 24: 83-157
70. TAKAHASHI T, D YOSHIDA 1960 Relation between the nutrition of tobacco plants and the accumulation of putrescine. J Sci Soil Manure (Japan) 31: 39-41
71. YOSHIDA D 1969 Formation of putrescine from ornithine and arginine in tobacco plants. Plant Cell Physiol 10: 393-397
72. YOSHIDA D 1969 Effect of agmatine on putrescine formation in tobacco plants. Plant Cell Physiol 10: 923-924
73. DELÉTANG J 1977 Alkaloid production and its relations to rhizogenesis in Nicotiana tabacum. Ann Tabac 14: (Sect 2) 5-110
74. DELÉTANG J 1974 Presence of caffeoylputrescine, caffeoylspermidine and di-caffeoylspermidine in Nicotiana tabacum. Ann Tabac 11:(Sect 2) 123-130
75. SMITH TA, J NEGREL, CR BIRD 1983 The cinnamic acid

amides of the di- and polyamines. Adv Polyamine Research 4: 347-370
76. PAPRSKÁŘOVÁ L, J MINÁŘ 1976 The effect of potassium deficiency on diamine oxidase activity in pea. Biol Plant 18: 99-104
77. SRIVASTAVA SK, V PRAKASH, BI NAIK. 1977 Regulation of diamine oxidase activity in germinating pea seeds. Phytochemistry 16: 185-187
78. MINÁŘ J 1979 The reasons for the different reactions of plants to potassium deficiency. Folia Biol 20:5-108
79. SMITH TA, GR BEST, AJ ABBOTT, ED CLEMENTS 1978 Polyamines in Paul's Scarlet rose suspension cultures. Planta 144: 63-68
80. KNOBLOCH K-H, G BEUTNAGEL, J BERLIN 1981 Influence of accumulated phosphate on culture growth and formation of cinnamoylputrescines in medium-induced suspension cultures of Nicotiana tabacum. Planta 153: 582-585
81. KNOBLOCH K-H, J BERLIN 1981 Phosphate mediated regulation of cinnamoylputrescine biosynthesis in cell suspension cultures of Nicotiana tabacum. Planta Medica 42: 167-172
82. SURESH MR, PR ADIGA 1978 Absence of parallelism between polyamine and nucleic acid contents during induced growth of cucumber cotyledons. Biochem J 172: 185-188
83. SURESH MR, S RAMAKRISHNA, PR ADIGA 1978 Regulation of arginine decarboxylase and putrescine levels in Cucumis sativus cotyledons. Phytochemistry 17: 57-63
84. CHO SC 1983 Effect of cytokinin and several inorganic cations on the polyamine content of lettuce cotyledons. Plant Cell Physiol 24: 27-32
85. HOFFMAN M, RM SAMISH 1971 Free amine content in fruit tree organs as an indicator of the nutritional status with respect to potassium. Recent Adv Plant Nutr 1: 189-206
86. SMITH GS, DR LAUREN, IS CORNFORTH, MP AGNEW 1982 Evaluation of putrescine as a biochemical indicator of the potassium requirements of lucerne. New Phytol 91: 419-428
87. MURTY KS, TA SMITH, C BOULD 1971 The relation between the putrescine content and potassium status of blackcurrant leaves. Ann Bot (London) 35: 687-695
88. FREEMAN GG 1967 Studies on potassium nutrition of plants. II Some effects of potassium deficiency on the organic acids of leaves. J Sci Food Agric 18: 569-576

89. LE RUDULIER D, G GOAS 1971 Mise en évidence et dosage de quelques amines dans les plantules de Soja hispida Moench., privées de leurs cotylédons et cultivées en présence de nitrates, d'urée et de chlorure d'ammonium. CR Hebd Séances Acad Sci Ser D 273:1108-1111
90. LE RUDULIER D, G GOAS 1975 Influence des ions ammonium et potassium sur l'accumulation de la putrescine chez les jeunes plantes de Soja hispida Moench. privées de leurs cotylédons. Physiol Veg 13: 125-136
91. HOHLT HE, DN MAYNARD, AV BARKER 1970 Studies on the ammonium tolerance of some cultivated Solanaceae. J Am Soc Hort Sci 95: 345-348
92. KLEIN H, A PRIEBE, H-J JÄGER 1979 Putrescine and spermidine in peas; effects of nitrogen source and potassium supply. Physiol Plant 45: 497-499.
93. LE RUDULIER D, G GOAS 1979 Contribution à l' etude de l'accumulation de putrescine chez des plantes cultivées en nutrition strictement ammoniacale. CR Hebd Séances Acad Sci Ser D 288: 1387-1390
94. PRIEBE A, H KLEIN, H-J JÄGER 1978 Role of polyamines in SO_2 polluted pea plants. J Exp Bot 29: 1045-1050
95. LE RUDULIER D, G GOAS 1974 Devenir de l'ornithine ^{14}C-5 dans les jeunes plantes de Soja hispida Moench. privées de leurs cotylédons et cultivées en présence de chlorure d'ammonium et de nitrates. CR Hebd Séances Acad Sci, Ser D 278: 1039-1042
96. LE RUDULIER D, G GOAS 1975 Etude de l'activité de l'arginine decarboxylase dans les jeunes plantes de Glycine max, privées de leurs cotylédons. Phytochemistry 14: 1723-1725
97. LE RUDULIER D, G GOAS 1974 Métabolisme azoté des jeunes plantes de Soja hispida Moench. privées de leurs cotyledons. Recherche des voies de formation de la putrescine. CR Hebd Séances Acad Sci, Ser D 279: 161-163
98. LE RUDULIER D, G GOAS 1977 Devenir de la putrescine-1,4-^{14}C chez Glycine max. Physiol Plant 40: 87-90
99. LE RUDULIER D, G GOAS 1977 La diamine oxydase dans les jeunes plantes de Glycine max. Phytochemistry 16: 509-512
100. YOUNG ND, AW GALSTON 1983 Putrescine and acid stress.

Induction of arginine decarboxylase activity and putrescine accumulation by low pH. Plant Physiol 71: 767-771

101. GALSTON AW, Y-R DAI, HE FLORES, ND YOUNG 1983 The control of arginine decarboxylase activity in higher plants. Adv Polyamine Res 4: 381-393

102. MOREL C, VR VILLANUEVA, O QUEIROZ 1980 Are polyamines involved in the induction and regulation of the Crassulacean acid metabolism? Planta 149: 440-444

103. JÄGER H-J, H KLEIN 1977 Die Bedeutung stoffwechselphysiologischer Reaktionen von Pflanzen als Kenngrössen für Immissionswirkungen. Phytopathol Z 89: 128-134

104. PIERRE M, O QUEIROZ 1981 Enzymic and metabolic changes in bean leaves during continuous pollution by sub-necrotic levels of SO_2. Environ Pollut, Ser. A 25: 41-51

105. STROGONOV BP 1964 Physiological basis of salt tolerance of plants. Pub. Acad. Sci. U.S.S.R. Translation by Israel Programme for Scientific Translations, Jerusalem

106. ASHOUR NI, AT THALOUTH 1971 Effect of putrescine on growth and photosynthetic pigments of broad bean plants grown under chloride salinization conditions. Biochem Physiol Pflanzen 162: 203-208

107. PRIKHOD'KO LS, LK KLYSHEV 1964 Nitrogen metabolism of pea shoots at different substrate salinity. Tr Inst Botan Akad Nauk SSR. 20: 166-182 (Chem Abst 62: 9474b)

108. ANON 1973 Structure and function of plant cells in saline habitats. Israel Programme for Scientific Translations. Wiley, New York, Toronto

109. ANDERSON JN, RO MARTIN 1973 Identification of cadaverine in *Pisum sativum*. Phytochemistry 12: 443-446

110. SHEVYAKOVA NI, NV ARUTYUNOVA, BP STROGONOV 1981 Disturbance in arginine and putrescine metabolism in cotton leaves in the presence of excessive Na_2SO_4. Fiziol Rast (Moscow) 28: 594-600

111. STROGONOV BP, NI SHEVYAKOVA, VV KABANOV 1972 Diamines in plant metabolism under conditions of salinization. Fiziol Rast (Moscow) 19: 1098-1104

112. BAGNI C, M CREMONINI 1973 Induction and repression of polyamine synthesis in *Oenothera biennis* by different salts. Inf Bot Ital 5: 103-104

113. FRIEDMAN R, A ALTMAN, U BACHRACH 1983 Polyamine involvement in plant response to salt stress. Plant Physiol (suppl.) 72: 137 abst 784

114. JÄGER H-J, A PRIEBE 1975 Zum Problem der durch Salinität induzierten Putrescinbildung in Pflanzen. Oecol Plant 10: 267-279

115. PRIEBE A, H-J JÄGER 1978 Effect of NaCl on the levels of putrescine and related polyamines of plants differing in salt tolerance. Plant Sci Lett 12: 365-369

116. LARHER F 1974 Métabolisme azoté des halophytes. Les amines non-volatiles de Limonium vulgare Mill.; évolution de la putrescine au cours du cycle de développement. CR Hebd Séances Acad Sci Ser D 279: 157-160

117. LARHER F 1973 Métabolisme azoté des halophytes. Recherche des voies de formation de la putrescine dans les jeunes rameaux de Limonium vulgare Mill. CR Hebd Séances Acad Sci, Ser D 277: 1333-1336.

118. LARHER F, M GOAS, G GOAS 1973 Métabolisme azoté des halophytes. Utilization de l'arginine $^{14}C(U)$ par les bourgeons de Limonium vulgare Mill. CR Hebd Séances Acad Sci, Ser D 276: 1429-1432

119. LARHER F 1974 Métabolisme azoté des halophytes. Rôle de l'agmatine dans la formation de la putrescine dans les rameaux et les racines secondaires de Limonium vulgare Mill. CR Hebd Séances Acad Sci, Ser D 279: 271-274

120. SHEVYAKOVA NI 1981 Metabolism and the physiological role of diamines and polyamines in plants. Fiziol Rast (Moscow) 28: 1052-1061

121. BRETELER H 1973 A comparison between ammonium and nitrate nutrition of young sugar-beet plants grown in nutrient solutions at constant acidity. 1 Production of dry matter, ionic balance and chemical composition. Neth J Agric Sci 21: 227-244

122. RAVEN JA, FA SMITH 1974 Significance of hydrogen ion transport in plant cells. Can J Bot 52: 1035-1048

123. RABEN JA, FA SMITH 1976 Nitrogen assimilation and transport in vascular land plants in relation to intracellular pH regulation. New Phytol 76: 415-431

124. RUBENSTEIN KE, E STREIBEL, S MASSEY, L LAPI, SS COHEN 1972 Polyamine metabolism in potassium deficient bacteria. J Bacteriol 112: 1213-1221

125. DUBIN DT, SM ROSENTHAL 1960 The acetylation of polyamines in Escherichia coli. J Biol Chem 235: 776-782

126. CHEN K, JS HELLER, ES CANELLAKIS 1976 The inhibition of the induction of ornithine decarboxylase by cations. Biochem Biophys Res Commun 70: 212-220

127. MUNRO GF, K HERCULES, J MORGAN, W SAUERBIER 1972 Dependence of the putrescine content of Escherichia coli on the osmotic strength of the medium. J Biol Chem 247: 1272-1280
128. BUSHELL ME, AT BULL 1974 Polyamine, magnesium and ribonucleic acid levels in steady-state cultures of the mould Aspergillis nidulans. J Gen Microbiol 81: 271-273
129. LOFTFIELD RB, EA EIGNER, A PASTUSZYN 1981 Polyamines and protein synthesis. In DR Morris, LJ Marton eds, Polyamines in Biology and Medicine. Marcel Dekker New York, Basel, pp 207-221
130. FLORES HE, AW GALSTON 1982 Polyamines and plant stress: Activation of putrescine biosynthesis by osmotic shock. Science 217: 1259-1261
131. GALSTON AW, HE FLORES, ND YOUNG 1983 Putrescine formation by arginine decarboxylase: a stress response in cereal leaves. Plant Physiol (suppl) 72: 105 abst 599
132. FLORES HE, AW GALSTON 1983 Osmotic stress-induced polyamine accumulation. Plant Physiol (suppl) 72: 106 abst 600
133. GÜNTHER T, HW PETER 1979 Polyamines and osmoregulation in Escherichia coli. FEMS Microbiol Lett 5: 29-31
134. FRIEDMAN Y, S PARK, S LEVASSEUR, G BURKE 1977 Activation of thyroid ornithine decarboxylase (ODC) in vitro by hypotonicity: a possible mechanism for ODC induction. Biochem Biophys Res Communs 77: 57-64
135. MUNRO GF, RA MILLER, CA BELL, EL VERDERBER 1975 Effects of external osmolality on polyamine metabolism in He La cells. Biochim Biophys Acta 411: 263-281
136. PERRY JW, T OKA 1980 Regulation of ornithine decarboxylase in cultured mouse mammary gland by the osmolarity in the cellular environment. Biochim Biophys Acta 629: 24-35
137. MITCHELL JLA, GE KOTTAS 1979 Osmotically-induced modification of ornithine decarboxylase in Physarum. FEBS Lett 102: 265-268
138. HARRIS WA, MJ NORTH 1982 Osmotically-induced changes in the ornithine decarboxylase activity of Dictyostelium discoideum. J. Bact 150: 716-721
139. CANELLAKIS ES, D VICEPS-MADORE, DA KYRIAKIDIS, JS HELLER 1979 The regulation and function of ornithine decarboxylase and of the polyamines. Current Topics Cell Regul. 15: 155-202
140. STEWART GR, F LARHER 1980 Accumulation of amino acids

and related compounds in relation to environmental stress. In BJ Miflin ed. The Biochemistry of Plants. Vol 5, Amino Acids and Derivatives, Academic Press, New York, pp 609-635

141. WYN JONES RG, R STOREY 1981 Betaines. In LG Paleg D Aspinall eds, The Physiology and Biochemistry of Drought Resistance in Plants. pp 171-204
142. ASPINALL D, LG PALEG 1981 Proline accumulation: physiological aspects. In LG Paleg, D Aspinall eds, The Physiology and Biochemistry of Drought Resistance in Plants. Academic Press, New York, London, pp 205-241
143. STEWART CR 1981 Proline accumulation. Biochemical Aspects. In LG Paleg, D Aspinall eds, The Physiology and Biochemistry of Drought Resistance in Plants. Academic Press, New York, London, pp 243-259
144. PALEG LG, TJ DOUGLAS, A VAN DAAL, DB KEECH 1981 Proline, betaine and other organic solutes protect enzymes against heat inactivation. Aust J Plant Physiol 8: 107-114
145. AHMAD I, F LARHER, AF MANN, SF McNALLY, GR STEWART 1982 Nitrogen metabolism of halophytes. IV Characteristics of glutamine synthetase from *Triglochin maritima* L. New Phytol 91: 585-595
146. WYN JONES RG, CJ BRADY, J SPEIRS 1979 Ionic and osmotic relations in plant cells. In DL Laidman, RG Wyn Jones eds, Recent Advances in the Biochemistry of Cereals. Academic Press, London, New York, San Francisco pp 63-103
147. GÖRING H, BH THIEN 1979 Influence of nutrient deficiency on proline accumulation in the cytoplasm of *Zea mays* L. seedlings. Biochem Physiol Pflanz 174: 9-16
148. MURUMKAR CV, BA KARADGE, PD CHAVAN 1982 Growth, mineral nutrition and nitrogen metabolism of potassium deficient *Sorghum*. Biovigyanam. 8: 37-42
149. BOZOVA L, D MIZEVA 1978 Effect of potassium on the content of free amino acids in young wheat plants. Rastenievud Nauki 15: 3-8 (Chem Abst 91: 210079)
150. MUKHERJEE I 1974 Effect of potassium on proline accumulation in maize during wilting. Physiol Plant 31: 288-291
151. EVANS HJ, RA WILDES 1971 Potassium and its role in enzyme activation. In Potassium in Biochemistry and Physiology. Int Potash Inst 8th Colloq. 13-39
152. TANDECARZ J, N LAVINTMAN, CE CARDINI 1975 Activation of

particulate starch synthetase from Zea mays embryo. Phytochemistry 14: 103-106

153. PETER HW, MR PINHEIRO, MS LIMA 1981 Regulation of the F_1 ATPase from mitochondria of Vigna sinesis (L.) Savi, cv. Pitiuba by spermine, spermidine and putrescine, Mg^{2+}, Na^+ and K^+. Can J Biochem 59: 60-66

154. TASHIMA Y, M HASAGAWA, LK LANE, A SCHWARTZ 1981 Specific effect of spermine on Na^+, K^+-adenosine triphosphatase. J Biochem 89: 249-255

155. SRIVASTAVA SK, TA SMITH 1982 The effect of some oligoamines and guanidines on membrane permeability in higher plants. Phytochemistry 21: 997-1008

156. ALTMAN A 1982 Polyamines and wounded storage tissues - Inhibition of RNase activity and solute leakage. Physiol Plant 54: 194-198

157. HUHTINEN OK, J HONKANEN, LK SIMOLA 1982 Ornithine- and putrescine-supported divisions and cell colony formation in leaf protoplasts of alders (Alnus glutinosa and A. incana). Plant Sci Lett 28: 3-9

158. GALSTON AW, R KAUR-SAWHNEY, A ALTMAN, HE FLORES 1980 Polyamines, macromolecular synthesis and the problem of cereal protoplast regeneration. In L Ferenczy, GL Farkas eds, Advances in Protoplast Research. Proc. 5th Int. Protoplast Symposium, Pergamon Press pp 485-497

159. KAUR-SAWHNEY R, HE FLORES, AW GALSTON 1980 Polyamine-induced DNA synthesis and mitosis in oat leaf protoplasts. Plant Physiol 65: 368-371

160. ABRAHAM AK, L PIHL 1981 Role of polyamines in macromolecular synthesis. Trends Biochem Sci 6: 106-107

161. GOREN R, NA PALAVAN, AW GALSTON 1982 Separating phytochrome effects on arginine decarboxylase activity from its effect on growth. J Plant Growth Regul 1: 61-73

162. GOREN R, N PALAVAN, HE FLORES, AW GALSTON 1982 Changes in polyamine titer in etiolated pea seedlings following red light treatment. Plant Cell Physiol 23: 19-26

Chapter Three

PHYTOCHEMICAL ASPECTS OF OSMOTIC ADAPTATION

R.G. WYN JONES

Department of Biochemistry and Soil Sciences
University College of North Wales
Bangor, Gwynedd, Wales, LL57 2UW

INTRODUCTION

The interactions between higher plants and excess salts in their growth environment are complex. Both ionic composition and total concentration of dissolved salts are important, as is the way in which these salts impinge upon the plant tissue. For example, the effects of salts in sprays falling on leaves either from sea-spray or overhead irrigation are somewhat different to those of salts in the root environment.[1,2] The degree of salt damage is also modified by factors such as relative humidity and ozone concentration.[3,4] In plants themselves a variety of anatomical, developmental, physiological and biochemical characteristics have been singled out as qualities to be associated with an ability to grow or survive in saline habitats.[5-8]

It would be difficult to include all of these characteristics in this chapter. Rather I will focus on certain cellular and subcellular aspects of adaptation related to phytochemistry; other facts such as membrane and cuticle composition will not be considered. Nevertheless, it is important to recognize that events at the organ and/or whole plant level such as xylem- and phloem-loading are at least as important as cellular and subcellular events in determining the overall response of a plant to a saline environment. While acknowledging these caveats, it can be fairly claimed that our understanding of the mechanisms of the adaptation of higher plant cells to salinity has been transformed in the last decade.

The leaves of halophytes contain large quantities of NaCl and maintain a roughly constant osmotic pressure differential between the leaf sap and the soil solution ($\pi_{sap} - \pi_{ex} \cong k$) (see ref. 4). Such osmotic adaptation is clearly essential if adequate cell turgor pressure is to be maintained and tissue dehydration avoided. Biochemical studies by Flowers, Greenway and others on enzymes from glycophytes and halophytes suggested that the sensitivity of these enzymes to salts (NaCl and KCl) were similar and not readily compatible with the levels of NaCl found in saps expressed from the halophyte leaves (see ref. 5). This led to the idea that the salts were largely occluded within the vacuoles away from salt-sensitive cytoplasmic enzymes. This basic concept has been refined and quantified into a specific hypothesis amenable to experimental testing.[9,10] In this chapter I shall briefly outline the hypothesis indicating its relationship to hyperosmotic adaptation in other eukaryotic and prokaryotic cells, before considering recent evidence giving firm support to the model. The need to modify aspects of the model will be discussed in the light of increasing evidence that certain solutes have an effect on the stability and activity of proteins and membranes as well as a direct osmotic role. In addition, the chemotaxonomic distribution of the cytosolutes themselves will be described.

MODEL OF HYPEROSMOTIC RESPONSE OF CELLS

The osmotic balance in cells of plants and indeed other eukaryotic and prokaryotic organisms (other than halobacteria) subjected to hyper-saline environments may be represented as in Figure 1. The basic features have been presented earlier,[9,10] and were derived in part from Steinbach.[11] In this model it is assumed:

(i) that the total ionic strength (dominantly univalent cations) of the cytoplasm is constrained in the range 100 to 200 mol m^{-3}. Making some allowance for other solutes, this is equivalent to an osmotic pressure of 260 to 440 mosmol kg^{-1} (0.65 - 1.0 MPa).

(ii) that the cytoplasm is highly selective for K^+ over Na^+.

(iii) that, where osmotic pressures in excess of those noted in (i) are required to maintain turgor, non-toxic organic cytosolutes are accumulated. Typical examples of such solutes in higher plants are glycinebetaine (betaine), sorbitol and proline.

(iv) that hyperosmotic adaptation in the vacuole may be achieved by the accumulation of a range of solutes but typically in the halophytes these are NaCl and other Na salts.

In our original hypothesis we noted the consistency in the responses of cytoplasms of plant and animal cells, especially those of marine animals, to hyperosmotic media as in all cases similar organic solutes were accumulated. We considered that the perturbation of protein stability by high salt (K or NaCl) concentrations was the major constraint on the use of these ions for cytoplasmic osmoregulation in non-halobacterial cells but offered no explanation for the K^+ selectivity of the cytoplasms. An almost identical hypothesis was presented recently by Yancey et al[12] again concentrating on solute-protein interactions as the underlying constraint on the ionic concentration in the cytoplasm leading to the accumulation of compatible solutes. These authors showed how interactions between solutes especially trimethylamine oxide/glycinebetaine and urea (a destabilizing solute) resulted in solute combinations causing a minimal perturbation of protein (enzyme) stability, thereby adding significantly to earlier ideas but without explaining the monovalent cation specificity.

Following Lubin[13] and Weber et al,[14] we later suggested that the close correlation between the ion specificity of protein synthesis and the ionic concentrations in cytoplasms could be a major factor in constraining the ionic composition of the cytosol[15] (Fig.

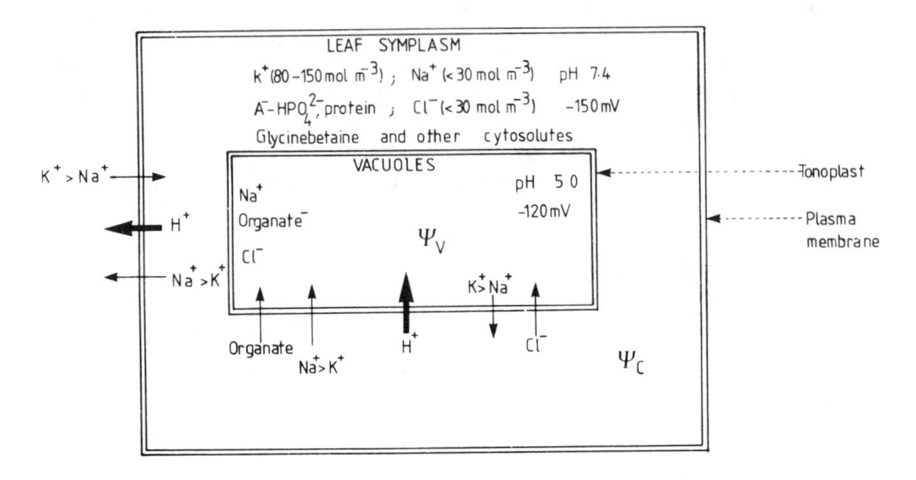

Fig. 1 Idealized model of solute compartmentation in higher plant cells accumulating sodium salts and glycinebetaine. The values quoted are intended to indicate a hypothetical 'norm' and not absolute unchangeable values. Thick arrow = active electrogenic transport. Thin arrow = selective transport (no energetic deduction).

$\psi_v = \psi_c$

$\psi_v = \psi^{\pi}$ Na salts

$\psi_c = \psi^{\pi}$ K salts $+ \psi^{\pi}$ cytosolutes $+ \psi^{\pi}$ other salts

Data from reference 7.

2). Since high K^+ concentrations are required for translation, a highly conserved process, this hypothesis also suggests a biochemical basis for the K^+ and Na^+ selectivity. These ideas and their relationship to other biochemical and physiological characteristics and integration of the various cytoplasmic functions of K^+ are explored in more detail elsewhere.[16]

In developing this model, particularly in its application to higher plant cells, considerable reliance was placed on comparative physiological and biochemical evidence although some direct evidence for subcellular solute compartmentation was available in higher plants

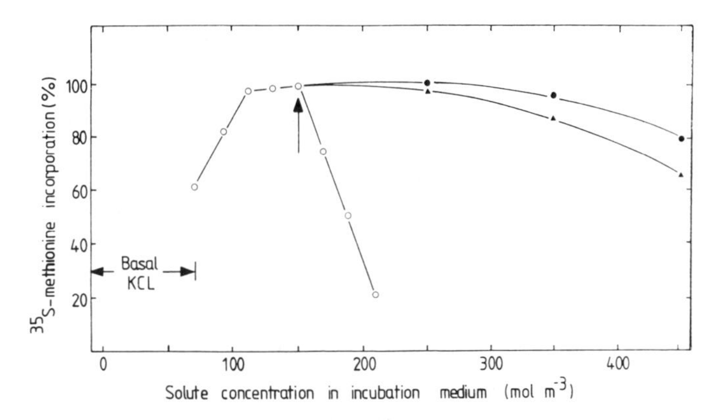

Fig. 2 Effect of potassium acetate, proline glycinebetaine on in vitro protein synthesis on a wheat germ system primed with wheat leaf mRNA. In addition to the basal KCl concentration (70 mol m^{-3}), potassium acetate was added (○). At the concentration of potassium salts which gave the maximum rate of incorporation, glycinebetaine (●) or proline (▲) was added to give the final solute concentration. Data from reference 7.

(see ref. 17). Efforts have been made to test the model in recent years. It has been established that glycinebetaine, proline and sorbitol are relatively nontoxic to metabolic functions at concentrations up to 300-500 mol m^{-3} (see for example Fig. 2) although inhibitory effects at very high concentrations cannot be ruled out.

It is worth noting that in a detailed study of the effects of glycinebetaine and proline on the translation of mRNA from Beta vulgaris, Chenopodium album, Pisum sativum and Triticum aestivum, proline at 500 mol m^{-3} allowed an incorporation rate about 30% of that found under optimum conditions. Glycinebetaine, less deleterious, allowed a rate of about 70% of the maximal. Sucrose was even more inhibitory than proline.[18] It is interesting also that Setter and Greenway[19] have tentatively attributed the inhibition of growth of Chlorella emersonii (which accumulates proline and sucrose) to metabolic damage caused by the imperfection of the so-called compatible solutes. Thus the undiscriminatory use of the term 'compatible solute' and the bland assumption that all the cytosolutes are equally benign and capable of being tolerated at very high concentrations is quite unjustified.

It has been more difficult to establish unequivocally the postulated solute distribution between cytoplasm and vacuole including the high K^+ selectivity of the former. In this chapter the cytoplasm is treated as a homologous entity. Although this is highly improbable, the contrast between the cytoplasm and vacuole is so striking that it may be accepted at this stage as a rather gross approximation.

SOME RECENT EVIDENCE FOR SOLUTE COMPARTMENTATION

A number of techniques have been employed to study solute compartmentation within higher plant cells, such as (1) X-ray microprobe analysis of sections of freeze-substituted tissue or freeze fractured/frozen samples, (2) microchemical analysis of tissues of different degrees of vacuolation, (3) efflux analysis and (4) isolation of purified vacuoles and their analysis. Each of these methods has its limitations and may be subjected to specific criticisms. For example, while X-ray microprobe analysis of freeze-substituted samples may be quantified, ion redistribution during sample preparation cannot be excluded. On the other hand, analyses of frozen samples minimize this error and probably produce accurate elemental ratios but are difficult to quantify and to convert to tissue concentrations of individual ions. Generally the low cytoplasm to vacuole ratio in mature cells means that a reliable separation of cell wall, cytoplasm and vacuole is only possible in certain cell types. While analyses (both X-ray microprobe and micro- chemical) of apical meristematic cells of relatively low vacuolation are possible, extrapolation of these data to the cytoplasmic concentration in mature vacuolated cells is clearly problematical. The criticism also applies to microchemical analyses of purified protoplasts and vacuoles where the possibility of the leakage of low molecular weight solutes is an ever present hazard. Despite these problems, evidence strongly indicates a marked degree of solute compartmentation in higher plants.[17,20-22]

With a few exceptions, the data agree quantitatively rather well with the concentrations assumed in the model (Fig. 1). However, the extent of K^+ selectivity and Na^+

and Cl^- exclusion from the cytoplasm in the most extreme halophytes has been somewhat uncertain.[20] Two recent studies on the halophytes Suaeda maritima and Atriplex spongiosa are revealing. Gorham and Wyn Jones[23] examined the elemental composition of apical cells and the vacuoles of mature leaf mesophyll cells using X-ray microprobe analysis of frozen/freeze-fractured samples combined with semi-microchemical analysis of K^+, Na^+ and glycinebetaine in sequential 2mm tissue sections from the shoot apical meristem (Tables 1 and 2). These demonstrated a great K^+ selectivity in the meristematic, cytoplasm-rich cells compared with the vacuoles of more mature cells. The total X-ray signal from the apical cells was 20 to 25% that of the vacuoles indicating the total ionic strength in the former was also substantially lower. Glycinebetaine analyses showed a steep inverse gradient, concentrations being highest in the smallest apical sample analyzed which nevertheless contained a large amount of vacuolated tissue. It is likely that the glycinebetaine concentration in the cytosol of meristematic cells exceeded 500 mol m^{-3}. An important study carried out by Storey and his colleagues on Atriplex spongiosa demonstrates in much greater detail the ion selectivity of both meristematic cells and the cytoplasms of mature cells.[24,25] In Table 3, the element ratios in apical cells are reproduced and reveal a 3:1 K:Na selectivity throughout a salinity range reaching 600 mol external NaCl. There is also a high selectivity for P and a constant K:P ratio but with selective exclusion of Cl,

Table 1. Solute content (nmol kg^{-1} fresh weight) of tissues of increasing age and size (leaf lengths in parentheses) from Suaeda maritima plants grown at 150 mol m^{-3} NaCl.[a] Values are the means of 3 replicates. Standard errors were <10%.

Tissue	K	Na	K/Na	Glycinebetaine
Leaf (2mm)	262	255	1.16	237
Axillary buds	211	379	0.56	168
Leaf (2-5 mm)	285	318	0.89	125
Leaf (5-10 mm)	174	301	0.58	86
Leaf (10-20 mm)	209	314	0.67	43
Leaf (20 mm)	189	490	0.40	20
Stem	50	412	0.12	14

[a] Data taken from Gorham and Wyn Jones.[23]

Table 2. X-ray microanalysis data for *Suaeda maritima* plants grown at 300 mol m^{-3} NaCl. Mean counts per 60 s (number of replicates in parentheses). Except for the leaf primordium, all the samples are vacuolar. Standard errors were less than 50%.[a]

Tissue	Na (uncorrected)	Na (corrected)	K	Cl	P	S	K/Na	K+Cl+P+S+Na (corrected)	Total + background (Na uncorrected)
Leaf primordium (13)	239	812	963	950	646	171	1.19	3,542	13,307
Young-leaf epidermis(6)	1,358	4,616	2,134	6,457	368	0	0.45	13,575	26,317
Young-leaf mesophyll (11)	1,359	5,232	2,866	5,747	297	760	0.53	14,902	30,831
Old-leaf epidermis (12)	1,639	5,572	583	4,650	0	0	0.11	10,805	24,093
Old-leaf pallisade (13)	2,113	7,184	746	4,995	0	0	0.10	12,925	25,352
Old-leaf mesophyll (8)	1,358	4,617	307	4,266	0	0	0.06	9,190	20,443

[a] Data from reference 23.

Table 3. Ratios of elements in meristematic cells of young leaf tissue from *Atriplex* plants grown on solutions containing varied levels of NaCl.[a]

Salt concentration (mM)	$\frac{Na}{P}$	$\frac{K}{P}$	$\frac{Cl}{P}$	$\frac{S}{P}$	$\frac{K}{Na}$	$\frac{Cl}{K+Na}$
0	<0.01	1.7	0.19	0.40	>10	0.11
200	0.39	1.8	0.68	0.46	4.5	0.31
400	0.59	1.7	0.81	0.46	2.8	0.35
600	0.47	1.6	1.00	0.42	3.4	0.48

[a] Data taken from Storey et al.[24]

although this gradually declines at the highest salinities.

The high K:Na ratio was also found in the cytoplasm but not in the vacuole of bundle sheath cells (Table 4). This observation indicates, at least to a reasonable approximation, that there is some validity in extrapolation from the young apical cells to the cytoplasms of mature cells. In this work, the total ionic strength of both the apical cells and the cytoplasm of the bundle sheath cells was much lower than that of vacuoles. No organic solute analyses were reported by Storey et al. Nevertheless, these studies together with the evidence for compartmentation noted earlier, provide compelling evidence for ion selectivity of cytoplasm in higher plants as well as other organisms and for solute distribution according to the proposed model. Even quantitative extrapolations from these data to cytoplasmic K^+ levels are in line with the idealized model.[24]

Interestingly, the phloem is modified cytoplasm[26] and consequently phloem-fed tissues e.g. grains, particularly the living embryonic aleurone cells, pollen and flower petals etc. also characteristically have low Na^+ and Cl^- contents and high levels of organic non-toxic solutes (see ref. 7).

It seems reasonable to conclude therefore that the model for solute distribution in higher plant cells is to a significant extent valid and is a useful guide to the processes involved in osmoregulation in halophytes as well as certain non-halophytes e.g. K^+-deficient cells. In the

Table 4. Ratio of K/Na in mature leaf of Atriplex spongiosa grown on culture solution with increasing concentrations of NaCl.[a]

Cell type	NaCl in culture solution (mM)			
	0	200	400	600
Bundle sheath				
Cytoplasm	>10	1.9	2.27	2.13
Vacuole	>10	0.14	0.13	0.1
Mesophyll				
Vacuole	>10	0.26	0.07	0.09
Hypodermis				
Vacuole	>10	0.14	0.09	0.08
Epidermis				
Vacuole	>10	0.22	0.125	0.07
Bladder				
Vacuole	>10	0.125	0.07	0.03

[a] Data taken from Storey et al.[24]

latter the preferential retention of K^+ in the cytoplasm has been demonstrated.[27] This is probably relevant to the interpretation of the K^+-nutrition e.g. 'luxury consumption' and putrescine accumulation in higher plants[28] and possibly to hypotheses on stomatal movement and CAM which involve substantial osmotic oscillations in vacuoles. Presumably these phenomena apply also in cytoplasms.

CHEMOTAXONOMY OF PUTATIVE CYTOSOLUTES

Extensive studies on the chemical distribution, physiological compatability and biochemistry of glycinebetaine and proline have been conducted and are summarized elsewhere.[29-31] Soon, it became evident that neither of these solutes was universally accumulated as a potential cytosolute. With proline the situation has been particularly confused. While proline is accumulated in higher plants as a near universal response to extreme water stress, the relation of this phenomenon to drought tolerance (or resistance) is very uncertain.[31] In terms of higher plant halophytism, adaptive proline accumulation is best characterized in Puccinellia and Agrostis species and in Triglochin maritima.[30] In other species such as barley and members of the Chenopodiaceae, the evidence clearly associates proline accumulation with growth inhibition at high salinities rather than on the adaptive responses involved in the growth of plants at modest salinities.[32]

Several other possible compatible cytosolutes have now been recognized in other species and their distribution is clearly related to taxonomy. While it is impossible to detail in this chapter all the evidence for the taxonomic distribution and physiological function of each solute, a general summary of the data is presented in Table 5. This listing is not comprehensive and other compounds such as γ-aminobutyric acid, alanine and some methylated inositols may have similar functions. The chemistry of compatible solutes remains unexplored in several taxa.

To date the organic cytosolutes fall into two chemical groups - polyols and their close derivatives, and small zwitterionic solutes, typically amino acids, betaines or their sulfur analogues. It is striking that very similar solutes are accumulated in all prokaryotic and eukaryotic cells with the exception of the halobacteria (Archebacteria).[10,12] Glycerol is accumulated in Dunaliella sp. and some fungi while mannitol is typically found in brown algae and in the unicellular Platymonas species. α- and β-galactosyl glycerols are osmoregulatory solutes in a number of algae.[6] Proline accumulates in bacteria, microalgae, some higher plants and invertebrates,[38,10,12] while glycinebetaine is found in the photosynthetic bacterium, Ectothiorhodospira halochloris,[39] in a number of halophytic blue green algae[40] as well as in marine invertebrates and plants.[33] This chemical consis- tency encourages the speculation that some fundamental physico-chemical phenomena underlie the interactions between these two very diverse groups of solutes and biopolymers. Presently, our understanding of such rela- tionships is limited as will be seen shortly.

The possible role of mono- and disaccharide sugars in cytoplasmic osmotic adjustment in higher plants is not clear. When exposed to drought, several species accumulate sugars to levels where they contribute significantly to the sap osmotic pressure.[41] Most of these sugars must be in the vacuole and their accumulation probably reflects

Table 5. Taxonomic distribution of possible cytosolutes in higher plants.

Trivial names	Structure	Family/Tribe	Comments	Ref
Glycinebetaine (betaine)	$(CH_3)_3N^+-CH_2-COO^-$	Chenopodiaceae	Characteristic of this family	33, 31
		Amaranthaceae		
		Asteraceae	Some members e.g. Aster tripolium	30
		Solanaceae	Some members e.g. Lycium sp.	
		Gramineae		
		Chlorideae	e.g. Spartina species	
		Triticeae	Modest levels in commercial cereals	
β-Alaninebetaine (homobetaine)	$(CH_3)_3N^+-CH_2-CH_2-COO^-$	Plumbaginaceae	Typical of this family, some evidence that it is partly esterified	33
Proline	$CH_2-CH_2-CH_2-CH(COO^-)-N^+H_2$ (ring)	Juncaginaceae	Triglochin species	30
		Gramineae		
		Festuceae	Seems to be typical of members of this tribe e.g. Puccinellia species	
		Asteraceae		

Table 5. (continued)

Trivial names	Structure	Family/Tribe	Comments	Ref
Prolinebetaine (stachycrine)	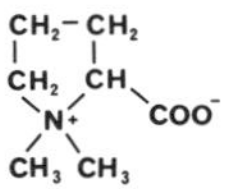 CH_2-CH_2, CH_2 $CH-COO^-$, N^+, CH_3 CH_3 (ring)	Labiatae Capparidaceae	Phytochemical reports but little physiological evidence	33
		Leguminoseae	Only reported in *Medicago sativa*	
β-dimethylsulphonio propionate (propiothetin)	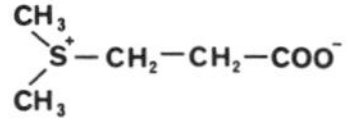 $(CH_3)_2S^+-CH_2-CH_2-COO^-$	Asteraceae	In *Melanthera biflora* only	33, 34
		Gramineae	Found in *Spartina* with glycinebetaine tentatively identified in *Posidonia* sp.	
D-Sorbitol	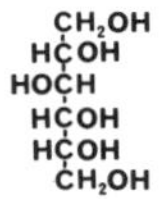 CH_2OH HCOH HOCH HCOH HCOH CH_2OH	Plantaginaceae	Characteristic of this family	35
D-Pinitol	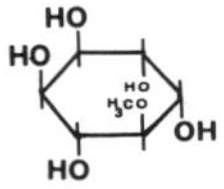 cyclohexane ring with HO, HO, H_3CO, OH, HO	Leguminoseae Caryophyllaceae	No evidence yet on biochemical compatibility	36, 37

a mechanism for osmotic adjustment when simple inorganic solutes K^+, Na^+, Cl^- and divalent ions are in short supply.[27] Under analogous circumstances, 'low salt' barley roots used monosaccharides as an alternative vacuolar osmotica,[27] while in beet root storage tissue, sucrose is localized in the vacuoles.[42]

I am not aware of any compelling evidence that sugars act as major cytoplasmic osmotica in higher plants, although sucrose, a major solute in the phloem, is, as noted earlier, a significant osmoticum in *Chlorella emersonii*. Similarly, galactosyl glycerols are accumulated in several species of marine and brackish water algae.

CYTOSOLUTES AS STABILIZERS OF BIOPOLYMERS

In the simple model for solute distribution and osmotic adjustment discussed so far in this chapter, the cytosolutes are visualized simply as benign osmotica which can be used to achieve hyperosmotic adjustment in the cytosol with a minimum of deleterious effects on biochemical functions. However, evidence is now emerging that, at least in certain circumstances, these solutes stabilize proteins and membranes against a range of perturbations.

In 1975 Skedy-Vinkler and Avi-Dor[43] showed that glycinebetaine and its sulfur analogue, dimethylsulfonioacetate, protected the respiratory activity of a bacterium against salt damage. The protective action diminished as the quaternary nitrogen group was selectively demethylated. A similar effect was observed by Pollard and Wyn Jones[44] when studying the effects of salt on the activity of malic dehydrogenase from barley (Table 6). On the basis of these interactions it was suggested that the stabilization of the enzyme was brought about not by direct 'binding' of the betaine to the enzyme (cf. Schobert[45]) but indirectly via solute-water-biopolymer interactions.

Several examples have now been published in which solutes such as sorbitol, glycinebetaine or proline are found to stabilize enzymes against perturbation by high temperature or non-optimum pH as well as salts.[46] An example of such evidence for glutamine synthetase from *Triglochin maritima* is presented in Table 7.[47] In this

Table 6. Comparison of the effects of glycinebetaine and close analogues on the salt inhibition of respiration and barley malic dehydrogenase

Compound	Formulae	% of inhibited value	
		MDHase[a]	Respiration[b]
Glycinebetaine	$(CH_3)N_3^+CH_2COO^-$	167	200
Dimethylglycine	$(CH_3)_2HN^+CH_2COO^-$	125	170
Sarcosine	$CH_3H_2N^+CH_2COO^-$	84	150
Glycine	$H_3N^+CH_2COO^-$	80	110
ß-Alaninebetaine	$(CH_3)_3N^+CH_2CH_2COO^-$	N.D.	150
Dimethylsulfonioacetate	$(CH_3)_2S^+CH_2COO^-$	N.D.	250
Tetramethylammonium chloride	$(CH_3)_4N^+Cl^-$	N.D.	100
Choline-O-sulphate	$(CH_3)_3N^+CH_2CH_2OSO_3^-$	121	N.D.

[a] Data from ref. 44: barley leaf malic dehydrogenase inhibited by addition of 300 mol m^{-3} NaCl and organic solute added at 500 mol m^{-3}.

[b] Data from ref. 43: respiration of halotolerant bacteria subjected to 1800 mol m^{-3} KCl and solutes added at 60 mol m^{-3}.

N.D. = not determined.

work sorbitol was the most effective 'stabilizer' while proline was the least effective. Second solute effects on the temperature-, pH- and salt-induced perturbations were noted.

An additional factor in this discussion is the realization that the solutes considered in this context as compatible solutes are either identical to or very close structural relatives of chemicals used in cryoprotection[48,49] and in the commercial stabilization of enzymes. Some data on the comparative efficacy of glycinebetaine and sucrose in cryoprotection is reproduced in Figure 3. Many enzymes are supplied commercially in glycerol - the classical 'compatible solute' of extremely halophytic micro algae!

These solutes also interact with membranes. The original report that glycinebetaine helped to stabilize

Table 7. Influence of temperature on pH on the stability of glutamine synthetase II

Treatment	Activity decay constant: Kd (h^{-1})[a]			
	Control	+Glycine betaine 600 mol m^{-3}	+Proline 600 mol m^{-3}	+Sorbitol 600 mol m^{-3}
PH 8.5 (30°C)	0.12	0.12	0.13	0.13
pH 6.0 (30°C)	1.64	1.00	1.19	0.88
pH 8.5 (35°C)	0.30	0.20	0.21	0.14
pH 8.5 (40°C)	1.09	0.83	0.71	0.42

[a] Data from ref. 47. An increasing numerical value indicates a more rapid loss of enzyme activity.

isolated chloroplasts has not been sustained in later experiments[50] but Jolivet et al.[51] have found that the integrity of the membranes of red beet discs subjected to temperature stress and oxalate destabilization was improved by glycinebetaine. Other work indicates rather more specific interactions between glycinebetaine and proline and the fluxes of specific ions across membranes. Early experiments on the effects of exogenously applied glycinebetaine on salt-stressed cereal seedlings showed that betaine-loading affected Na^+ and Cl^- transport.[52] Later work indicated that there was a rather specific effect on Na^+ fluxes across the tonoplast.[52] Recent studies on cultured barley embryos subjected to NaCl and other ionic stresses revealed that proline fed to the embryos from agar media significantly effected Na^+ and Cl^- loading at the xylem-symplasm boundary. However it was not possible to show whether this was a direct effect of proline on the mechanism of ion transport or an indirect effect on solute distribution in the roots, which caused a lower symplasmic Na^+ and Cl^- level.[53]

CONCLUDING REMARKS

This chapter has sought to cover only certain aspects of the phytochemistry of osmotic adaptation; other equally important areas such as membrane lipid composition and the nature of ion pumping mechanisms have been neglected. Biophysical topics related to cell wall elasticity and water relations have been ignored altogether. Nevertheless, it is important to emphasize that both the mainte-

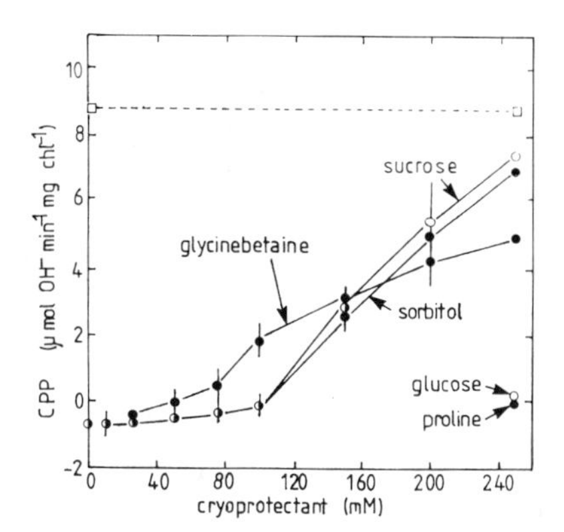

Fig. 3 Effect of freeze/thawing on thylakoid photophosphorylation (CPP). Thylakoids were frozen for 3h at -20°C in 100 mM NaCl, 0-250 mM cryoprotectant. Afterwards the vials were thawed at 20°C and stored on ice. Cyclic photophosphorylation was estimated as OH^- formation according to

$$ADP^{3-} + HPO_4{}^{2-} \longrightarrow ATP^{4-} + OH^-$$

Data from reference 49.

nance of the correct ionic and solute environment for biochemical activity and the adjustment of water and carbon balances and cellular osmotic and turgor pressures are equally important in successful adaptation.

The role of organic cytosolutes is of interest for several reasons. There is by now little doubt that the accumulation of solutes such as glycinebetaine is part of the mechanism of salt tolerance in specific taxa. Furthermore, it is a factor that can potentially be manipulated either by conventional breeding techniques or by the newer methods of genetic engineering.[54] The presence of two separate biosynthetic pathways - a soluble pathway in the typically salt-tolerant family, the Chenopodiaceae, and one that is dependent on membrane turnover in the less tolerant barley - invites further study.[31]

The stabilizing effects of these cytosolutes are attracting increasing interest from the viewpoint of ecophysiology,[55] as probes of enzyme structure and because of their possible commercial importance. In assessing this development it must be remembered that in the two studies[44,56] where the effects of glycinebetaine on the range of salt-inhibited enzymes were considered, only four out of six enzymes showed appreciable protection. In the crucial process of protein synthesis there is no evidence that inhibition of translation of supra-optimal ion concentrations is ameliorated by glycinebetaine or other solutes.[18] While studying the temperature-sensitivity of bovine glutamate dehydrogenase Pollard[57] observed stabilization by glycinebetaine but an examination of the solute-specificity of the effect showed simple salts, e.g. KCl, to be more effective. Care must therefore be taken in interpreting the results of these experiments and it cannot be assumed that all enzymes will react in the same way.

Although there is much to learn about the physico-chemical reactions underlying compatibility, some important clues have emerged. Schobert[45] postulated that solutes such as proline and glycinebetaine did not owe their physiological function to an ability to generate a high osmotic pressure or low water activity (a_w) without concomitant damage to the biochemical machinery as discussed in this chapter. Rather she suggested that proline 'bound' to protein and in turn bound on further water thus more effectively maintaining the hydration of the proline (or glycinebetaine) - enzyme complex and preventing water loss by non-osmotic forces. It is now clear that, although betaine and proline can stabilize some enzymes against 'perturbations', the mechanism proposed by Schobert is mistaken.

Timasheff et al[58] showed for a number of structure-stabilizing solutes e.g. glycerol, sucrose, and structure-stabilizing amino acids and ammonium sulfate, that they are excluded from the hydration sphere of proteins as defined by a negative preferential interaction parameter $(\delta m_3/\delta m_2)m_3$. Destabilizing solutes, urea etc. have a positive value. Similarly glycinebetaine does not bind to proteins (Fig. 4) and is also probably excluded from their hydration sphere.[59] It appears therefore that compatible solutes although interacting themselves with solvent water and lowering a_w do so in such a way as <u>not</u> to compete with the hydration shell of the proteins and presumably also of membranes. This emphasizes that it is

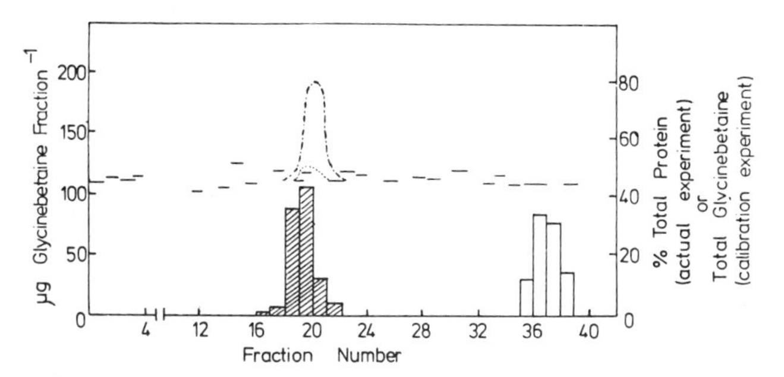

Fig. 4 The equilibrium of filtration profile from a Sephadex G-25 column, equilibrated with 1mM glycinebetaine, during elution of 1 mg BSA in 1 mM glycinebetaine. BSA elution peak (▨), glycinebetaine elution peak (□) when solute (0.1 mmole) run independently. The lines above protein peak represent the hypothetical glycinebetaine concentrations that would result from the binding of 1 mol (.....) and 10 moles (·—·—·—·) of glycinebetaine per mole of BSA. Actual glycinebetaine values from column eluate shown as (-). Data from reference 59.

not the activity of the cytoplasmic water itself that is crucial to the maintenance of biochemical integrity[60] but the nature of the thermodynamic state of that water and its interactions with the biopolymers. Such interactions are obviously important in a wide range of plant responses to stress and their definition in unequivocal physico-chemical terms would be a major advance.

NOTES ADDED IN PROOF

Recent work by Popp[61,62] on Australian mangroves has identified pinitol as a major solute in leaves of all members of the Rhizophoraceae while quebrachitol was found in *Excoecaria agollocha* (Euphorbiaceae). Mannitol was the dominant organic solute in *Aegiceras corniculatum* (Myrsinaceae), in *Lumnitzera* species (Combretaceae) and *Scyphiphora hydrophyllacea* (Rubiaceae). Glycinebetaine was accumulated in *Avicennia* species (see also 33), *Heritiera littoralis* (Sterculiaceae) and *Hibiscus tiliaceus* (Malvaceae). Elevated proline levels were found in *Xylocarpus* species (Meliaceae) while both nitrogenous solutes were found in *Acanthus ilicifolius* (Acanthaceae).

ACKNOWLEDGEMENTS

I am grateful to Drs. John Gorham and Elizabeth McDonell for valuable discussions on this topic, to Professor George Stewart, Dr. Marianne Popp and Dr. Richard Storey for preprints of their manuscripts and to the Overseas Development Administration for their support of our work on this topic.

REFERENCES

1. AHMAD I, SJ WAINWRIGHT 1976 Ecotype differences in leaf surface properties of Agrostis stolonifera from salt marsh, spray zone inland habitats. New Phytol 76: 361-366
2. HUMPHREYS MO 1982 The genetic basis of tolerance to salt spray in populations of Festuca rubra L. New Phytol 91: 287-296
3. HOFFMAN GH, EV MAAS, SL RAWLINS 1975 Salinity-ozone interactive effects on alfalfa yield and water relations. J Environ Qual 4, 326-331
4. WAISEL Y 1972 Biology of Halophytes. Academic Press, New York
5. FLOWERS TJ, PF TROKE, AR YEO 1977 The mechanism of salt tolerance in halophytes. Annu. Rev. Plant Physiol. 28: 89-121
6. MUNNS R, H GREENWAY, GO KIRST 1983 Halotolerant eukaryotes. In OL Lange, PS Nobel, CB Osmond, H Ziegler, eds, Encyclopedia of Plant Physiology, New Series. Vol 12C Springer-Verlag, Berlin pp 60-135
7. WYN JONES RG 1981 Salt Tolerance. In CB Johnson, ed, Physiological Processes Limiting Plant Productivity. Butterworth Press, London, pp 271-292
8. YEO AR 1983 Salinity resistance: Physiologies and prices. Physiol. Plant 58, 214-222
9. WYN JONES RG, R STOREY, A POLLARD 1977 Ionic and osmotic regulation in plants particularly halophytes. In M Thellier, A Monnier, M Demarty, J Dainty, eds, Transmembrane Ionic Exchange in Plants. Colloques Internat. C.N.R.S. Paris, pp 537-544
10. WYN JONES RG, R STOREY, RA LEIGH, N AHMAD, A POLLARD 1977 A hypothesis on cytoplasmic osmoregulation. In E Marrè, O Ciferri, eds, Regulation of Cell Membrane Activities in Plants. North Holland, Amsterdam pp 121-136

11. STEINBACH HB 1962 The importance of potassium. Perspect Biol Med 5, 338-355
12. YANCEY PH, ME CLARKE, SC HAND, RD BOWLES, GN SOMERO 1982 Living with water stress: evolution of osmolyte systems. Science 217, 1214-1222
13. LUBIN M, HL ENNIS 1964 On the role of intracellular potassium in protein synthesis. Biochim Biophys Acta 80: 614-631
14. WEBER LA, ED HICKEY, PA MARONEY, C BAGLIONI 1977 Inhbition of protein synthesis by Cl-. J Biol Chem 252, 4007-4010
15. WYN JONES RG, CJ BRADY, J SPEIRS 1979 Ionic and osmotic relations in plant cells. In DL Laidman, RG Wyn Jones, eds, Recent Advances in the Biochemistry of Cereals. Academic Press, London, New York pp 63-104
16. WYN JONES RG, A POLLARD, 1983 Proteins, enzymes and inorganic ions. In A Läuchli, RL Bieleski eds, Encyclopaedia of Plant Physiology, New Series Vol 15, Springer Verlag, Berlin, pp 528-562
17. FLOWERS TJ, A LÄUCHLI 1983 Sodium versus potassium: substitution and compartmentation. In A Läuchli, RL Bieleski, eds, Encyclopaedia of Plant Physiology, New Series Vol. 15, Springer Verlag, Berlin, pp 651-681
18. GIBSON TS 1983 Stability and Function of Plant Ribosomes in Cytoplasmic Solutes. Ph.D. Thesis, Macquarie University Sydney
19. SETTER TL, H GREENWAY 1979 Growth and osmoregulation of *Chlorella emersonii* in NaCl and neutral osmotica. Aust J Plant Physiol 6: 47-60 and corrigendum in Aust J Plant Physiol 6: 569-572
20. HARVEY DMR, JL HALL, TJ FLOWERS, B KENT 1981 Quantitative ion localization with *Suaeda maritima* leaf mesophyll cells. Planta 151: 555-560
21. LEIGH RA, N AHMAD, RG WYN JONES 1981 Assessment of glycinebetaine and proline compartmentation by analysis of isolated beet vacuoles. Planta 153: 34-41
22. JESCHKE WD 1980 Cation selectivity and compartmentation; involvement of protons and regulation. In RM Spanswick, WJ Lucas, J Dainty, eds, Plant Membrane Transport: Current Conceptual Issues. Elsevier Amsterdam, pp 17-28
23. GORHAM J, RG WYN JONES 1983 Solute distribution in *Suaeda maritima*. Planta 157: 344-349

24. STOREY R, MG PITMAN, R STELZER, D CARTER 1983 X-ray microanalysis of cells and cell compartments of Atriplex spongiosa. 1. Leaves. J Expt Bot 34: 778-794
25. STOREY R, MG PITMAN, R STELZER 1983 X-ray microanalysis of cells and cell compartments of Atriplex spongiosa. 2. Roots. J Expt Bot 34: 1196-1206
26. RAVEN JA 1977 H^+ and Ca^{2+} in phloem and symplast: relation of the immobility of the ions to the cytoplasmic nature of the transport paths. New Phytol 79: 465-480
27. PITMAN MG, A LÄUCHLI, R STELZER 1981 Ion distribution in roots of barley seedings measured by electron probe X-ray microanalysis. Plant Physiol 68: 673-679
28. LEIGH RA, RG WYN JONES 1984 A hypothesis relating critical potassium concentrations for growth to the distribution and functions of this ion in the plant cell. New Phytol In press
29. PALEG LG, D ASPINALL eds 1981 Physiology and Biochemistry of Drought Resistance in Plants. Academic Press New York
30. STEWART GR, F LARHER 1980 Accumulation of amino acids and related compounds in relation to environmental stress. In BJ Miflin ed, Biochemistry of Plants Vol. 5 Amino Acids and Derivatives. Academic Press, London, New York, pp 609-635
31. HANSON AD, WD HITZ 1982 Metabolic responses of mesophytes to plant water deficits. Annu Rev Plant Physiol 33, 163-203
32. STOREY R, RG WYN JONES 1979 Responses of Atriplex spongiosa and Suaeda monoica to salinity. Plant Physiol 63, 156-162
33. WYN JONES RG, R STOREY 1981 Betaines. In ref. 29, pp 171-204
34. GORHAM J, R STOREY unpublished data
35. AHMAD I, F LARHER, GR STEWART 1979 Sorbitol, a compatible organic solute in Plantago maritima. New Phytol 82: 671-678
36. FORD CW 1982 Accumulation of O-methyl inositols in water stressed Vigna species. Phytochemistry 21: 1149-1151
37. GORHAM J, LL HUGHES, RG WYN JONES 1981 Low molecular weight carbohydrates in some salt-stressed plants. Physiol Plant 53: 27-33

38. KUSHNER DJ 1978 Life in high salt and solute concentrations: halophilic bacteria. In DJ Kushner, ed, Microbial Life in Extreme Environments Academic Press, London New York pp 318-368

39. GALINSKI EA, HG TRÜPER 1982 Betaine, a compatible solute in the extremely halophilic phototrophic bacterium Ectothiorhodospira halochloris. FEMS Microbiol Lett 13: 357-360

40. MOHAMMAD FAA, RH REED, WDP STEWART 1983 The halophilic cyanobacterium, Synechocystis DUN52: its osmotic responses. FEMS Microbiol Lett 16: 287-290

41. TURNER NC, MM JONES 1980 Turgor maintenance by osmotic adjustment: A review and evaluation. In NC Turner, PJ Kramer, eds, Adaptation of Plants to Water and High Temperature Stress, Wiley, New York pp 87-104

42. LEIGH RA, T AP REES, WA FULLER, J BANFIELD 1979 The location of acid invertase and sucrose in vacuoles of storage roots of beetroot (Beta vulgaris L.). Biochem J 178: 539-547

43. SKEDY-VINKLER C, Y AVI-DOR 1975 Betaine: Induced stimulation of respiration of high osmolarities in halotolerant bacterium. Biochem J 150: 219-226

44. POLLARD A, RG WYN JONES 1979 Enzyme activities in concentrated solutions of glycinebetaine and other solutes. Planta 144: 291-298

45. SCHOBERT B 1977 Is there an osmotic regulatory mechanism in algae and higher plants? J Theor Biol 68: 17-26

46. PALEG LG, TJ DOUGLAS, A VAN DAAL, DB KEECH 1981 Proline, betaine, and other organic solutes protect enzymes against heat inactivation. Aust J Plant Physiol 8: 107-114

47. AHMAD I, F LARHER, AF MANN, SF McNALLY, GR STEWART 1982 Nitrogen metabolism of halophytes. IV. Characteristics of glutamine synthetase from Triglochin maritima. New Phytol 91: 585-595

48. WITHERS LA, PJ KING 1979 Proline: A novel cryoprotectant of the freeze preservation of cultured cells of Zea mays L. Plant Physiol 64: 675-678

49. COUGHLAN SJ, U HEBER 1982 The role of glycinebetaine in the protection of spinach thylakoids against freezing stress. Planta 156: 62-69

50. LARKUM AWD, RG WYN JONES 1979 Carbon dioxide fixation in chloroplasts isolated in glycinebetaine: a putative cytoplasmic osmoticum. Planta 145: 393-394
51. JOLIVET Y, J HAMELIN, F LARHER 1983 Osmoregulation in halophytic higher plants: the protective effects of glycinebetaine and other related solutes against oxalate destabilization of membranes in beet root cells. Z Pflanzenphysiol 109: 171-180
52. AHMAD N 1978 Glycinebetaine phytochemistry and metabolic functions in plants. Ph.D Thesis University of Wales, Cardiff
53. LONE IM 1983 Salt relations in cultured embryos and seedlings of barley. Ph.D. Thesis University of Wales, Cardiff
54. LE RUDULIER D, RC VALENTINE 1982 Genetic engineering in agriculture osmoregulation: Trend Biochem Sci 7: 431-433
55. SMIRNOFF N, GR STEWART 1984 Stress metabolites and their role in coastal plants. Vegetatia, in press
56. FLOWERS TJ, JL HALL, ME WARD 1978 Salt tolerance in halophyte *Suaeda maritima* (L.) Dum. Properties of malic enzymes and PEP carboxylase. Ann Bot (London) 42: 1065-1074
57. POLLARD AS 1979 Glycinebetaine and enzyme activity. Ph.D. Thesis University of Wales, Cardiff
58. TIMASHEFF SN, T ARAKAWA, H INOUE, K GEKKO, MJ GORBUNOFF, JC LEE, GC NA, EP PITTZ, V PRAKASK 1982 The role of solvation in protein structure stabilization and unfolding. *In* F Franks, S Mathias eds, The Biophysics of Water Wiley, London pp 48-51
59. WYN JONES RG, A POLLARD 1982 Towards the physical chemical characterisation of compatible solutes. *In* F Franks, S Mathias eds, The Biophysics of Water Wiley, London pp 335-339
60. FRANKS F 1982 Water activity as a measure of biological viability and quality control. Cereal Foods World, 403-407
61. POPP M 1984 Chemical composition of Australian mangroves II Low molecular weight carbohydrates. Z. Pflanzenphysiol In press
62. POPP M, F LARHER, P WEIGEL ibid III Free amino acids, total methylated onium compounds and total nitrogen. Z Pflanzenphys In press

Chapter Four

THIGMOMORPHOGENESIS: CALLOSE AND ETHYLENE IN THE HARDENING OF MECHANICALLY STRESSED PLANTS

MORDECAI J. JAFFE AND FRANK W. TELEWSKI*

Biology Department
Wake Forest University
Winston-Salem, North Carolina 27109

INTRODUCTION

Thigmomorphogenesis encompasses the group of phenomena which includes the growth and developmental responses of plants to various mechanical perturbations (MP) such as the wind. In general, most plants that react to MP have their elongation retarded and their stems thickened (Fig. 1).[1] This is usually accompanied by an increase in respiration and a decrease in photosynthesis.[2,3] This laboratory has previously shown that of the morphological events, the thickening response is directly mediated by, and the decrease in elongation is indirectly mediated by the phytohormone ethylene.[4,5] We have also been able to demonstrate that when plants are subjected to MP, they are thereby hardened to mechanical injury by other, subsequent mechanical stress (Fig. 2).[6] It is also of important adaptive advantage to the plant that when MP-plants are subjected to other stresses such as freezing or drought, they are less susceptible to injury than control plants.[7,8] In other words, MP can harden plants to reduce strains resulting from other kinds of stresses.

*Current Address: Laboratory of Tree Ring Research, University of Arizona, Tucson, Arizona.

Fig. 1 Thigmomorphogenesis in young plants of Cherokee wax beans (left) and Loblolly pine (right). In each case the non-perturbed controls (C) are the taller plants, and the mechanically perturbed (MP) plants are the shorter plants.

One of the responses of plants to MP that has only been reported anecdotically, is the deposition of the 1,3-β-glucan, callose.[9] This polysaccharide is known to occur in the phloem in response to some stresses.[9] In this paper, we shall develop the thesis that callose and ethylene are both necessary and causal mediators of thigmomorphogenesis, and that via the production of thigmomorphogenetic ethylene, the plants can be hardened to resist other stresses.

THE ROLE OF ETHYLENE

In order to show that a physiologically active growth substance mediates a particular plant response, several requirements must be met. First, the native titer of the substance should be shown to change during, or better still, before the reaction takes place. Second, exogenous application of the growth substance should mimic the physiological effect of the environmental cue. Third, known inhibitors of the synthesis or action of the growth substance should also inhibit the reaction. Fourth, known promotors of the synthesis or action of the growth substance should also promote the reaction. Of course,

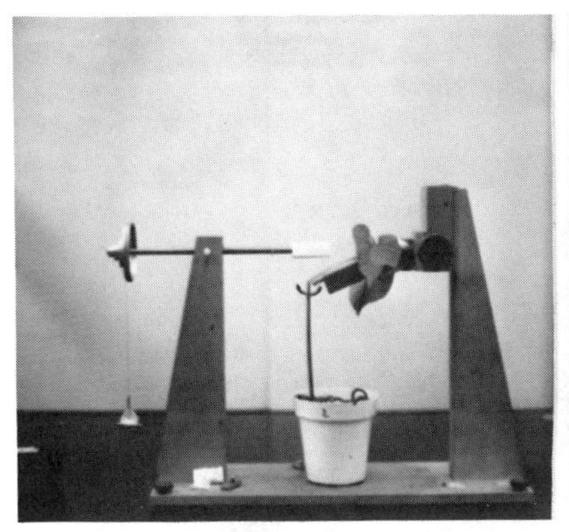
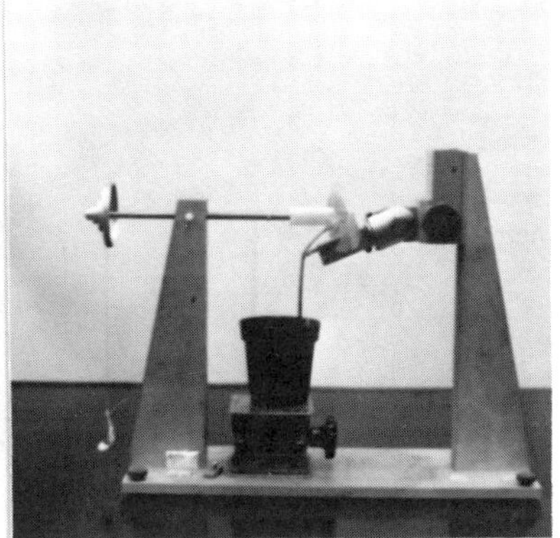

Fig. 2. Mechanical rupture of control bean plants (left) but not of MP-plants (right) subjected to a lateral force load of 1.41 N in the thigmostimulator.

the experimental implementation of these requirements may be very difficult, but if they cannot be met, the probability must be faced that the growth substance does not mediate the reaction. In the case of thigmomorphogenetic stem thickening in beans, all four requirements have been met. Figure 3 shows that bean plants that are given MP begin to evolve ethylene after a 40 to 60 min lag period, that ethylene production peaks after 2 to 3 h, and that it decays to ground level by 5 to 6 h. 1-Aminocyclopropanecarboxylic acid (ACC), the biosynthetic precursor to ethylene, also is induced in the bean plant by MP, and is measurable after a briefer lag period. If exogenous ethephon ((2-chloroethyl)phosphonic acid) or ACC are added to the bean stem, they cause both of the aspects of thigmomorphogenesis (Table 1). Certain varieties of tomato, which elongate less, but do not thicken following MP, also produce little or no ethylene (Table 1). Half-sibs of loblolly pine that produce MP-induced ethylene, show a much longer time course than do beans, (with a 2-4 h lag compared to a 30-60 min lag for beans) (Fig. 3). Certain half-sibs of loblolly pine, which do not thicken following MP, also produce no ethylene (Table 1). The inhibitors of ethylene synthesis and action, aminoethoxyvinylglycine (AVG) and $CoCl_2$, both block MP induced stem thickening, but have much less an effect on the retardation of stem elongation (Table 1). High concentrations of exogenous auxin, which are known to activate the enzyme

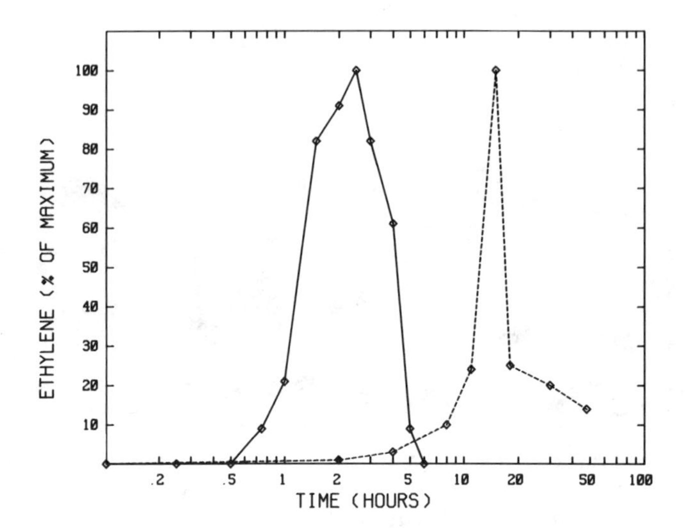

Fig. 3. Time courses of MP-induced ethylene production in stems of bean (◇——◇) and pine (◇---◇). The data points represent percentages of the maximal ethylene evolution (100%) for each plant.

ACC synthase and cause the production of ethylene, mimic the effect of rubbing and of exogenous ethylene on growth (Table 1). Together, these observations strongly indicate that ethylene is the native mediator of the stem thickening effect of thigmomorphogenesis but the evidence is not strong enough to conclude that it directly mediates the retardation of elongation.

The ethylene that is produced as the result of MP is not "wound" ethylene[10] but represents an increase in endogenous ethylene. We have shown this in two ways. First, MP-induced ethylene has the same time course, but is about twice as great in amount as "wound" ethylene when measured in excised tissue (Table 2).[4] Second, when the wound ethylene arising from the cut ends of an excised segment is accounted for, the ethylene produced by that segment exceeds that amount (Table 2). We have previously shown by grafting experiments that when the lower part of a bean plant is given MP, elongation of the upper, unstimulated internodes is also retarded.[11] We have now been able to demonstrate that when the first internode, for example, is rubbed, both the subtending hypocotyl and the

Table 1. Species and varietal differences in growth and ethylene evolution responses to MP. The effects of various inhibitory and promotive conditions in beans are also given.

			% of Control					
			Elongation		Thickening		Ethylene Evolution	
Plant	Variety	Additive	Control	MP	Control	MP	Control	MP
Bean	Cherokee Wax	None	100	48	100	143	100	195
		IAA (10μM)	64	77	63	--	--	--
		ACC (5mM)	56	51	165	161	364	--
		Hypobaria	93	71	--	--	--	--
		AVG (0.5 mM)	91	44	100	104	35	--
		$CoCl_2$ (0.5 mM)	88	47	108	118	42	--
Tomato	Hosen	None	--	--	100	103	100	111
	Alcobaca	None	--	--	100	150	100	253
Loblolly Pine	Half-sib #8-61	None	100	72	100	97	100	35
	Half-sib #8-27	None	100	70	100	110	100	231

supertending second internode produce first ACC and then ethylene (Table 2). These experiments suggest that integrative communication takes place within the plant following MP.

The nature of this communication is, at present, unknown. However, another kind of experiment may suggest the possible mode of action. Dr. Moshe Huberman, in this laboratory, has shown that within 30 sec of the time that MP is given to the first internode, the titer of reducing sugars in the roots decreases by 50%. Since the roots are 6 to 8 cm from the rubbed internode, phloem or xylem transport does not seem fast enough to account for the communication. Action potentials have been recorded in plants, but not having that kind of velocity.[12] Hence another explanation should be sought. It seems possible that if hydrostatic pressure in the vascular bundles were to be suddenly changed in the rubbed internode, then that change must be instantaneously transmitted to every other level of the vascular column. It is such a possibility that will be explored in the next section.

THE ROLE OF CALLOSE

As we have mentioned above, it is part of the botanists anecdotal lore that when a plant is handled, or otherwise mechanically perturbed, the 1,3-β-glucan, callose, is formed, probably in the phloem sieve cells.[9] Callose can be visualized in sectioned or squashed tissue because of its bright yellow-green fluorescence when reacted with alkaline aniline blue.[9] We have developed a method for quantifying callose using the TOPOGRAPHER II program of our digital image processing system, "DARWIN".[13] We have studied this effect carefully in beans and confirmed its occurrence.[14] One of the important aspects of this study is a detailed time course of callose deposition. Figure 4 shows the early time courses of callose appearance for both bean and pine. The production of callose begins well before the production of ethylene and its precursor. The compound 2-deoxy-D-glucose (DDG) is a known inhibitor of protein glycosylation.[15] In our hands, DDG also inhibits the deposition of callose (Table 3). In addition, it also blocks thigmomorphogenetic stem thickening and partially blocks MP-induced ethylene formation (Table 3). Since DDG does not seem to block stem elongation, it seems possible that it does not inhibit cellulose synthesis in this system.

Table 2. The location of ethylene evolution as affected by MP in beans

Treatment	Ethylene Evolution (% of Control) Control	MP
First internode MP'd, ethylene evolution measured in 5 mM segments (all wound ethylene)	100	113
or in 15 mM segments (wound ethylene + endogenous ethylene	113	142
Difference between wound ethylene and total ethylene = endogenous ethylene	13	29
First internode MP'd, ethylene measured evolution at maximum production ("N" minutes)		
First internode MP'd (45 min)	--	143
Petioles Supertending first internode (60 min)	--	167
Second internode (100 min)	--	146

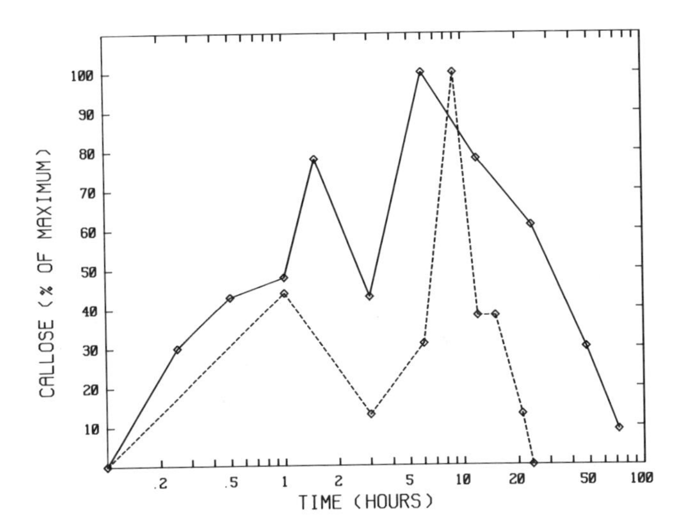

Fig. 4. Time courses of MP-induced callose deposition in stems of bean (◇——◇) and pine (◇---◇). The data points represent percentages of the maximal ethylene evolution (100%) for each plant.

MP also induces callose deposition in a pine half-sib that undergoes MP-induced stem thickening, but not in one that does not (Table 3). In this plant, as in beans, callose deposition precedes ethylene production and both are inhibited by DDG (Fig. 3 and 4, Table 3). These experiments suggest that callose deposition may be a necessary causal part of the thigmomorphogenetic syndrome in beans. If it is, we must ask how it is involved.

There is some evidence that in some plants, callose may block phloem transport at the sieve plates.[9,16] However, in just as many cases, it does not seem to be involved since callose may be formed but phloem transport not effected,[17] or phloem transport may be inhibited but callose may not be deposited.[18] Where it has been examined, callose does not seem to be directly involved in blocking the basipetal transport of photosynthate in beans.[17] However, MP does cause a decrease in phloem transport in beans. Recently, we have employed the integrated tracer kinetics system of Magnuson et al.[19] to observe the effect of MP on the basipetal translocation of [$^{11}CO_2$]photosynthate in bean stems. Figure 5 shows that

Table 3. The effect of 2-deoxy-D-Glucose (DDG) on stem thickening in beans and on callose deposition and ethylene production in beans and pine

	% of Non-DDG Control			
	Bean		Pine	
Measurement	C	MP	C	MP
Stem thickening	100	10	--	--
Callose deposition	108	26	103	82
Ethylene Evolution	138	46	110	64

immediately following MP, the rate of translocation below the point of MP decreases, and above, increases (probably due to accumulation caused by blockage at the point of MP). It may be that the deposition of callose in the phloem, which happens immediately upon MP, may affect the hydrostatic pressure in the phloem. Some evidence for this has been obtained by Dr. Huberman, who found that exudation from cut stems was drastically increased in MP-plants (Table 4). Much research remains to be done in this area before we can be sure if such a causal relationship exists.

Table 4. The effect of MP or ethephon on exudation from the hypocotyl at 2 cm above the root-hypocotyl interface after 5 hours. The plants were MP'd at zero hours

Treatment	Exudation (% of zero h)
Control	93
MP	125
Ethephon	147

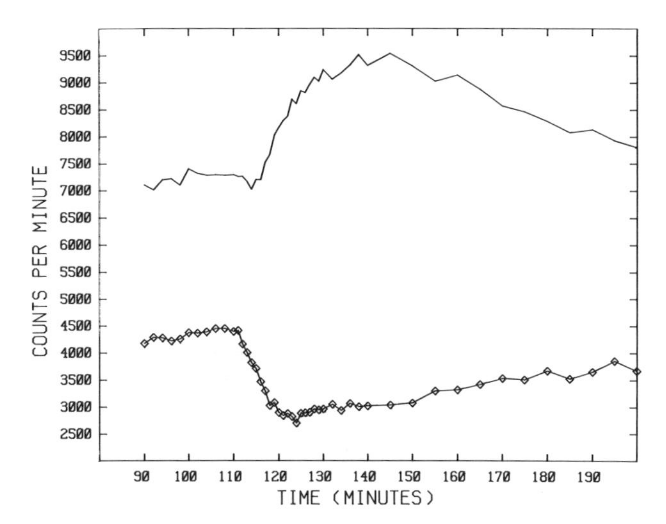

Fig. 5. The effect of brief rubbing (MP) a previously unperturbed first internode on basipetal transport of [$^{11}CO_2$] photosynthate in the first internode above the MP region (———) and in the subtending hypocotyl (◇——◇). The MP was given between 111 and 112 min. The previous level of radioactivity represents the pre-MP baseline.

The fact that callose deposition occurs before ethylene production and that the callose inhibitor DDG also seems to block ethylene production, is not inconsistent with the hypothesis that callose deposition is part of the causal mechanism of ethylene biosynthesis. The arguments for this model are as follows. Callose is deposited on the inside of the cell wall of mature cells, adjacent to the plasma membrane.[9] In many of the systems that have been studied, it is apparent that the callose, as it is deposited, presses against the plasma membrane. The enzyme system that converts ACC to ethylene is known to be found at the cell surface. In fact, it is probable that this enzyme system requires both the plasma membrane and the callose which is being laid down in contact with it. The reason for this is as follows. A.K. Mattoo (personal communication) has shown that isolated protoplasts are incapable of synthesizing ethylene. However, when the protoplasts begin to produce new cell walls, they become capable of synthesizing ethylene. This finding is very pertinent, because several workers have found that when protoplasts begin to synthesize new cell walls, the first polysaccharide that they make is callose.[20] Thus, it is possible that there is something about the callose-plasma membrane interface that is necessary for the production of ethylene from ACC. We are following this line of reasoning, and future experiments will be planned to prove or disprove this hypothesis.

MECHANICAL PERTURBATION-INDUCED HARDINESS

Stress is an environmentally induced change in a plant which causes an injurious strain in the plant, or which causes a strain which is potentially injurious. The nomenclature of stress and strain is derived from the realm of mechanical engineering.[21] When plants are considered in that sense, mechanical stresses may be caused by high winds or by plants being pushed or rubbed by animals or machinery. Such stresses may set up mechanical strains in the plant tissues which may end when the rupture point is reached and the plant breaks. However, slight to moderate mechanical perturbations usually do not result in injurious strains. They do, however, produce changes in the patterns of growth and development of the plant which may, in fact, harden it against injury by subsequent severe stresses.

Mechanical Stress Hardiness

In most vascular plants, thigmomorphogenesis includes a decrease in stem elongation and an increase in stem thickening, caused by some mechanical perturbation.[1] It has been assumed that in nature, the perturbation is usually the wind, and the effect is to harden the plant to further wind induced injury.[1,22] In other words, a plant which has undergone thigmomorphogenesis might be less likely to be injured by subsequently stressful windy conditions. There have been several reports in the literature which suggest that this hypothesis may be correct. Thus, Jacobs[23] found that trees protected from the wind by guying were likely to be blown over by the wind when the wires were removed. It is also well known to foresters that when a stand of trees is protected by other trees that are exposed to the wind, and it loses that protection by logging or other means, a strong subsequent wind can cause their trunks to break. This results in a "windfall".[24] Similarly, Steucek and Gordon[25] reported that wheat plants that were mechanically perturbed by rubbing were less likely to fall over.

This laboratory has studied thigmomorphogenesis in beans and showed that this phenomenon occurs in plants grown in windy areas in the field[26] as well as in the laboratory.[26,27] In order to explain the strengthening effect of MP, it was considered necessary to examine the mechanical properties of MP-plants, and to see how they differed from those of non-perturbed, control plants. The experiments described below are designed to provide that information.

Bean plants were grown for up to 13 days after the first internode reached 2 cm in length. The first internodes of half of the plants were rubbed for a few seconds once daily. The unrubbed plants were the controls. Each day 15 plants of each type were selected. Each plant was placed in the thigmostimulator (Fig. 2),[28] and a lateral force was applied to the upper 2 cm of the internode. The deflection of the internode at each force, together with its reflection when force was removed were measured. A higher force was then applied, and the process was repeated. Force increments were increased until the rupture point of the stem was reached and the stem broke. Figure 6 shows that previously rubbed plants were much more resistant to rupturing by lateral force loading than were the controls. Further computations showed that the increase in mechanical hardiness was due to both a 40% increase in flexibility and a 60% increase in the second moment of area (i.e. inertia) of the stem (Table 5).[6]

There is very little data from other plants to compare to the present observations. S.A. Wainwright and Hebrank (personal communication) have studied the way in which coconut palm fronds escape rupture in gale or hurricane force winds. In this case, the petiole not only bends, but has extraordinarily low torsional stiffness, which allows all of the fronds to twist around to the lee side of the plant and stream in the wind. However, no comparison has been made of sheltered plants with previously MP-plants. Grace and Russell[29] found that MP of the grass *Festuca arundinacea* Schreb. cv. S170 induced

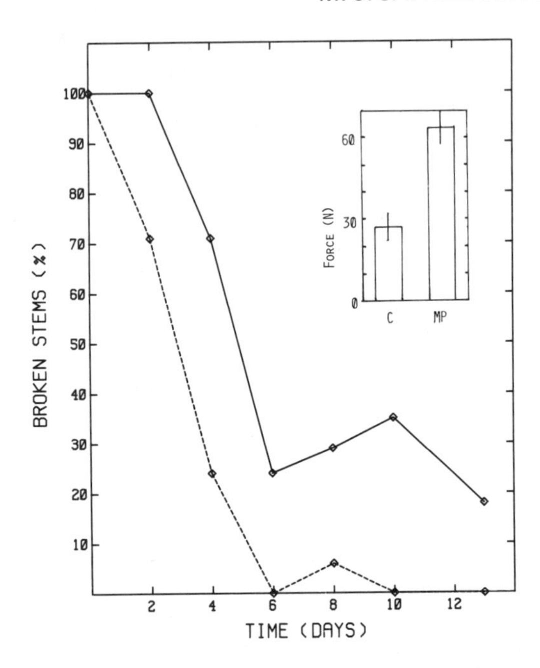

Fig. 6. The effects of previous MP on mechanical stress-induced stem rupture in beans (main graph) and pine (inset). The main graph shows the time courses of the amount of bean stems broken due to a laterally applied force of 1.76 N in the thigmostimulator. Plants were rubbed (◇----◇) or not (◇——◇) for up to 13 days. The inset bar graph shows the lateral force, in a 3-point test, needed to rupture pine stems that had been given MP or not for one growing season (about 7 months).

a 64.5% increase in the modulus of elasticy (i.e. a decrease in flexibility). Using their morphological data for computation, we calculate a 34.8% increase in flexural stiffness of the MP plants over the controls. Hence, the MP-induced hardening to wind in this grass can be accounted for not by an increase in flexibility, but by an increase in flexural stiffness. Using an Instron Universal Tensiometer, Heuchert[30] found that the modulus of elasticity of tomato stem sections also increases (by 7.9%) due to MP. We calculate from this data that the flexural stiffness of the tomato stem decreases by 22.1%. Thus wind hardening cannot be accounted for by either increased flexural stiffness or flexibility of the MP-plants. Heuchert reports that the MP plants had a 10.4% greater modulus of rupture than the control plants. In this case the rupture was caused by lateral shear force (not the kind of force found in nature), but no explanation can yet be offered for this difference on the basis of the mechanical properties of the stem. Thus, this analysis of the MP-induced changes in the mechanical parameters of three species suggests an important caveat: "The changes responsible for hardening during thigmomorphogenesis should not be extrapolated from one species to another." All of these studies suggest that in nature, stems growing taller in a windy environment will become more and more capable of surviving in that environment.

Resistance to Pithiness

It is, perhaps, to be expected that plants growing in a mechanically stressful environment should adapt to that environment. However, several studies have also shown that, at least in some instances, such plants are more resistant to other stresses. For example, Jaffe and Biro[7] have shown that bean plants exposed to drought stress or to freezing stress are better able to recover if they were previously mechanically perturbed (Table 6). Similarly, Suge[31] demonstrated that mechanically perturbed beans were more resistant to drought stress injury than were controls.

Table 5. Summary of the means of the time course values of the various mechanical parameters[21] of bean first internodes and of pine hypocotyls

Parameter	(Control - MP)/Control x 100	
	Bean	Pine
Length (mm)	-32	- 6
Diameter (mm)	+ 8	+34
Plants broken under a force load (% of total)	-65	-55
Elastic resiliance (reflection/ deflection) x 100	+22	--
Plasticity (deflection-reflection/ deflection x 100	-38	
Elastic modules (E)	-39	-36
2d Moment of area (I)	+53	+109
Flexural stiffness (EI)	-30	+18

Table 6. The effects of MP (rubbing and flexing) on the hardiness to non-mechanical stress in bean and pine

Plant	Treatment	Control	MP
Bean	Withold irrigation, then measure new growth 5 days after re-irrigation (weight of new growth, grams)	0.36 ± 0.12	0.77 ± 0.12
Bean	Expose to -20°C for 5 min, then incubate 4 days at 26°C and measure new growth (mm)	2.9 ± 0.8	7.1 ± 0.8
Pine	Expose to overnight frost, then allow to grow for 1 season and compute % of plants with live candles (%)	80 ± 4	94 ± 2

When tomato plants are subjected to drought stress, and then re-irrigated, the stem pith parenchyma undergoes autolysis, and the resulting stem becomes a hollow tube.[8,32] This syndrome is known as "pithiness". We have used the AREA program of DARWIN[13] to quantify the amount of pithiness in this and other systems. We have found that if the internodes of the plants are pretreated with MP, they become resistant to drought induced pithiness (Table 7). Furthermore, pretreatment with ethephon can substitute for MP in inducing hardiness (Table 7). Since rubbed plants produce ethylene, and AVG blocks MP-induced hardening (Table 7), we conclude that the MP-induced ethylene formation mediates the hardening mechanism.

CONCLUSION

From the description of these observations, it is apparent that when a plant is perturbed by some mechanical effector, such as the wind, the ecological effects are many and broad. In general, the effects of such perturbations are beneficial. The plant is not only better adapted to mechanically stressful environments, but may become hardier in a general sense. The effect of wind and other mechanical perturbations, acting through the mediation of endogenous ethylene, should be a factor to be considered when evaluating stress-strain relationships in plants.

Table 7. The effect of MP on stem elongation, ethylene evolution and pithiness and of 5 ppm ethephon (E) on pithiness in alcobaca tomatoes

Treatment and measurement	Control	MP	E
Stem elongation after 6 days of MP or 2 days E (cm)	8.5±0.8	4.3±0.4	---
Peak ethylene production (pica m/g. fr. wt/h)	45	57	---
Pithy Internodes (%)	95±3	6±2	18±1

ACKNOWLEDGEMENTS

This research was supported by National Science Foundation grant PCM 8206560, National Aeronautics and Space Administration grant NSG 7352, and Binational Agric. Res. and Develop. grant I-127-79 to M.J.J. We thank Drs. J.E. Goeschl and B.R. Strain for allowing us to use the CO Integrated Tracer Kinetics System. Special thanks are due to Dr. C.H. Jaeger for expert assistance in the CO tracer experiment.

REFERENCES

1. JAFFE MJ 1973 Thigmomorphogenesis: The response of plant growth and development to mechanical stimulation: with special reference to _Bryonia dioica_. Planta 114: 143-157
2. FERREE DC, FR ALL 1981 Influence of physical stress on photosynthesis and transpiration of apple leaves. J Amer Soc Hort Sci 106: 348-351
3. TODD GW, DL CHADWICK, S-D TSAI 1972 Effect of wind on plant respiration. Physiol Plant 27: 342-346
4. BIRO R 1980 Thigmomorphogenesis: Ethylene biosynthesis and its role in the changes observed in mechanically perturbed bean internodes. Doctoral Dissertation, Ohio University
5. JAFFE MJ, R BIRO 1977 Thigmomorphogenesis: The role of ethylene in wind induced growth retardation. Proc 1977 Plant Growth Regulator Working Group 4: 118-124
6. JAFFE MJ, FW TELEWSKI, P COOKE 198? Thigmomorphogenesis: On the mechanical properties of mechanically perturbed bean plants. Physiol Plant In press

7. JAFFE MJ, R BIRO 1979 Thigmomorphogenesis: The effect of mechanical perturbation on the growth of plants, with special reference to anatomical changes, the role of ethylene, and interaction with other environmental stresses. In H. Mussell, R Staples, eds, Stress Physiology in Crop Plants, John Wiley & Sons Inc., N.Y.
8. PRESSMAN E, M HUBERMAN, B ALONI, MJ JAFFE 1983 Thigmomorphogenesis: The effect of mechanical perturbation and ethrel on stem pithiness in tomato (Lycopersicon esculentum Mill) plants. Ann Bot 52: 93-100
9. ESCHRICH W 1975 Sealing systems in plants. In MH Zimmerman, JA Milburn, eds, Encyclopedia of Plant Physiology New Series Vol 1(I) Springer-Verlag, New York pp 39-56
10. ABELES FB 1973 Ethylene in plant biology. Academic Press, New York London 302 p
11. ERNER Y, R BIRO, MJ JAFFE 1980 Thigmomorphogenesis: Evidence for a translocatable thigmomorphogenetic factor induced by mechanical perturbation of beans (Phaseolus vulgaris). Physiol Plant 50: 21-25
12. ASHER WC 1968 Response of pine seedlings to mechanical stimulation. Nature 217: 134-136
13. TELEWSKI FW, AH WAKEFIELD, MJ JAFFE 1983 Computer-assisted image analysis of tissues of ethrel-treated Pinus taeda seedlings. Plant Physiol 72: 177-181
14. JAFFE MJ, M HUBERMAN, J JOHNSON, FW TELEWSKI 1983 Thigmomorphogenesis: The induction of callose and ethylene by mechanical perturbation in bean stems. Physiol Plant In press
15. DATEMA R, RT SCHWARZ 1979 Interference with glycosylation of glycoproteins Inhibition of formation of lipid-linked oligosaccharides in vivo. Biochem J 184: 113-123
16. MCNAIRN RB, HB CURRIER 1968 Translocation blockage by sieve plate callose. Planta 82: 369-380
17. PETERSON CA, WE RAUSER 1979 Callose deposition and photoassimilate export in Phaseolus vulgaris exposed to excess cobalt, nickel and zinc. Plant Physiol 63: 1170-1174
18. VAN DE VENTER HA, HB CURRIER 1977 The effect of boron deficiency on callose formation and ^{14}C translocation in bean (Phaseolus vulgaris L.) and cotton (Gossypium hirsutum L.). Mer J Bot 64: 861-865

19. MAGNUSON CE, Y FARES, JD GOESCHL, CE NELSON, BR STRAIN, CH JAEGER, EG BILPUCH 1982 An integrated tracer kinetics system for studying carbon uptake and allocation in plants using continuously produced CO. Radiat Environ Biophys 21: 51-65

20. TAKEUCHI Y, A KOMAMINE 1981 Glucans in the cell walls regenerated from *Vinca rosea* protoplasts. Plant Cell Physiol 22: 1585-1594

21. WAINWRIGHT SA, WD BIGGS, JD CURREY, JM GOSLINE 1976 Mechanical Design in Organisms. Princeton Univ Press NJ (ISBN 0-691-08308-8 pbk)

22. LAWTON RO 1982 Wind stress and elfin stature in a montane rain forest tree: An adaptive explanation. Am J Bot 69: 1224-1230

23. JACOBS MR 1954 The effect of wind sway on the form and development of *Pinus radiata* D Don. Aust J Bot 2: 35-51

24. SOMERVILLE A 1981 Wind-damage profiles in a *Pinus radiata* stand. N Z J For Sci 11: 75-78

25. STEUCEK GL, LK GORDON 1975 Response of wheat (*Triticum aestivum*) seedlings to mechanical stress. Bot Gaz 136: 17-19

26. HUNT ER JR, MJ JAFFE 1980 Thigmomorphogenesis: The interaction of wind and temperature in the field on the growth of *Phaseolus vulgaris* L. Ann Bot 45: 665-672

27. JAFFE MJ 1976 Thigmomorphogenesis: A detailed characterization of the response of beans (*Phaseolus vulgaris* L.) to mechanical stimulation. Z Pflanzenphysiol 77: 437-453

28. JAFFE MJ, R BIRO, K BRIDLE 1980 Thigmomorphogenesis: Calibration of the parameters of the sensory function in beans. Physiol Plant 49: 410-416

29. GRACE J, G RUSSELL 1977 The effect of wind on grasses. III Influence of continuous drought or wind on anatomy and water relations in *Festuca arundinacea* Schreb. J Exp Bot 28: 268-278

30. HEUCHERT JC 1981 Characterization of growth response and stem strengthening in mechanically-stressed tomato plants. Master's Thesis Purdue University

31. SUGE H 1980 Dehydration and drought resistance in *Phaseolus vulgaris* as affected by mechanical stress. Rep Inst for Agric Res Tohoku Univ 31: 1-10

32. ALONI B, E PRESSMAN 1981 Stem pithiness in tomato plants: The effect of water stress and roll of abscisic acid. Physiol Plant 51: 39-44

Chapter Five

STRESS AND SECONDARY METABOLISM IN CULTURED PLANT CELLS

FRANK DiCOSMO

Department of Botany
University of Toronto
Toronto, Ontario
Canada, M5S 1A1

and

G. H. N. TOWERS

Department of Botany
University of British Columbia
Vancouver, British Columbia
Canada, V6T 2B1

INTRODUCTION

Grime[1] defines "stress" in plants as the external constraints which limit the rate of dry matter production of all or part of the vegetation, e.g. shortages of water, light, mineral nutrients and suboptimal temperatures. These shortages may be an inherent characteristic of the environment, or they may be induced or intensified by the vegetation itself. The use of the word "stress" to describe an external constraint on dry matter production differs from that of many plant physiologists however, who have used the word to describe the physiological state of the plant. Grime's definition could be applied to cultured plant cells; stress in this case would be any type of constraint which reduces the dry matter of the cultures. This otherwise excellent and precise definition is not very useful, however, to phytochemists who are generally not interested in dry matter production. We will define stress as the external constraints on cultured cells which limit the normal production of secondary metabolites. Let us call this phytochemical stress to be absolutely specific. The stress may involve nutrient, light and temperature regimes or chemical and microbial insult or genetic manipulation.

The use of cultured plant cells should be a powerful tool in providing novel insights into our understanding of phytochemistry, including physiology, enzymology and enzyme regulation. The environment of the culture can be strictly monitored, the effects of microbial associations and edaphic parameters eliminated, nutritional and hormonal levels can be strictly controlled. Limitations in the technique relate to difficulties in initiating cultures of certain species or of obtaining good growth of others, but, most important of all is the fact that cultured plant cells often fail to show the normal pattern of secondary metabolites characteristic of the intact plant. In recent years cell cultures have been used to study phytochemical stress. For example, NAA (naphthalene-acetic acid) has been shown to increase the synthesis of PAL (phenylalanine ammonia-lyase) and of the stress-metabolite (phytoalexin) phaseollin in cell suspensions of _Phaseolus vulgaris_. The addition of 2,4-D (2,4-dichloro-phenoxy-acetic acid), on the other hand, suppressed both PAL and phaseollin synthesis.[2] Sulfydryl agents, including mercuric chloride, induce medicarpin accumulation in _Trifolium repens_ callus[3] and glyceollin accumulates in ultraviolet-irradiated callus of _Glycine max_.[4] A thio-disulfide of _Ricinus communis_ cell suspension inhibits kaurene synthetase, an enzyme in the pathway leading to diterpenoids viz. gibberellins.[5]

The altered metabolism of plant cell cultures with respect to their secondary metabolism when maintained under optimal growth conditions implies that some kind of stress is being applied to the cells. This stress usually results in "switching" off of secondary metabolite production. Production may be "switched" on again by alterations in nutritional or hormonal levels, altered light or temperature regimes and by techniques which do not appear to have as profound effects in the intact plant. By careful manipulation of genetic selection and environmental factors it is possible, however, to obtain yields of secondary products in amounts comparable to, or even exceeding those of the intact plant (Fig. 1).[6-18] Sometimes novel compounds not detected in intact plants are produced in culture (Fig. 2).[19-25]

The usual experience, however, is that many characteristic types of compounds are produced in low amounts, if at all, by cultured plant cells.[26-35] Synthesis of the compounds can sometimes be induced in various ways, some of which involve the application of further stress superimposed on the phytochemical stress of the cell culture system.[28,36-42]

The ability or apparent inability of cultured cells to produce normal patterns of phytochemicals are manifestations of stress induced regulation, such as induction, catabolite repression, and feed-back regulation. The addition of specific activators, elicitors, removal of regulatory repressors, genetic and epigenetic manipulation of enzyme pathways to alter the effects of metabolic regulation may increase the concentrations of enzymes of secondary metabolism many-fold. Unfortunately, the phytochemical effects of a particular manipulation or stress have most often been determined empirically and many diverse growth media have been devised in which to study secondary metabolism; this makes integration of these many various studies exceedingly difficult. The discrepancies observed in various culture-systems with respect to secondary phytochemicals are so inconsistent as to make it obvious that regulation of secondary metabolism is not well understood.

Various aspects of secondary product synthesis by cultured plant cells have been extensively reviewed.[26,31,33,43-51] It is not our intention to repeat that information here, but, some overlap is necessary. In this review we assess the nature of, and some of the causes and effects of various stress-phenomena affecting secondary

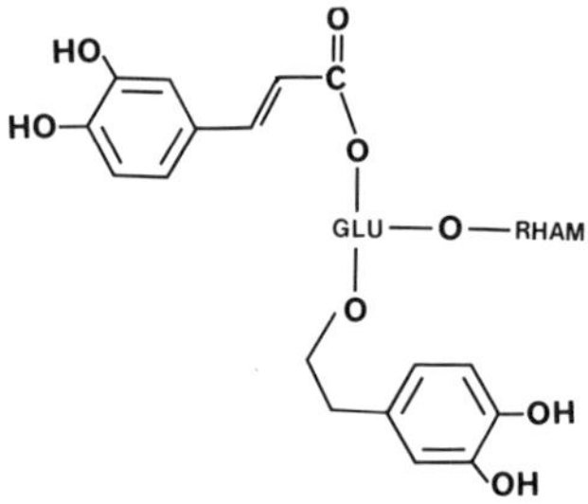

HO
HO
O
C
O
GLU — O — RHAM
O
OH
OH

ACTEOSIDE from Syringa vulgaris

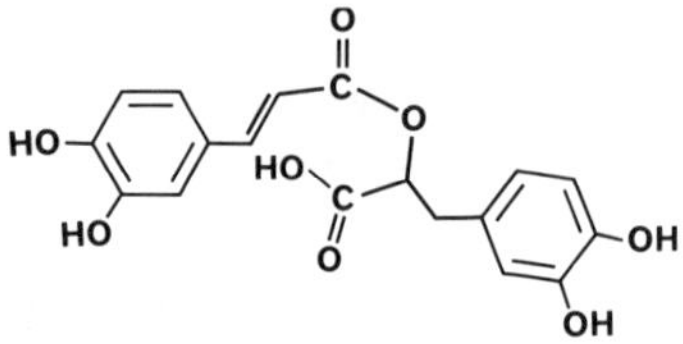

O
C
O
HO
HO
HO
C
O
OH
OH

ROSMARINIC ACID from Coleus blumei

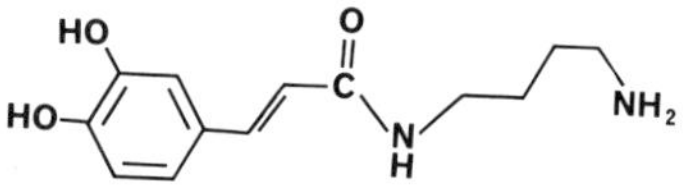

HO
HO
O
C
N
H
NH_2

CAFFEOYL PUTRESCINE from Nicotiana tabacum

N^+
N
H
H
O
H
O=C
OCH_3

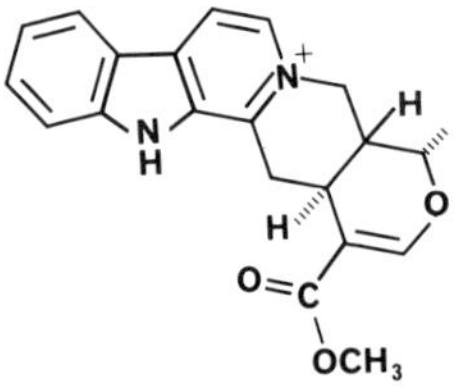

SERPENTINE from Catharanthus roseus

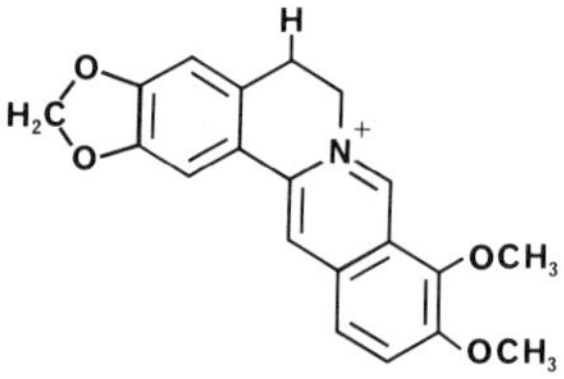

H
O
H_2C
O
N^+
OCH_3
OCH_3

BERBERINE from Thalictrum minus

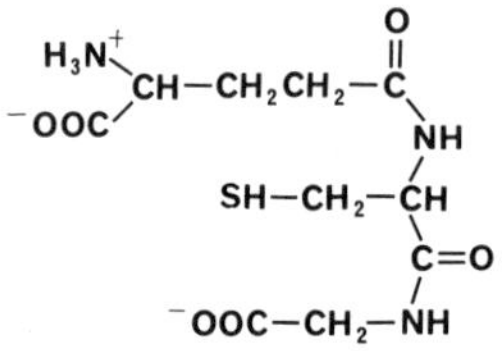

H_3N^+
^-OOC
CH—CH_2CH_2—C
O
NH
SH—CH_2—CH
C=O
^-OOC—CH_2—NH

GLUTATHIONE from Nicotiana tabacum

O
O
HO

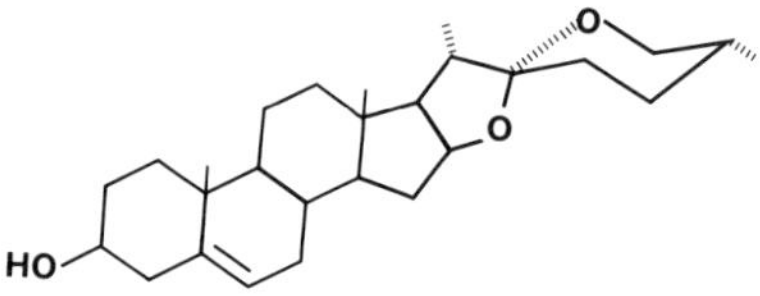

DIOSGENIN from Dioscorea deltoidea

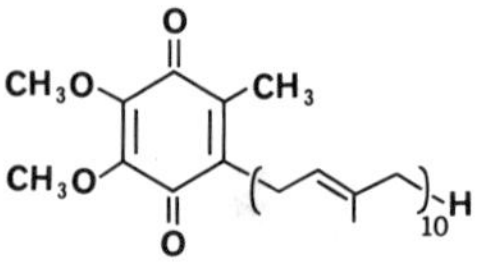

O
CH_3O
CH_3
CH_3O
O
$)_{10}$H

UBIQUINONE-10 from Nicotiana tabacum

O
OH
OH
O

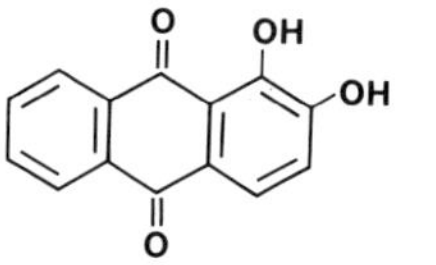

ALIZARIN from Morinda citrifolia

Fig. 1. Compounds produced by cultured plant cells in amounts comparable or exceeding those of the intact plant.

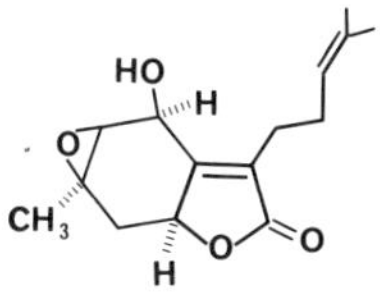

A 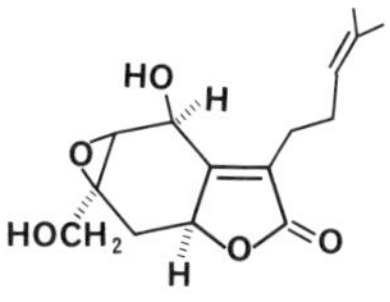 B 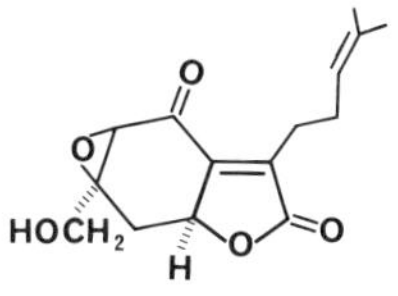 C

PANICULIDES A, B, C from Andrographis paniculata

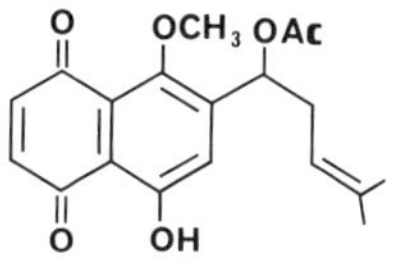

ECHINONE from Echium lycopsis

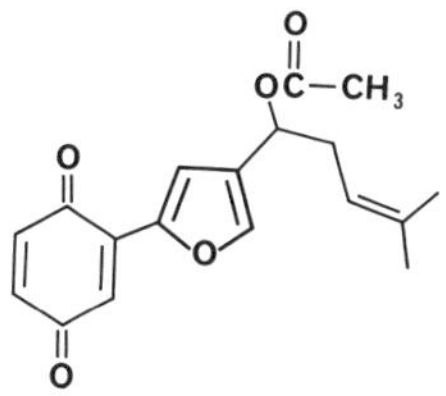

ECHINOFURAN from Echium lycopsis

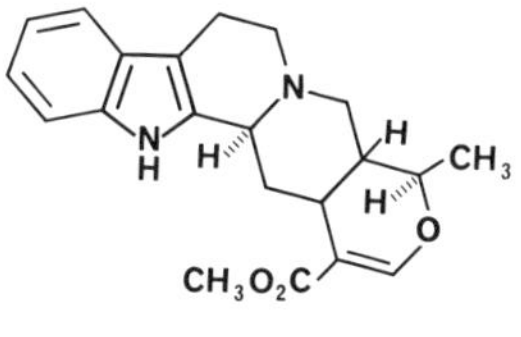

19-epi-AJMALICINE fom Catharanthus roseus

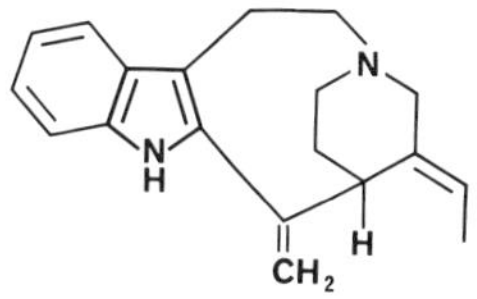

PERICINE from Picralima nitida

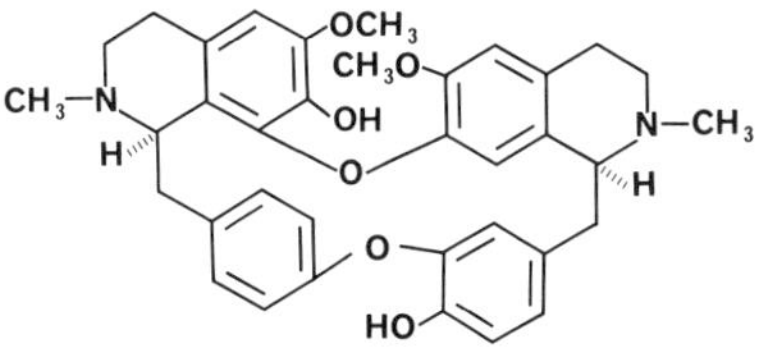

AROMOLINE from Stephania cepharantha

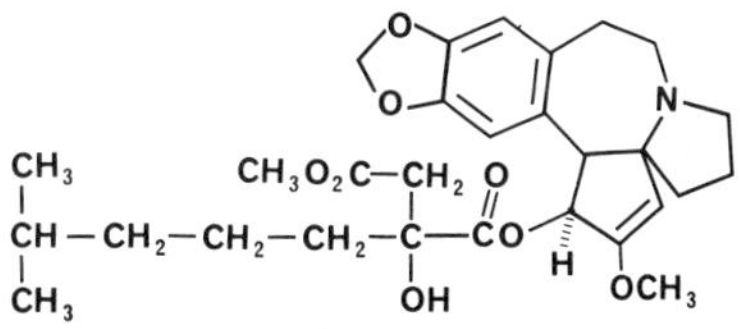

HOMODEOXYHARRINGTONINE from Cephalotaxus harringtonia

Fig. 2. Compounds produced by cultured cells but not detected in the respective intact plant.

metabolism in cultured plant cells, and point to areas of plant biochemistry that offer exciting new areas of research.

EFFECTS OF NUTRITION ON PHYTOCHEMICAL SYNTHESIS

Cultured plant cells are usually supplied with all essential minerals, vitamins, and carbohydrate sources for vigorous growth and active primary metabolism. These cells possess all the biochemical capabilities of the intact plant from which they were derived.

Effects of Carbohydrate Stress on Phytochemical Synthesis

Carbohydrate sources, for example sucrose, are normally supplied in a 2 to 3% concentration to cultured plant cells, and have been shown repeatedly to influence significantly the production of phytochemicals. For example, Zenk et al.[11] found the alkaloid level in cultured cells of *Catharanthus roseus* to fluctuate in response to the growth medium used, especially when the concentration of sucrose was varied. Knobloch et al.[52] found anthocyanin and indole alkaloid synthesis in *C. roseus* cell suspensions to increase when the sucrose concentration was increased from 4% to 10%. The stimulating effect of sucrose was less significant on total phenolic and alkaloid synthesis. However, Tal et al.[53] observed that the nature and concentration of the carbohydrate source had a significant effect on diosgenin production by *Dioscorea deltoidea* cell suspensions. They found that 1.5% sucrose-amended media resulted in greatest diosgenin production relative to fructose-, galactose-, lactose-, or starch-amended media. The cultured cells with the greatest diosgenin productivity were those grown on 3% sucrose-containing media.

Phenolic metabolite accumulation is also influenced by levels of sucrose supplied. Increasing sucrose from 2% to 4% stimulated synthesis of polyphenols in cell suspensions of Paul's Scarlet rose.[54] Westcott and Henshaw[55] reported a three-fold increase in sucrose-stimulated phenolic metabolite accumulation in cell suspensions of *Acer pseudoplatanus* when the carbohydrate concentration was increased from 2% to 4%. Constabel[56] found 5% sucrose to enhance the production of tannins in tissue cultures of

Juniperus communis, Mizukami et al.[57] found the yield of shikonin derivatives in callus cells of Lithospermum erythrorhizon to increase with increasing sucrose from 1% to 5%, and attain constant levels when the sucrose concentration varied from 7% to 10%. However, cell proliferation declined at higher sucrose concentrations. Increasing sucrose levels from 0.3% to 5% stimulated PAL and anthocyanin production in callus of Populus sp., and cell growth was not affected.[58]

Increasing glucose concentration from 0.95% to 3.6% supplied to cell suspensions of Paul's Scarlet rose stimulated phenolic metabolite synthesis.[59] Matsumoto et al.[58] reported that sucrose, glucose, fructose, and raffinose supplied as 1% glucose equivalents were all metabolized for anthocyanin synthesis in Populus sp. cell suspensions. Galactose, lactose, maltose, melezitose, sorbitol, trehalose, and starch had no net anthocyanin-stimulating effect, but reduced growth. In contrast, Ikeda et al.[60] observed insignificant variation in ubiquinone production in Nicotiana tabacum cell suspensions grown on either glucose-, or sucrose-amended media, and found that increased sucrose concentrations (2%-5%) resulted in decreased ubiquinone synthesis.

Altered carbohydrate levels and sources therefore obviously affect secondary metabolism (Table 1).[10,11,27,55-58,60-77] It is now beginning to be recognized that carbon source specificity may be important to the synthesis of certain classes of secondary compounds. Sucrose had the greatest stimulating effect on diosgenin production by Dioscorea deltoidea.[53] Sucrose, glucose, fructose, and raffinose were metabolized for anthocyanin synthesis by Populus sp., but sucrose had the greatest stimulating effect. Starch, however, inhibited growth and had no stimulatory effect,[58] although cultured plant cells are capable of growing well on starch-amended media.[78-79] Lowered sucrose concentration stimulated coumarin synthesis in N. tabacum but synthesis was decreased in response to glucose and fructose.[76]

Edelman and Hanson[77] found sucrose to inhibit chlorophyll synthesis by Daucus carota cells; glucose and fructose alone or in concert had no inhibitory effect. These authors also reported a reduction of carotenoid content in cells grown in the presence of sucrose.

Table 1. Effects of carbohydrate source and level on phytochemical synthesis

Carbohydrate	Phytochemical effect	Species	Reference
Increase glucose	Increase alkaloids	*Choisya ternata*	61
	Increase anthraquinones	*Morinda citrifolia*	10
Remove glucose	Decrease lunularic acid	*Marchantia polymorpha*	62
Add glucose	Increase phylleblin	*Emblica officinalis*	63
	Decrease podophyllotoxin	*Podophyllum peltatum*	64
Add lactose	Induce β-galactosidase	*Nemesia strumosa*	65
Add ribose	Decrease alkaloids	*Peganum harmala*	66
Increase sucrose	Increase phenolics	*Acer pseudoplatanus*	55
	Increase virus inhibitor	*Agrostemma githago*	67
	Increase alkaloids, phenolics	*Catharanthus roseus*	11,52,68,69
	Decrease alkaloids		27
	Decrease anthocyanins		70
	Increase rosmarinic acid	*Coleus blumei*	71
	Increase diosgenin	*Dioscorea deltoidea*	53
	Increase anthraquinones	*Galium mollugo*	72
	Increase tannins	*Juniperus communis*	56
	Increase shikonin	*Lithospermum erythrorhizon*	57
	Decrease ubiquinone	*Nicotiana tabacum*	60
	Increase cinnamoyl putrescines		73
	Increase anthocyanin	*Populus* sp.	58
	Increase solasodine	*Solanum laciniatum*	74
	Decrease ecdysterone	*Trianthema portulacustrum*	75
Decrease sucrose	Increase scopoletin, scopolin	*N. tabacum*	76
	Decrease ecdysterone	*T. portulacustrum*	75
Add sucrose	Inhibit carotenoids, chlorophyll	*Daucus carota*	77

The general consensus is that sucrose is better than all other carbohydrate sources for growth,[80] and concentrations above 3% often enhance the biosynthesis of phytochemicals produced by heterotrophic plant cells. In some systems increased product synthesis can be traced to greater cell proliferation as a consequence of increased sucrose, but this alone cannot explain fully the net phytochemical accumulations observed. Several examples clearly show that this general opinion, i.e. that increased sucrose concentration results in higher product yields, is not tenable. It should be expected, however, that distinct cultured cells grown in the presence of different carbon sources and levels will show altered phytochemical patterns. No broad generalizations can yet be inferred from the many interesting studies surveyed. Several studies have shown that β-galactosidase is induced in cells grown with lactose.[65,81-83] These observations suggest that catabolite repression (which involves the repression of enzymes by the metabolic intermediates of a rapidly consumed carbon source) may exert considerable regulatory control on secondary metabolism. Sacher and Glasziou[84] showed that invertase was repressed by its catabolic products. Demain[85] has provided an excellent overview of catabolite repression of fermentation products in microorganisms, and Drew and Demain[86] reviewed the effects of primary metabolites on secondary metabolism. Carbon source has been shown to influence fatty acid composition in the fungus _Microsporum gypseum_,[87] and the production of secondary metabolites by _Cephalosporium acremonium_.[88] Catabolite repression had strong control in the latter system. Certainly, similar regulatory mechanisms may alter secondary metabolism in cultured plants cells.

The metabolism of carbohydrates by plant cells includes the oxidative pentose phosphate pathway, glycolysis, and the tricarboxylic acid cycle, which ultimately yield the precursors of phytochemicals.[98-91] The pentose phosphate cycle is a source of part of the carbon skeleton of aromatic amino acids and phenolic compounds, and provides a source of reducing potential (NADPH) necessary for biosynthetic reductions.[92,93] The pentose phosphate pathway, glycolysis, and the tricarboxylic acid cycle are interrelated and are feed-back regulated.[89-91,94,95] In plant cells, glycolysis predominates, and the pentose phosphate pathway probably accounts for no more than 30% of the total carbohydrates metabolized.[89] Several reports suggest that there is a close relationship between the pattern of carbohydrate metabolism, and nitrate and phosphate assimilation in plants. For example, although cell proliferation and protein content are limited by the availability of nitrogen, dry and fresh weight yields may also be affected by a complex interaction via carbohydrate availability and catabolism.[76,96] This relationship can lead to increased respiration and a shift in the balance of carbon flow from glycolysis to the pentose phosphate pathway.[97,98]

We conclude that the effects of carbohydrate stress on phytochemical synthesis by cultured plant cells are complex, and partially dependent on the other nutritional factors present in the culture system. More penetrating studies should be directed toward study of the regulation of secondary metabolism by carbohydrates. Such studies will undoubtedly uncover novel regulatory controls.

Effects of Nitrogen Stress on Phytochemical Synthesis

Standard plant cell culture media usually provide nitrogen in the form of a mixture of nitrate, and ammonia. However, some cultured plant cells cannot tolerate the often excessive levels of ammonia provided.[99,100] It has been shown repeatedly that the composition of cultured cells grown with nitrate differs from that of cells grown on ammonia containing media.[101-103]

Nitrogen metabolism proceeds essentially through the phases reviewed by Miflin & Lea[104,105] and outlined by the simple model provided (Figure 3).

Davies[54] found increasing levels of nitrate from 10 mM to 20 mM resulted in decreased polyphenol synthesis in suspension cultures of *Rosa* sp. Similarly, increased nitrate concentrations resulted in decreased total phenolic synthesis in *Rosa* cells when 0.2 M glucose, but not 0.1 M glucose, was supplied to the cells, and then only between 8 and 12 days of growth. Leucoanthocyanin accumulation was also decreased by increased nitrate levels.[59] Mizukami et al.[57] however, found synthesis of 1,4-naphthoquinones by callus cells of *Lithospermum erythrorhizon* to increase when total nitrogen in the medium was increased from 67 mM to 104 mM. Further increases resulted in decreased yields. Zenk et al.[10] showed that anthraquinone production by *Morinda citrifolia* cells decreased when KNO_3 levels were varied either above or below the range 2.0-4.5 $g \cdot l^{-1}$. Ikeda et al.[60] found altered total ubiquinone production in *N. tabacum* suspension cultures when the ammonia:nitrate ratio was altered from 3:1 to 1:3, but total nitrogen was kept constant. The biosynthesis of indole alkaloids in *Peganum harmala* decreased when ammonia or glutamine was substituted for nitrate.[66] In contrast, the caffeine content of cells of *Camellia sinensis* increased about 4-fold in response to ammonia.[106] Knobloch et al.[52] found relatively high levels of alkaloids (ajmalicine and serpentine), anthocyanins and phenolics in the absence of phosphate and mineral nitrogen. The addition of KNO_3 and NH_4NO_3 inhibited anthocyanin accumulation by 90% and alkaloid synthesis by 80%.

Zielke and Filner[106] showed induction of nitrate reductase to occur in cultured cells of N. tabacum when nitrate was added to nitrate-deficient media. The nitrate uptake system can be induced by both nitrate and nitrite[107] and may be inhibited by some amino acids, but not ammonia.[108] Amino acids are end-products of nitrate metabolism and repress nitrate reductase in tobacco,[109] and other cultured cells.[108,110,111] Tyrosine is an effective feedback inhibitor of arogenate dehydrogenase.[112] Jordan and Fletcher[113] found accumulation of nitrite was increased 30-fold in cells of Rosa sp. grown on nitrate-containing media relative to cells grown with both ammonia and nitrate. Evidence suggests that induced nitrate reductase activity is independent of nitrate in N. tabacum cells.[109] The effect depends on RNA and protein synthesis. Nitrate or its metabolic products may provide positive feedback regulation controlling nitrate reductase. Halhbrock[114] has suggested that correlation between nitrate levels and nitrate reductase activity

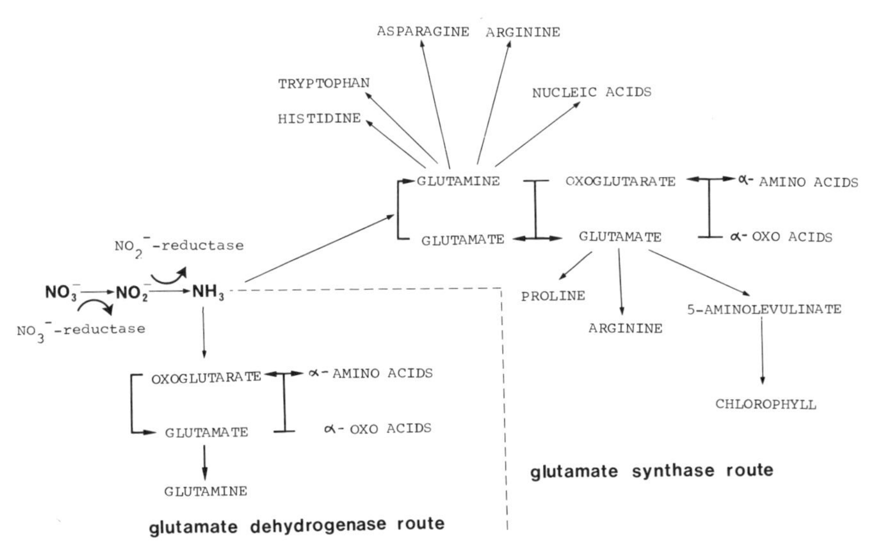

Fig. 3. Metabolism of nitrogen in cells of higher plants.

indicates a strict regulation of enzyme activity by its substrate, and possibly by other nitrogenous metabolites. However, enzymes related to general phenylpropanoid metabolism show dramatically increased activities in coordinated fashion in response to nitrate depletion from the growth media.

The available evidence suggests that nitrogen assimilation in higher plants from either nitrate or low concentrations of ammonia proceeds through the glutamate synthase route,[100,104,105] and during conditions of high ammonia load, or low energy levels a switch from the glutamate synthase route to the glutamate dehydrogenase pathway could occur. The activity of glutamate dehydrogenase has been shown to increase at low ammonia concentration.[115] The greatest portion of mineral nitrogen assimilated by cultured cells is used for the biosynthesis of amino acids, proteins, including enzymes, and nucleic acids. It is to be expected that nitrogen stress will have pronounced effects on the pattern of nitrogenous metabolites produced by cultured plant cells, but the effects will not always be predictable. We have outlined the unpredictable nature of mineral nitrogen stress in Table 2.[10,52-54,57,59,60,62,66,69,74,76,100,106,116-121]

Numerous studies have indicated that feeding amino acid precursors to cultured cells may increase the formation of specific secondary metabolites. Some selected examples illustrating the variable effects of added organic nitrogen and amino acid precursors are documented in Table 3.[6,10,11,57,66,68,70,71,76,117,122-139] Clearly, increased synthesis of phytochemicals is not always observed, and these results too are not predictable, in fact they are often contradictory.

Margna[140] suggested that the availability of phenylpropanoid intermediates, e.g. phenylalanine, exert greater control than PAL over phytochemical synthesis. However, L-phenylalanine, or L-tyrosine when added to cell cultures of *Coleus blumei* had no effect on the synthesis of rosmarinic acid.[126] In contrast, Zenk et al.[71] found cell cultures of *C. blumei* to produce 100% more rosmarinic acid in response to added phenylalanine. Cell cultures of *Catharanthus roseus* produced increased levels of indole alkaloids in response to added tryptophan,[11,131,

Table 2. Effects of mineral nitrogen source and level on phytochemical synthesis

Nitrogen source	Phytochemical effect	Species	Reference
Increase nitrogen as NH_4NO_3:KNO_3	Increase diosgenin	Dioscorea deltoidea	53
	Decrease naphthoquinones	Lithospermum erythrorhizon	57
	Decrease nicotine	Nicotiana tabacum	116
	Increase scopoletin and scopolin		76
	Inhibit phenolics	Rosa sp.	cited in 117
	Increase plasmin inhibitor	Scopolia japonica	118, 119
Add KNO_3 and NH_4NO_3	Inhibit anthocyanins and alkaloids	Catharanthus roseus	52,69
Only KNO_3 or NH_4NO_3	Inhibit diosgenin	D. deltoidea	53
Alter ammonia:nitrate from 3:1 to 1:3	Increase ubiquinone	N. tabacum	60
Substitute ammonia for nitrate	Decrease indole alkaloids	Peganum harmala	66
Decrease KNO_3 below 0.2%, or increase KNO_3 above 0.45%	Decrease anthraquinones	Morinda citrifolia	10
Increase nitrate	Decrease polyphenols	Rosa sp.	54
	Decrease leucoanthocyanin		59
Only nitrate	Increase amides		100
Increase ammonia	Increase caffeine	Camellia sinensis	106
	Increase glutathione	N. tabacum	120
Delete nitrate, ammonia	Decrease indole alkaloids	P. harmala	121
	Decrease lunularic acid	Marchantia polymorpha	62
Delete mineral nitrogen	Increase alkaloids, anthocyanins, phenolics	C. roseus	52
Decrease mineral nitrogen	Increase solasodine	Solanum laciniatum	74

[132] but, other studies noted decreased alkaloid synthesis.[6,68] Tryptophan had no influence on alkaloid synthesis by Camptotheca acuminata.[118] Surprisingly, when L-tryptophan was fed to cultured cells of Phaseolus vulgaris indole alkaloids were recovered; P. vulgaris does not normally produce indole alkaloids.[134,135] Krueger and Carew[131] found tryptamine to enhance alkaloid synthesis by cells of C. roseus if added 2-3 weeks after inoculation. However, if tryptamine was added at the time of inoculation alkaloid synthesis in C. roseus is repressed.[6,11,68]

Table 3. Effects of organic nitrogen source on phytochemical synthesis

Nitrogen source	Phytochemical effect	Species	Reference
Add cadaverine	Stimulate lupinine	Lupinus polyphyllus	122
Add 4-hydroxy-2-quinolone	Increase dictamnine	Ruta graveolens	123
Add ornithine	Increase alkaloids	Datura stramonium	cited in 117
	No effect on tropane alkaloids	Scopolia parviflora	124
	Slight increase in tropane alkaloids	D. innoxia	125
Add phenylalanine	Increase anthocyanins	Catharanthus roseus	70
	Increase rosmarinic acid	Coleus blumei	71
	No effect on rosmarinic acid	C. blumei	126
	Slight increase in tropane alkaloids	D. innoxia	125
	Increase alkaloids	D. stramonium	cited in 117
	Increase alkaloids	D. tatula	127
	Increase naphthoquinones	Lithospermum erythrorhizon	57
	Increase thebaine	Papaver bracteatum	128
	Increase alkaloids	Scopolia acutangula	129
	No effect on tropane alkaloids	S. parviflora	130
Add phenylalanine at early growth phase	Decrease scopoletin, scopolin	Nicotiana tabacum	76
Add phenylalanine and casamino acids at early growth phase	Increase scopoletin, scopolin		
Add phenylalanine, casamino acids at inoculation	Decrease scopoletin, scopolin		
Add tryptamine	Increase indole alkaloids	C. roseus	131
Add tryptamine and secologanin	Synthesis of new alkaloids		131
Add tryptophan	No effect on alkaloids	Camptotheca acuminata	118
	Increase indole alkaloids	Catharanthus roseus	11,136
	Decrease indole alkaloids	C. roseus	6,68
	Decrease cinnamoyl putrescines	N. tabacum	133
	Increase harmaline, harmine	P. harmala	66
Add tryptophan	Synthesis of indole alkaloids	Phaseolus vulgaris	134,135
Add tyrosine	No effect on rosmarinic acid	C. blumei	126
	Increase alkaloids	Papaver somniferum	136,137
Glutamine for nitrate	Decrease indole alkaloids	Peganum harmala	66,138
Add peptone or casein hydrolysate	Decrease naphthoquinones	L. erythrorhizon	57
Increase casein hydrolysate	Inhibit anthraquinones	Morinda citrifolia	10
Add yeast extract	Increase diosgenin extract	Dioscorea deltoidea	139
	Increase phenolics	Rosa sp	cited in 117
Add peptone, yeast extract, NZ-amine, casein hydrolysate	Decrease plasmin inhibitor	Scopolia japonica	118,119

The addition of other organic nitrogen sources, such as casein hydrolysate, peptone and yeast extract added alone or in concert with amino acids has also produced conflicting results in various systems. Cultures of *Lithospermum erythrorhizon* showed decreased synthesis of 1,4-naphthoquinones in response to casein hydrolysate,[57] *Scopolia japonica* showed decreased synthesis of plasmin inhibitor.[118,119] Increased levels of casamino acids resulted in increased levels of scopoletin and scopolin in cells of *N. tabacum*.[76] However, casamino acids and L-phenylalanine added in concert at inoculation, or phenylalanine added alone during the early growth phase inhibited the synthesis of scopoletin and scopolin. But, phenylalanine and casamino acids added together during the early growth phase resulted in increased synthesis of the compounds.[76] A relatively high concentration of yeast extract resulted in increased synthesis of diosgenin, but reduced cell growth in cultures of *Dioscorea deltoidea*.[139]

The evidence suggests that the effects of nitrogen stress on cultured plant cells are variable, and depend in part on the nutritional status of the cells at the time the stress is applied. Clearly, any nitrogen-containing compound may be affected by nitrogen stress of any type; these will include amino acids, proteins, non-protein amino acids, alkaloids, amides and pseudoalkaloids. However, any metabolic perturbation affecting protein metabolism will surely influence the formation of non-nitrogenous compounds also because enzyme pathways may be altered.

Feed-back regulation can dramatically affect the synthesis of nitrogen metabolites by influencing both primary metabolism and pathways leading to secondary compounds. For example, tryptophan regulates its own biosynthesis by feed-back inhibition of anthranilate synthetase.[141-143] There is no activity of the enzyme in the absence of glutamine or Mg^{2+}.[142] Other studies indicate that tryptophan may inhibit synthesis of phenylalanine, tyrosine and itself by feed-back regulation at some point between shikimate and chorismate.[144,145] Some metabolites which may be altered by varied nitrogen flux are illustrated in Figure 3.

Nitrogen catabolite regulation observed in fungi,[146] which involves the repression of enzymes acting on nitrogenous substrates by the catabolic products of rapidly assimilated nitrogen sources, e.g. nitrate and ammonia, may also be a regulatory mechanism in cultured plant cells. For example, Mehta and Shailaga[117] found accumulation of phenolics by cell cultures of Rosa sp. to be inhibited by increased levels of nitrogen. However, maximum phenolic accumulation was observed when the medium was supplemented with yeast extract. The effect of yeast extract is probably due to avoidance of catabolite repression since it may be metabolized slowly into amino acids and ammonia which could become repressive. Here too the results can be contradictory (Table 3).

Another source of regulation could be the loss of activity of an active enzyme; this has been reviewed with respect to microorganisms.[147] In the unicellular alga Chlorella vulgaris nitrate reductase is inactivated by the addition of nitrate and ammonia to cells grown in the presence of light.[148] This regulatory mechanism is most likely to occur in cultured plant cells during conditions of altered metabolic flux experienced with an altered nitrogen source. Under such stressful conditions enzyme inactivation could divert a metabolite from one pathway to another and so affect the pattern of phytochemicals observed. The selective inactivation of enzymes in response to metabolic stress is an important regulatory system in microorganisms. Such a mechanism must certainly operate in cultured plant cells, but its relation to and integration with secondary phytochemical synthesis remains relatively unexplored.

Effects of Phosphate Levels on Phytochemical Synthesis

Inorganic phosphate (Pi) participates as a regulator in plant cells. It is important in metabolic processes such as photosynthesis,[149] respiration,[150] glycolysis,[151] and is essential to nucleic acid[152] and phospholipid synthesis.[153] It seems reasonable to expect that altered phosphate levels in growth media may affect profoundly the biosynthesis of phytochemicals by cultured plant cells. Many secondary phytochemicals are synthesized through phosphorylated intermediates, e.g.terpenes, terpenoids, and phenylpropanoids, with subsequent dephosphorylation. Thus, phosphate-cleaving steps must occur in the synthesis of such compounds. The inhibition of some secondary phytochemicals, e.g. scopoletin,[76] cinnamoyl putrescine,[73,154] may involve the inhibition of phosphatases by Pi. Phosphate levels have been shown to alter dramatically both the synthesis and net accumulation of phytochemicals.[10,27,52,73,76] For example, altered phosphate levels had no effect on the synthesis of protoberberine alkaloids by cultured cells of *Berberis* spp.[155] Other studies have shown that cultured cells of *Catharanthus roseus* and *Nicotiana tabacum* grown in media depleted of Pi showed increased synthesis of indolic alkaloids,[52,66,69] and coumarins respectively.[76] In contrast, increased Pi concentrations caused increased anthraquinone production in *Morinda citrifolia*,[10] and increased synthesis of indolic alkaloids in *C. roseus*,[27] and *Ipomoea violacea*.[156]

Consider the biosynthesis of elymoclavine, (Figure 4) an ergoline alkaloid found in *I. violacea*, some other members of the Convolvulaceae, and in the fungus *Claviceps*. Elymoclavine is formed in *Claviceps* spp. as outlined in Figure 4. Inorganic phosphate represses the first enzyme,[157] and chanoclavine cyclase.[158] The addition of tryptophan or other compounds that reduce inhibition derepress the enzyme. However, increased Pi

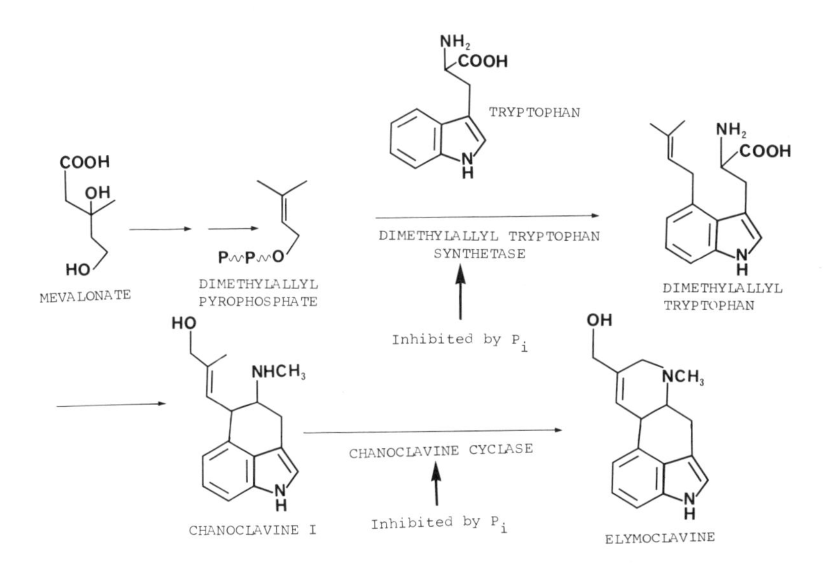

Fig. 4. Biosynthesis of elymoclavine; enzymes inhibited by P_i are indicated.

concentrations resulted in increased total alkaloid synthesis in _I. violacea_, but reduced alkaloid synthesis in _Argyreia nervosa_ and _Rivea corymbosa_.[156]

In both eucaryotic and procaryotic systems secondary pathways are often inhibited by Pi levels which appear to be optimal for growth, and low Pi concentrations are often beneficial for an active secondary metabolism.[85,86,159,266] Table 4 [10,27,52,62,66,68,69,73,76,121,154-156,160,161] outlines the effects of altered Pi levels on secondary metabolism in cultured plant cells. No firm conclusions can yet be made regarding the effects of Pi stress on secondary metabolism in plant cells. It is apparent that plant cells possess some of the regulatory mechanisms found in other organisms.[162] However, relatively little is known about the regulatory relation between Pi levels and scondary phytochemicals in cultured plant cells. This represents a challenging area not fully explored.

Effects of Mineral Nutrients other than Nitrate and Phosphate on Phytochemical Synthesis

Much attention has been devoted to the study of physiology, differentiation and optimization of cell growth in plant cell cultures. However, there is a conspicuous lacuna on the effects of altered concentrations of nutrients other than nitrate and phosphate on phytochemical synthesis by cultured cells. Various compilations of metabolic processes affected by essential mineral nutrients in intact plants are available.[163,164]

Dobberstein and Staba[156] found that the alkaloid levels produced by _Argyreia nervosa_ were increased by growth on low K^+-containing media. However, _Rivea corymbosa_ showed decreased synthesis of alkaloids. Glutathione biosynthesis required Mg^{2+} and was stimulated by K^+.[165] Croteau and Karp[166] found that the enzyme catalysing the synthesis of the monoterpenes camphor and borneol required Mg^{2+} for activity. Activity of ent-kaurene synthetase which is involved in the synthesis of diterpenoids, viz. gibberellins, is dependent on divalent ions, Mg^{2+}, Mn^{2+}, and Ni^{2+}.[167] Other enzymatic reactions involved in the synthesis of other diterpenoids are affected by the divalent ions Fe^{2+}, Mg^{2+}, and Mn^{2+}.[168] It seems reasonable to expect diterpenoids, steroids,

Table 4. Effect of phosphate level on phytochemical synthesis

Phosphate level	Phytochemical effect	Species	Reference
Increase P_i	Increase alkaloids	*Catharanthus roseus*	27
	No effect on alkaloids		68
	Increase alkaloid	*Ipomoea violacea*	156
	Increase anthraquinones	*Morinda citrifolia*	10
	Increase diosgenin	*Dioscroea deltoidea*	160
Decrease P_i	Increase alkaloids, anthocyanins, phenolics	*C. roseus*	52,69
	Increase cinnamoyl putrescine	*Nicotiana tabacum*	73,154
	Increase scopoletin		76
	Increase alkaloids	*Peganum harmala*	66,121
	Increase solasodine	*Solanum laciniatum*	74
	No effect on alkaloids	*Tylophora indica*	161
Increase or decrease P_i	No effect on protoberberine alkaloids	*Berberis* spp.	155
Remove P_i	Increase lunularic acid	*Marchantia polymorpha*	62

carotenoids and higher isoprenoids to be regulated by nutrient ion stress.

Sulfur is also important for active metabolism in plant cells.[169,171] Sulfate ultimately provides the sulfur required for the biosynthesis of sulfur-containing phytochemicals, such as amino acids,[172] some polyacetylenes, mustard oil glycosides and flavor components, e.g. S-trans-propenyl-L-cysteine of *Allium cepa*. Sulfur-containing compounds are important in the synthesis of various intermediates of secondary phytochemicals. S-adenosylmethionine functions in the methylation of a multitude of compounds,[173] such as the N-methylation of ergoline alkaloid precursors[143] and C-24 alkylation of sterol precursors.[174]

Table 5 outlines some of the phytochemical effects of mineral nutrient stress in cultured plant cells;[10,57,66,69,120,121,175-180] unfortunately, these data have been determined empirically and there seems to be no general concensus as to the interpretation of these results.

EFFECTS OF PHYTOHORMONES ON PHYTOCHEMICAL SYNTHESIS

Experience indicates that increased levels of phytochemicals in cultured plant cells coincide with differentiation, or organogenesis. It is therefore not

Table 5. Effects of mineral nutrients other than nitrate and phosphate on phytochemical synthesis

Nutrient and level	Phytochemical effect	Species	Reference
Increase calcium	Decrease naphthoquinones	Lithospermum erythrorhizon	57
	Supress chlorophyll	Nicotiana tabacum	175
Decrease calcium	Increase lignin	Daucus carota	176
		N. tabacum	176
		Parthenocissus tricuspidata	176
	Decrease serotonin	Peganum harmala	121
Increase iron	Decrease naphthoquinones	L. erythrorhizon	57
Increase potassium	Supress chlorophyll	N. tabacum	175
Decrease potassium	Increase alkaloids	Argyreia nervosa	156
	Decrease serotonin	P. harmala	
	Supress chlorophyll	N. tabacum	175
Increase sulfate	Increase amides	N. tabacum	177
Decrease sulfate	Decrease glutathione		120
	Decrease serotonin	P. harmala	121
Increase $MgSO_4$	Decrease alkaloids	P. harmala	66
Add $MgSO_4$	No effect on alkaloids	Catharanthus roseus	69
Add NaCl	Increase proline	N. tabacum	178
	Accumulate amino acids		179
	Stimulate ethylene and ethane		180
Add MgCl	No effect on alkaloids	C. roseus	69
Add KCl	Decrease alkaloids	P. harmala	66
Add micronutrients	No effect on alkaloids	C. roseus	69
Remove micronutrients	Decrease anthraquinones	Morinda citrifolia	10

surprising that the majority of studies of secondary metabolism in cultured plant cells have focused on the effects of altered phytohormone levels. Two types of hormones are required by cultured plant cells; these are auxins and cytokinins. For example, auxins, IAA (indole-acetic acid), IBA (indolebutyric acid), NAA, and 2,4-D, are required by some cultured plant cells. Cytokinins, for example, kinetin and GA (gibberellins) are sometimes required in addition to auxins for vigorous growth of plant cells. There are numerous reports that the type and concentration of auxin, or cytokinin, or auxin:cytokinin ratio, or gibberellin in the growth medium alters dramatically both the growth and secondary product accumulation of cultured plant cells (Table 6).[10,11,18-20,27,35,52,54,58,60,64,66,67,75,121,129,181-230] For example, GA and IAA may enhance RNAase activity with a resultant decrease in RNA level.[231] Such an effect would surely affect secondary metabolism in plant cells.

Carew and Krueger found that increased concentration of 2,4-D resulted in a slight increase in indole alkaloid production by C. roseus cell suspensions.[27] Decreased levels of 2,4-D resulted in suppressed growth and decreased alkaloid recovery from the medium. However, the low 2,4-D concentrations were not as repressive to alkaloid synthesis as were the higher levels. The optimal concentration for alkaloid production was found to be 1 $mg.l^{-1}$. The addition of IAA to the growth medium resulted in qualitative and quantitative differences in alkaloid production.[27] Knobloch et al.[52] noted 2,4-D to supress the synthesis of the alkaloids ajmalicine and serpentine by C. roseus cell suspensions, but growth with IAA resulted in increased alkaloid levels. Benzyladenine (BA) in the presence of auxin gave high alkaloid yields but low growth rates. Gibberellin suppressed alkaloid synthesis.[11]

Nettleship and Slaytor[66] found indole alkaloid synthesis by P. harmala callus to increase in response to decreased 2,4-D concentration. Callus grown on media containing 2,4-D produced only gentisate glucoside.[66,183] and ruine.[232] However, harmine, harmalol and ruine are the major alkaloids synthesized by callus grown on auxin-deficient media.[66] Harmine and harmalol were detected in one year-old callus of P. harmala grown with NAA. (DiCosmo and Towers unpublished). Recently, Sasse et al.[121] found harmine and harmalol to be the main alkaloids detected in cultured cells of P. harmala grown in the presence of 2,4-D. Figure 5 illustrates the altered pattern of alkaloid biosynthesis presented by P. harmala grown under different conditions of auxin stress.[66, 121,183,233] Barz et al.[212] found that photoautotrophic cells of P. harmala on reversion to heterotrophy in the presence of 2,4-D and kinetin synthesized harmalol, harmol, and 5-hydroxytryptamine; but in the presence of NAA these cells produced harmine, harmaline and harmol. It was

Table 6. Effects of phytohormones on phytochemical synthesis

Phytohormone	Phytochemical effect	Species	Reference
Increase 2,4-D	Increase ethylene	Acer pseudoplatanus	181
	Increase alkaloids	Catharanthus roseus	27
	Decrease polyphenols	Cassia fistula	182
	Inhibit indole alkaloids	Peganum harmala	66,183,121
	Stimulate ethylene	Nicotiana tabacum	185
	Decrease anthocyanins in dark-grown cells	Rosa sp.	54,184
	Increase anthocyanins in light-grown cells	Rosa sp.	54
Add 2,4-D	Increase malate	Acer pseudoplatanus	186
	Increase anthraquinones	Cassia tora	187
	Decrease alkaloids	Catharanthus roseus	11,52
	Inhibit cinchoninone, quinidinone, cinchophylline, dihydrocinchonidine	Cinchona pubescens	188
	Increase alkaloids	Coptis japonica	189
	Inhibit anthocyanins	Daucus carota	190
	Esterification of tropine with acetate	Datura innoxia	191,192
	Increase anthocyanins	Dimorphotheca auriculata	193
	Increase diosgenin	Dioscorea deltoidea	122
	Depress protoberberine and aporphine alkaloids	Dioscoreophyllum cumminsii	35
	Inhibit naphthoquinones	Echium lycopsis	195
	Supress anthraquinones	Galium mullugo	72
	Stimulate hemigossypol, gossypol	Gossypium hirsutum	194a
	Inhibit anthocyanin, chlorogenic acid	Haplopappus gracilis	196
	Accumulate anthocyanins		197
	Stimulate cyanidin-3,5-diglucoside	Linum usitatissimum	198
	Inhibit anthraquinones	Morinda citrifolia	10
	Inhibit alkaloids	Nicotiana tabacum	199
	Decrease nicotine		200
	Increase ubiquinone		60
	Increase scopolin in cells, Decrease scopoletin in media		201
	Inhibit alkaloids	Papaver bracteatum	202
	Synthesis of codeine alone	Papaver somniferum	203
	Increase podophyllotoxin	Podophyllum peltatum	64
	Inhibit anthocyanins	Populus sp.	58
	Increase diosgenin	Solanum xanthocarpum	203
	Inhibit anthocyanins	Strobilanthes dyeriana	204
	Synthesis of ecdysterone	Trianthema portulacustrum	75
	Stimulate hydroxyanthracene, hydroxybenzene, hydroxy-naphthalene derivatives	Rumex alpinum	205
Decrease 2,4-D	Increase phenolics, lignin	Acer seudoplatanus	206,207
	Increase phthalides	Apium graveolens	208
	Increase diosgenin	Dioscorea deltoidea	139,209
	Decrease carotenoids, steroids	Daucus carota	210,211
Add 2,4-D + kinetin	Inhibit alkaloids	Cinchona pubescens	188
	Increase berberine, but decrease jatrorrhizine	Coptis japonica	189
	Synthesis of harmol, harmalol and 5-hydroxytrytamine	Peganum harmala	212
	Inhibit anthocyanins	Populus sp.	58
	Stimulate alkaloids	Stephania cepharantha	19

Table 6. (continued)

Phytohormone	Phytochemical effect	Species	Reference
Add IAA	Altered alkaloid pattern	*Catharanthus roseus*	27
	Increase alkaloids		11
	Inhibit alkaloids	*Cinchona pubescens*	188
	Esterification of tropine and acetate	*Datura innoxia*	191,192
	No effect on cardiac glycosides	*Digitalis purpurea*	213
	Increase alkaloids	*Dioscoreophyllum cumminsii*	18
	Increase sterols	*Euphorbia tirucalli*	214
	Synthesis of naphthoquinones	*Lithospermum erythrorhizon*	215
	Synthesis of anabasine, anatabine, nicotine	*Nicotiana tabacum*	199
	Decrease nicotine		200
	Decrease podophyllotoxin	*Podophyllum peltatum*	64
	Increase diosgenin, decrease β-sitosterol	*Solanum xanthocarpum*	203
	Increase anthocyanins	*Strobilanthes dyeriana*	204
Add IAA + BA	Increase scopoletin and scopolin in cells, decrease scopoletin in media	*Nicotiana tabacum*	201
Add IAA + kinetin	Stimulate alkaloids	*Stephania cepharantha*	216
Add IBA	Inhibit cinchinonine, quinidinone,cinchophylline, dihydrocinchonidine	*Cinchona pubescens*	188
	Esterification of tropine and acetate	*Datura innoxia*	191,192
	Inhibit solasodine	*Solanum xanthocarpum*	203
Add IBA + GA + kinetin	Increase linoleic acid	*Glycine max*	217
Add indolecaproic acid	Increase saponins	*Panax ginseng*	cited in 218
Add NAA	Decrease alkaloids	*Catharanthus roseus*	11
	Increase alkaloids	*Chiosya ternata*	67
	Inhibit cinchoninone, quinidinone,cinchophylline, dihydrocinchonidine	*Cinchona pubescens*	188
	Esterification of tropic and tropic acid (increase atropine)	*Datura innoxia*	191
	Increase alkaloids	*Dioscoreophyllum cumminsii*	18
	Increase deoxyisoflavones	*Glycine max*	219
	Stimulate hemigossypol, gossypol	*Gossypium hirsutum*	194
	Stimulate anthocyanins	*Haplopappus gracilis*	197
	Stimulate naphthoquinones	*Juglans major* *Juglans microcarpa*	20
	Increase anthraquinones	*Morinda citrifolia*	10
	Synthesis of harmine, harmol harmaline	*Peganum harmala*	212
	Decrease phenolics	*Perilla ocymoides*	221
	Decrease podophyllotoxin	*Podophylum peltatum*	64
	Inhibit anthocyanins	*Populus* sp.	58
	Supress hydroxyanthracene, hydroxybenzene, hydroxy-napthalene derivatives	*Rumex alpinus*	222

Table 6. (continued)

Phytohormone	Phytochemical effect	Species	Reference
Add NAA + BA	Inhibit cinchoninone, quinidinone,cinchophylline, dihydrocinchonidine	Cinchona pubescens	188
	Stimulate anthocyanins	Haplopappus gracilis	197
Add NAA + kinetin	Inhibit cinchoninone, quinidinone, cinchophylline, dihydrocinchonidine	Cinchona pubescens	188
	Synthesis of codeine, morphine, thebaine	Papaver somniferum	203
	Inhibit anthocyanins	Populus sp.	58
	Inhibit ecdysterone	Trianthema portulacustrum	75
Add GA	Inhibit alkaloids	Catharanthus roseus	11
	Inhibit anthocyanins	Daucus carota	223
	Increase chlorogenic acid		224
	Increase digoxin	Digitalis lanata	225
	Inhibit cardiac glycosides	Digitalis purpurea	213
	Stimulate ecdysterone	Trianthema portulacustrum	75
Add GA + kinetin	Increase linoleic acid, Decrease linoleic acid	Glycine max	217
Add kinetin	Increase polyphenols	Cassia fistula	182
	Inhibit nicotine	Nicotiana tabacum	200
	Decrease ubiquinone		60
	Increase lignin		226,227
	Synthesis of codeine	Papaver somniferum	203
	Increase phenolics	Perilla ocymoides	221
	Inhibit anthocyanins	Populus sp.	58
Add 6-N-(2-isopentenyl)-aminopurine	Inhibit codeine, morphine but synthesis of thebaine	Papaver somniferum	203
Add abscissic acid	Induce suberin	Solanum tuberosum	228,229
Add cyclic AMP	Decrease campesterol, Increase sitosterol	Corchorus olitorius	230

suggested that growth on 2,4-D and kinetin favored the preferential synthesis of hydroxylated compounds, but growth on NAA favoured methoxylated alkaloids. This is provocative, and should be examined futher because it offers a technique whereby the preferential synthesis of either hydroxylated or methoxylated compounds could be realized.

Mulder-Krieger et al.[188] found the alkaloids quinamine and cinchonamine in callus cultures of Cinchona pubescens grown in the presence of 1 μM of the auxins IAA, IBA, 2,4-D, NAA and 1 μM of zeatin. In contrast, the

Compounds from intact plants		Compounds from cultured cells		
		2,4-D	NAA	Auxin-deficient
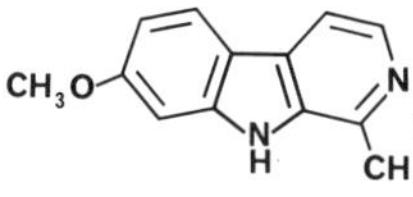 HARMINE	+	- +	+	+
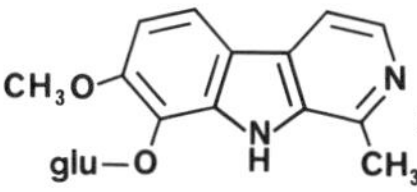 RUINE	+	- -	-	+ DIHYDRORUINE
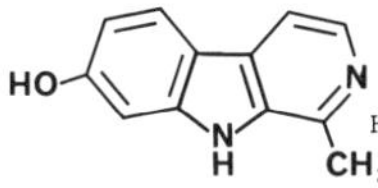 HARMOL	+	- +	- +	+
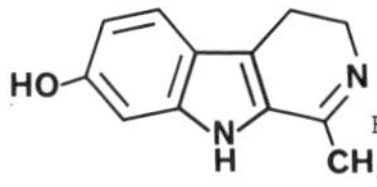 HARMALOL	+	- +	-	+
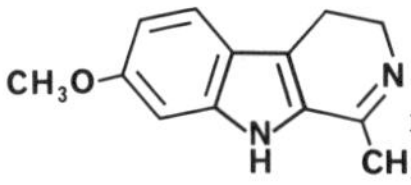 HARMALINE	+	- -	- +	+
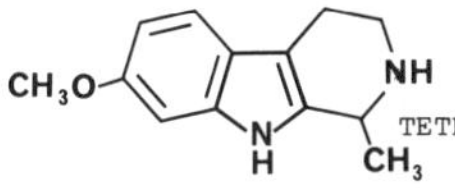 TETRAHYDROHARMINE	+	- -	-	-

Fig. 5. Pattern of β-carboline alkaloids found in intact Peganum harmala and cultured cells grown on auxin-containing and auxin-deficient media.

compounds were not detected in cells grown with IAA, IBA,2,4-D and NAA, and the cytokinins BA or kinetin, except when NAA and BA were present together. All of the auxin and cytokinin combinations tested supressed synthesis of cinchoninone, quinidinone, 3α,17β-cinchophylline and 3α,17α-dihydrocinchonidine. The greatest production of alkaloids was found to occur in cells cultured in the presence of IBA and zeatin, or 2,4-D and zeatin present at 1μM concentrations. Akasu et al.[19] isolated two biscolaurine alkaloids berbamine and aromoline from callus cells of Stephania cepharantha cultured with IAA and kinetin; NAA and kinetin also supported alkaloid synthesis at lower levels. Surprisingly, IAA-kinetin amended media allowed synthesis to levels surpassing those of the intact plant.

The synthesis of the phenanthrene alkaloids, papaverine, morphine, codeine and isothebaine by *Papaver bracteatum* cell suspensions is suppressed by 0.1 $mg.l^{-1}$ 2,4-D.[202] In contrast, Hodges and Rapoport[203] found thebaine, codeine, and morphine to be produced by callus of *P. somniferum* grown in the presence of 2,4-D, NAA, 6-N-(2-isopentenyl) aminopurine and kinetin. When 6-N-(2-isopentenyl) aminopurine was supplied alone only thebaine accumulated. Codeine was the major product found in callus grown in the presence of 2,4-D, or kinetin, lesser amounts of morphine and thebaine were also produced. Not surprisingly, it was found however, that the callus cultures failed to maintain the observed pattern and level of alkaloid production after repeated subculturings.

Ikuta et al.[189] found callus of *Coptis japonica* to produce all the main alkaloids found in the rhizome of the intact plant. Berberine and jatrorrhizine predominated in callus cells, but the amounts were lower than those found in the intact plant. Alkaloid content of cells grown in the presence of 2,4-D and kinetin increased with prolonged culture periods. Jatrorrhizine content and growth were both reduced when 2,4-D was excluded from the growth medium. However, kinetin added in the presence of 2,4-D resulted in increased berberine content, but reduced jatrorrhizine synthesis. Jatrorrhizine and related alkaloids in callus cultures of *Thalictrum minus* were shown to be regulated by the concentration of 2,4-D. The spectrum of alkaloids was altered in response to increased 2,4-D concentration; columbamine, thalifendine, thalidastine and deoxythalidastine were not detected in callus cells grown in the presence of 5 $mg.l^{-1}$ 2,4-D. However, berberine was present in callus grown with 2,4-D and kinetin in amounts greater than those found in the stems and leaves of intact plants.[234]

Callus of *N. tabacum* repeatedly subcultured for up to 5 years in the presence of 2,4-D produced no alkaloids; but the alkaloids anabasine, anatabine and nicotine were synthesized rapidly when cells were grown on media with IAA.[199] Ohta et al.[116] found nicotine synthesis by *N.*

tabacum was best on media with 0.15 and 0.2 $mg.l^{-1}$ NAA; alterations of auxin level above or below that range reduced nicotine content. However, nicotine content in callus of N. tabacum was found to decrease as the concentrations of 2,4-D, IAA or NAA were increased.[200] The suppressive effect on nicotine synthesis was greatest in the presence of 2,4-D, NAA and IAA respectively, and 1 $mg.l^{-1}$ 2,4-D, or 10 $mg.l^{-1}$ NAA or 100 $mg.l^{-1}$ IAA almost inhibited completely nicotine synthesis. The inhibitions were found to be reversible. Kinetin added to the growth medium at 5 $mg.l^{-1}$ also inhibited nicotine synthesis completely.[200]

The synthesis of anthocyanins by Daucus carota cell suspensions occurred in the absence of 2,4-D. Addition of 2,4-D to the growth medium suppressed synthesis, and concentrations above 10^{-7}M inhibited completely anthocyanin synthesis.[190] However, Haplopappus gracilis cell suspensions accumulate anthocyanin in the presence of 4.5 x 10^{-6}M 2,4-D; transfer of cells to kinetin- or BA-amended media with no auxin stimulated anthocyanin.[197] Schimtz and Seitz[223] found 10^{-12}M GA to suppress anthocyanin synthesis by D. carota, 10^{-4}M inhibited synthesis completely. GA_3 added to H. gracilis at 10^{-6} $g.ml^{-1}$ inhibits PAL formation.[235] This may explain, in part, the decreased synthesis of anthocyanins observed in cells exposed to GA. Sugano and Hayashi[236] found anthocyanin in D. carota cell cultures grown with 2,4-D instead of IAA. Harborne et al.[193] found the anthocyanins cyanidin-3-glucoside and delphinidin-3-glucoside in callus cells of Dimorphotheca auriculata when grown in the presence of 2,4-D; these compounds were not detected in intact stems. In contrast, Linum usitatissimum callus cultures produced cyanidin-3,5-diglucoside, but not malvidin and hirsutidin glucosides which occur in the normal plant.[198] Smith et al.[204] isolated cyanidin-3,5-diglucoside and peonidin-3,5-diglucoside from callus cultures of Strobilanthes dyeriana grown in IAA, but not in 2,4-D. Other studies have shown that anthocyanin synthesis is inhibited by auxins, especially 2,4-D in cultured cells of Populus sp.,[58] H. gracilis,[196] and Rosa sp.[54,237] However, 2,4-D has been shown to stimulate anthocyanin synthesis by H. gracilis,[197] and Rosa sp. in the presence of light.[54] and high levels of 2,4-D do not inhibit the activity of PAL in vitro.[237]

Okazaki _et al_[201] found 2,4-D, IAA and BA to increase the synthesis of the coumarins scopoletin and its glucoside scopolin. IAA and BA increased the scopoletin and scopolin concentrations in the cells, and scopoletin level in the medium. On the other hand, 2,4-D caused increased accumulation of the glucoside in the cells and decreased the level of the aglycone in the medium. Ethylene, abscissic acid or GA did not affect the synthesis of either compound.[210]

Callus cultures of _Trianthema portulacustrum_ grown on 2,4-D-amended media produced ecdysterone.[75] Ecdysterone production was stimulated in the presence of 2,4-D and kinetin which stimulated root formation. GA was found to stimulate ecdysterone synthesis. Biesboer and Mahlberg[214] found sterol production by callus cultures of _Euphorbia tirucalli_ to be enhanced by the addition of IAA and depressed by BA added to the growth medium.

Waller et al.[238] showed that callus cells of _Delphinium ajacis_ produced much more sterols than intact plants.

Rucker et al.[239] observed the synthesis of cardiac glycosides in callus tissue of _Digitalis purpurea_ to be influenced by IAA and unaltered by the ratio 2,4-D:IAA. GA was found to inhibit glycoside biosynthesis. Kaul and Staba[209] and Kaul et al.[139] reported diosgenin production from undifferentiated callus of _Dioscorea deltoidea_ grown with 0.1 to 1.0 $mg.l^{-1}$ 2,4-D. However, removal of 2,4-D resulted in differentiation and rather unexpectedly, a decreased level of diosgenin. Heble et al.[203] found callus of _Solanum xanthocarpum_ grown with IAA to show a 2-fold increase in diosgenin levels above controls, but the level of β-sitosterol was reduced. When 2,4-D was substituted for IBA similar results were observed, and diosgenin synthesis was stimulated a further 0.01%. Stearns and Morton[217] found _Glycine max_ cell suspensions grown in the presence of IBA, with or without GA contained more palmitic acid and less polyunsaturated acids. Oleic acid levels were reduced in response to IBA. Linoleic acid synthesis was increased by IBA, GA and kinetin, and was stimulated 4-fold over controls with kinetin and GA which decreased synthesis. Tabata et al.[215] found increased synthesis of 1,4-naphthoquinones in callus tissue of _L. erythrorhizon_ grown in dark on media with IAA; production decreased when 2,4-D was substituted for IAA, or cultures were irradiated with blue light (380-560 nm). However, 2,4-D and irradiance with red light (660 nm) yielded maximum synthesis of podophyllotoxin by Podophyllum peltatum callus.[64]

The majority of studies related to phytohormone-stress have been concerned with optimization of secondary phytochemical yield using either callus cells or cell suspensions in batch cultures. With such studies it is not possible to evaluate the effects of auxins and cytokinins independently of the effects of changing nutritional parameters intrinsic to these systems. Using a continuous culture technique Balague et al.[240] found that 2,4-D-starved cells of Pyrus communis consumed ammonia, K^+ and Ca^{2+} at high rates, but nitrate and Mg^{2+} were not used, and carbohydrate uptake was lower relative to cells entering a stationary phase. Later, an efflux of serine, threonine and aspartic acids occurred with serine representing 52% of the total. Such an efflux of serine would probably have some effects on the biosynthesis of phytochemicals derived from it, e.g. lupinic acid.[241] The evidence is clear that synthesis of all types of phytochemicals are altered by phytohormone stress, but that this change is transient. Although general opinion suggests that increased concentrations of auxins suppress secondary metabolism and low concentrations may allow synthesis of phytochemicals, some studies contradict this notion.[27,181,186,189,216]

EFFECT OF LIGHT ON PHYTOCHEMICAL SYNTHESIS

Light has pronounced effects on the growth and development of plant cells in culture,[242,243] and influences the formation of phytochemicals. Phytochemical responses are affected by both irradiance and light quality. For example, two groups of enzymes involved in flavonoid biosynthesis are recognized on the basis of activity changes induced by illumination.[244] Blue light induced maximum anthocyanin formation in Haplopappus gracilis cell suspensions.[245-247] White light induced anthocyanin synthesis in C. roseus,[52] and Populus sp.[58] In contrast, Tabata et al.[215] found either white or blue light inhibited completely 1,4-naphthoquinones in callus cultures of Lithospermum erythrorhizon.

The production of chlorogenic acid in *H. gracilis* was stimulated by white, blue and red light; blue light was most effective.[196,246] However, Lackmann[245] found that light-dependent anthocyanin induction in cell suspensions of *H. gracilis* was controlled by blue light only. Anthocyanin synthesis was not affected by red or far-red radiation even after induction with high intensity blue light. In contrast, podophyllotoxin production by callus cultures of *Podophyllum peltatum* is stimulated by red light (660 nm) and inhibited by irradiation with blue light (371,420,460 nm).[64] Anthocyanin synthesis in callus cultures of *Daucus carota*, *Helianthus tuberosus*, *L. usitatissimum*, *M. pumila* and *R. multiflora* was found to have a white light requirement.[248] Takeda and Katoh found a 4-fold increase in yields of the sesquiterpenoid azulene in cell suspensions of the liverwort *Calypogeia granulata* grown in white light (300-800 nm).[249] Cultured cells of *Digitalis lanata* synthesized increased levels of glycosides in response to increased irradiance.[216]

Ultraviolet light (UV 280-320 nm) was shown to stimulate flavone glycoside synthesis in *Petroselinum hortense* cell suspension cultures.[250,251] Flavone glycoside formation induced by UV radiation was increased by subsequent irradiation with red light, but irradiation with far-red light reduced the UV-induced effect.[250] Ample evidence shows that the activities of the enzymes responsible for flavone glycoside synthesis in cultured cells of *P. hortense* increase dramatically upon UV irradiation.[244,252-254] Continuous irradiation with red, far-red and blue light also increased the UV-mediated flavonoid synthesis, if administered before UV irradiation.[251] UV (254 nm) irradiation of *G. max* cells resulted in rapid synthesis of glyceollin,[4] and such irradiation resulted in increased synthesis of γ-aminobutyric acid in cell suspensions of *Rosa damascena*.[255]

Cells of *Ruta graveolens* produce several coumarins and alkaloids when grown in continuous white light.[34] *R. graveolens* produces 2-nonanone, 2-nonanyl acetate and 2-nonanol preferentially in dark, and their concentrations decreased with increasing white light intensity. In contrast, the levels of 2-undecanone, 2-undecanyl acetate and 2-undecanol increased in response to increased light intensity.[256] Dark-grown callus of *Centaurea ruthenica* synthesized very low concentrations of two aliphatic polyacetylenes when exposed to light for six weeks.[257] Callus tissue of _Bidens pilosa_ grown in dark, and subsequently irradiated with continuous white light for three weeks failed to produce any aromatic polyacetylenes.(DiCosmo and Towers unpublished). Various aliphatic polyacetylenes were isolated from callus of _Carthamus tinctorius_ incubated 14 days in light. However, overall quantities of acetylenic compounds declined with further subculturing.[258] Callus cells of _Parthenium argentatum_ synthesized ten times the amount of _cis_-1,4-polyisoprene when illuminated with white light as compared to dark-grown cells.[259]

Callus cultures of _Scopolia acutangula_ produce more alkaloids in light than in dark.[129] Incubation of _P. harmala_ callus in light favors the production of the indole alkaloid harmine.[183] White light influences the synthesis of indole alkaloids by _C. roseus_.[57,70,260] Decreased synthesis of nicotine by _N. tabacum_ was observed in response to illumination. The inhibitory effect increased with increased light intensity and duration of illumination, and no differences in nicotine synthesis were observed between the effects of blue and red light.[261]

Hahlbrock isolated the flavone apigenin from illuminated cell suspensions of _G. max_, and observed correlation between apigenin formation and increased activities of PAL and p-coumarate coA ligase.[262] Sinensetin, nobiletin and other methoxylated flavones were found in callus cultures of _Citrus aurantium_ and _C. medica_ exposed to light, but not in dark. A rapid increase in PAL activity preceded flavonoid synthesis.[263] Irradiation of _H. gracilis_ callus with blue-light increased PAL activity and anthocyanin production.[264,265]

It has been already shown that nitrate assimilation by plant cells is regulated mainly at the level of nitrate reductase, and visible light is one parameter controlling its synthesis and activity.[266] Nitrate reductase is an enzyme complex containing FAD (flavin adenine dinucleotide), cytochrome and molybdenum as prosthetic groups, and can be inactivated by ammonia in light, or NADPH. NADPH-inactivated nitrate reductase can however be reactivated by blue light,[267] and flavin nucleotides dramatically enhance blue light-mediated reactivation.[268] Red light is known to enhance the levels of the GA diterpenoids.[269] Madyastha et al.[270] showed that _C. roseus_ contains a monooxygenase associated with cytochrome P-450, which hydroxylates C-10 methyl moieties of geraniol and nerol. This hydroxylation is an important step in the biogenesis of certain indole alkaloids. Grunwald[271] indicated that irradiance was an important factor in controlling sterol synthesis; light-grown plants generally show lower sterol contents (mainly stigmasterol) than dark-grown plants.

Light affects several metabolic processes during chloroplast development in plants.[272] Blue light-induced responses include lipid biosynthesis, increased carbon:nitrogen assimilation and oxidative phosphorylation. Significant increases in either total lipid, carotenoid or RNA occur only after adequate photophosphorylation.[272] These light-induced changes are thought to be regulated through phytochrome. Carotenoid synthesis is also photoregulated,[273,274] a process also mediated by phytochrome. Phytochrome may be the only clearly defined pigment in vascular plants which is able to translate illumination into metabolic regulation.[272,273,275] Schopfer[275] reviewed phytochrome control of enzymes and listed 52 enzymes whose activity has been reported to be regulated by light. For example, enzyme activities involved in the reductive and oxidative pentose phosphate cycle are modulated by light in vivo, light and nitrate are needed to maintain a high level of nitrate-reductase activity in plant cells, and peroxidase activity is modulated by light. A remarkable range of enzymes may be regulated by light.[276-279]

Although various aspects of light-mediated regulation of enzymes related to photosynthesis and chloroplast development have been reviewed,[272,273,275,278,280] relatively little information is available regarding the effects of irradiance on the enzymes of secondary metabolites by cultured plant cells, except for those of general phenylpropanoid metabolism.[244,281] This is largely because the enzymes involved with the majority of secondary metabolites remain to be characterized. Also most of the studies concerned with the effects of irradiation on phytochemical synthesis have been concerned mainly with the altered phytochemical pattern. Examples of

Table 7. Effects of light on phytochemical synthesis

Light quality	Phytochemical alteration	Species	Reference
White	Increase sesquiterpene, paniculide B	Andrographis paniculata	20
	Increase polyphenols, but inhibit polymerization of leucoanthocyanins	Camellia sinensis	282
	Decrease caffeine		106
	Increase anthocyanins, indole alkaloids	Catharanthus roseus	52,70,260
	Increase digalactosyldiacyl-glycerols	Chenopodium rubrum	283
	Increase flavonoids	Citrus aurantium C. medica	263
	Increase anthocyanins	Daucus carota	248
	Increase cardiac glycosides	Digitalis purpurea	213
	Increase diosgenin	Dioscorea spp.	242
	Inhibit naphthoquinones	Echium lycopsis	195
	Increase apigenin	Glycine max	262
	Increase chlorogenic acid	Haplopappus gracilis	196,246
	Increase sulpholipids	Kalanchoe crenata	284
	Increase anthocyanin	Linum usitatissimum	248
	Increase anthocyanin	Malus pumila	248
	Decrease nicotine	Nicotiana tabacum	261
	Favors harmine synthesis	Peganum harmala	183
	Increase lignin	Pinus strobus	44
	Increase anthocyanins	Populus sp.	58
White	Increase anthocyanin	Rosa multiflora Rosa sp.	248
	Increase alkaloids, coumarins	Ruta graveolens	34
	Decrease 2-nonanone, 2-nonanyl acetate, 2-nonanol; increase 2-undecanyl acetate, 2-undecananol, 2-undecanone		256
	Increase alkaloids	Scopolia acutangula	129
	Increase glycoalkaloids	Solanum acculeatissimum	242
Red	Increase chlorogenic acid	H. gracilis	196,246
	Increase podophyllotoxin	Podophyllum peltatum	64
Green	Increase anthocyanins	Populus sp.	58
Blue	Inhibit naphthoquinones	E. lycopsis	195
	Increase chlorogenic acid	H. gracilis	196,246
	Increase anthocyanin		245,264,265
	Inhibit podophyllotoxin	P. peltatum	64
	Increase anthocyanin	Populus sp.	58
UV	Increase glyceollin	G. max	4
	Increase flavonoid glycosides	Petroselinum hortense	244,250-254
UV + far-red	Decrease flavonoid glycosides		250
Far red, or red,or blue + UV	Increase flavonoid glycosides		251

effects of light on secondary metabolite formation are illustrated in Table 7.[4,20,34,44,52,58,64,70,106,129,183,195,242,244-246,248,250-254,256,262-265,283,284]

EFFECTS OF TEMPERATURE ON PHYTOCHEMICAL SYNTHESIS

Plant cells are usually cultured in the temperature range of 25-28°C. Altered temperature regime affects both the accumulation of specific compounds and the type of phytochemicals synthesized. For example, the amount of saturation of fatty acids can be affected.[285,286] Thus, the composition of fatty acids is related to temperature; saturated fatty acids increase in response to increased temperature, and unsaturated acids increase in response to decreased temperatures.[287]

Growth of *Catharanthus roseus* cells at temperatures below 27°C (to 16°C) results in increased synthesis of indole alkaloids, and growth at temperatures above 27°C results in decreased synthesis of alkaloids.[288] Either a decrease, or increase in the incubation temperature of *Camellia sinensis*[106] or *Nicotiana tabacum*,[116] results in decreased synthesis of caffeine, and nicotine respectively. Increased temperature inhibits carotenoid synthesis in tomato fruit.[289]

Plant cells incur biochemical and metabolic changes in response to temperature-stress. Intact plants show altered protein content,[290-292] and altered hormonal fluxes.[293-297] Abscissic acid may be a mediator in plant cell temperature-stress phenomena.[297] Specific proteins are synthesized in response to lowered temperature regime, and abscissic acid is able to induce certain protein species.[297] Increased temperatures, likewise, result in altered protein metabolism.[298,299] The effects of altered temperatures on phytochemical synthesis are outlined in Table 8.[10,31,66,116,285,286,300-303]

EFFECTS OF pH ON PHYTOCHEMICAL SYNTHESIS

Plant cells are usually cultured in media having a pH range of 5-6. Several studies have shown that the pH of the growth medium can dramatically influence the production of phytochemicals by cultured cells, including anthocyanins,[58,300,301] anthraquinones,[10,304] and alkaloids.[304] Table 9 illustrates the phytochemical

Table 8. Effects of temperature on phytochemical synthesis

Temperature regime (°C)	Phytochemical alteration	Species	Reference
26-30	Increase anti-plant virus substance	Agrostemma githago	66
Increase 25 to 28 at pH 4.2	Decrease anthocyanin	Daucus carota	300,301
Increase 25 to 28	Increase chlorophyll	Mentha arvensis	302
Increase above 25	Decrease nicotine	Nicotiana tabacum	116
Increase above 26	Decrease caffeine	Camellia sinensis	106
Increase 27 to 38	Decrease indole alkaloids	Catharanthus roseus	288
Increase to 30	Increase anthraquinones	Morinda citrifolia	10
Increase 30 to 35	Decrease arbutin	Juglans major J. minor	220
Decrease	Increase linolenic acid Increase linoleic	Brassica napus Tropaeolum majus	285,286 285,286
Decrease to 10	Increase phospholipid	Rauwolfia serpentina	303
Decrease to 15	Increase linoleic, linolenic	Glycine max	cited in 36
Decrease below 25	Decrease nicotine	N. tabacum	116
Decrease below 26 to 16	Increase indole alkaloids	C. roseus	288
Decrease below 26	Decrease caffeine	Camellia sinensis	106
Decrease below 30 to 25	Decrease arbutin	J. major, J. minor	220

effects of altered pH.[106,300,301,304,306] For example, Daucus carota grown at pH 5.5 produced less anthocyanin than when cultivated at pH 4.5.[300,301] It was suggested that the lower level of anthocyanin was due to increased degradation of the compound at the higher pH. When the pH was increased from 4.5 to 5.5 the yield constant for anthocyanin production decreased by 90%.[300,301] Changes of either temperature or pH when glucose, ammonia and phosphate were limiting led to altered yield constants.[301] Altered pH affected the formation of tryptophol from tryptophan, at controlled pH of 6.3 tryptophol synthesis was stimulated 71% over control cultures at neutral pH, but at pH 4.8 tryptophol synthesis was inhibited.[305]

Murphy et al.[255] found that UV-irradiation (254 nm)

of *Rosa damascena* cell suspensions caused K^+ leakage from cells, as well as a drop in cytoplasmic and vacuolar pH. In these cells there was a reduction of nitrate and a large increase in γ-aminobutyric acid synthesis. It was suggested that the conversion of nitrate to amino acids limited the pH change induced by UV-irradiation.

EFFECTS OF ALTERED AERATION ON PHYTOCHEMICAL SYNTHESIS

The supply of oxygen has been shown to affect the phytochemical composition of cultured plant cells,[286,307] but there seems to be little information on the effects of oxygen-stress on secondary compound synthesis. Metabolic alterations produced by anaerobiosis in plants have been extensively reviewed.[308] The activity of alcohol dehydrogenase, malic enzyme and nitrate reductase increase under anoxic conditions.[309-311] The best known effect of oxygen depletion is the increased activity of glycolysis,[308] which leads to accumulation of ethanol.

Under aerobic conditions, a large portion of *cis*-9-octadecenol was oxidized to oleic acid and incorporated into phospholipid; up to 30% was esterified to wax in heterotrophic cell suspensions of *Glycine max* and photomixotrophic cells of *Brassica napus*, but the oxidation rate was increased in the latter. However, in

Table 9. Effects of pH on phytochemical synthesis

pH	Phytochemical alteration	Species	Reference
6.0	Increase serpentine	*Catharanthus roseus*	304
7.2	Increase caffeine	*Camellia sinensis*	106
Increase 4.5 to 5.5	Decrease anthocyanin	*Daucus carota*	300,301
Increase 4.6 to 5.0	Increase anthraquinone	*Morinda citrifolia*	304
Decrease 7 to 6.3	Increase tryptophol	*Ipomoea* sp.	305
Decrease to 4.8	Inhibit tryptophol		
Decrease 4.8 to 4.0	Increase anthraquinones	M. *citrifolia*	305
Decrease to 3.5	Increase tropane alkaloids	*Hyoscyamus muticus*	306

anoxic cultures little octadecenol was converted to oleic acid; most was found as wax esters.[312] Lipids in well-aerated cultures contain more unsaturated fatty acids than poorly aerated cultures which contain relatively high proportions of long-chain saturated fatty acids.[307,313] After one generation in O_2-enriched, and O_2-deprived systems the amounts of volatile oils in Ruta graveolens callus decreased. But after two generations the oil content was greater in O_2-enriched systems.[256]

Free amino acids were found to increase in cultures of Acer pseudoplatanus in response to increased CO_2 and decreased air supply.[314]

Glycosylation of the isoflavonoid daidzein is favored when mungbean cell suspensions are grown in the absence of oxygen (cited in 315).

A range of metabolites including shikimic acid, lactate, and malate, are known to be produced by plants in response to partial or total anoxia.[316-318] It has been indicated that a major portion of ethanol-carbon is converted into glutamine.[319] Acetaldehyde has been shown to be an intermediate in ethanol metabolism by pea cotyledons.[320]

Recently, it was shown that 10% CO_2 in the ambient gas inhibits the production of CO_2 via respiration by 65%, and 5% CO_2 inhibits it by 85% in pea tissue. Also malate, α-oxoglutarate and succinate decreased in content in 10% CO_2. The data suggested that in 10% CO_2 the activity of the TCA cycle was reduced by a reduced supply of pyruvate. One of the effects of high CO_2 tension on pea is reduced glycolytic activity.[321] Oxygen (2% O_2 in N_2) reduced the uptake of glucose by 20%, and the activity of glucose-6-phosphate by 50%. Large increases in pyruvate and α-oxoglutarate were noted, suggesting a decreased rate of oxidation and subsequent decreased activity of the TCA cycle. High CO_2 and low O_2 increased the proportion of the pentose phosphate pathway relative to the TCA cycle.[321]

Many enzymes are known to have molecular oxygen (dioxygen) as a substrate, e.g. oxygenases catalyze the incorporation of oxygen into organic substrates,[322,323] and hypoxic conditions would be expected to affect the biosynthetic rates of oxygenated phytochemicals.

Obviously, much more needs to be done in this interesting area of metabolism.

EFFECTS OF ANTIBIOTICS ON PHYTOCHEMICAL SYNTHESIS

There are several reports that antibiotics, including protein synthesis inhibitors and amino acid analogs, alter the pattern of phytochemicals synthesized by cultured plant cells. For example, streptomycin sulfate, which inhibits protein synthesis, promoted the synthesis of naphthoquinones by callus of *Lithospermum erythrorhizon.*[57] Cycloheximide and puromycin inhibited the production of daidzen in cell suspensions of *Glycine max.*[324]

However, actinomycin D stimulated the synthesis of daidzen, and related glucosides as well as a lignan.[325] Several RNA-synthesis-inhibitors, on the other hand, reduced the production of daidzen.[324] Puromycin and actinomycin promoted accumulation of phenanthrene alkaloids in cell suspensions of *Papaver somniferum*; cycloheximide reduced alkaloid synthesis. However, when antibiotics were added at incubation, both puromycin and actinomycin enhanced codeine synthesis, but when inhibitors were added on the third day of incubation, actinomycin depressed and puromycin increased akaloid synthesis.[137]

Hahlbrock and Ragg[326] found that light-induced flavonoid glycoside-producing enzyme activities in parsley cells were inhibited by low concentrations of actinomycin D and cycloheximide, but the activities of other enzymes were unaffected. Actinomycin D and puromycin inhibit blue light-induced enzyme activity in *Halopappus gracilis.*[327] Chloramphenicol caused a 60% decrease in cytochrome oxidase and blocked synthesis of other components in mitochondrial electron transport in soybean cell suspension cultures.[327]

The data presented in Table 10 show phytochemical alterations in response to various antibiotics.[23,137,324,326,330-332]

Amino acid analogues which inhibit cultured cell growth when the endogenous pools of free protein-amino acids are low have found broad applications in selecting for cells with altered biochemical virtuosities. The use

Table 10. Effects of antibiotics on phytochemical synthesis

Antibiotic	Phytochemical alteration	Species	Reference
Aminooxyphenyl-propionic acid	Decrease anthocyanins	Daucus carota	328,329
Actinomycin D	Decrease deoxyisoflavone	Glycine max	324
	Inhibit flavonoid glycosides	Petroselinum hortense	326
	Induce lignan, isoflavone glycosides	Vigna angularis	325
Add Actinomycin D at inoculation	Increase phenanthrene alkaloids		
Add Actinomycin D at 3 days growth	Decrease alkaloids	Papaver somniferum	137
Chloramphenicol	Stimulate alkaloids	Nicotiana tabacum	23
Cycloheximide	Decrease deoxyisoflavone, Increase unknown compound	G. max	324
	Induce novel alkaloids	Nicotiana tabacum	137
	Decrease alkaloids	P. somniferum	
	Inhibit flavonoid glycosides	Petroselinum hortense	326
Puromycin	Decrease deoxyisoflavone, Increase unknown compound	G. max	324
Add Puromycin at 3 days growth	Increase alkaloids	Papaver somniferum	137
Increase Puromycin	Inhibit codeine	P. somniferum	137
Streptomycin	Increase naphthoquinones	Lithospermum erythrorhizon	57
p-Fluorophenylalanine	Increase phenolics	Nicotiana tabacum	330,331 332

of amino acid analogues in studies of plant metabolism has been reviewed.[333,334] Cells of Nicotiana tabacum which proliferate in the presence of p-fluorophenylalanine, an analogue of phenylalanine, produced increased amounts of PAL and showed enhanced phenylalanine synthesis.[13,330] This combination of increased enzyme activity and substrate titre may allow much more phenylalanine to be directed into phenylpropanoid metabolism.[331,335-338] However, p-fluorophenylalanine-resistant Daucus carota cells may accumulate phenylalanine but show no increased PAL activity nor increased polyphenol synthesis.[339,340] Resistance to 5-methyl-tryptophan may result in increased levels of tryptophan.[341-343]

The various phytochemical effects caused by antibiotic agents have been derived empirically and it is difficult to interpret the results meaningfully, or extrapolate to alternate systems. This is mainly due again to the fact that the enzymes involved in many biosyntheses are not characterized with respect to their synthesis, compartmentalization or substrate.

EFFECTS OF MICROORGANISMS AND VIRUSES ON PHYTOCHEMICAL PRODUCTION

It has been shown repeatedly that plants often synthesize additional levels of antibiotic compounds, and even novel ones in response to microbial insult.[343-352]

Effects of Ti-Plasmids on Phytochemical Synthesis

The effects of introduction of microbial DNA (Ti-plasmid), on phytochemical synthesis by cultured cells is outlined in Table 11.[218,353-358] Brown and Tenniswood[354] showed that normal tobacco callus cells produce the coumarins, scopoletin, esculetin, umbelliferone and bergapten. However, crown gall-derived callus tissue showed altered patterns of phytochemical synthesis and produced scopoletin and esculetin, but no umbelliferone, or bergapten. Differentiated teratoma tissues of tobacco contain predominantly n-C_{29}, 2-methyl C_{30}

Table 11. Effects of Ti-plasmids on phytochemical synthesis in cultured plant cells

Species transformed with Ti plasmid	Altered secondary metabolism	References
Bidens alba	Synthesis of polyacetylenes	Norton & Towers unpublished
Catharanthus roseus	Synthesis of vindoline	353
Coptis japonica	Synthesis of berberine	cited in 218
Nicotiana tabacum	Inhibition of umbelliferone, bergapten	354
N. tabacum (teratoma)	Synthesis of ubiquinone 10 Synthesis of polyamines	355 356
Panax ginseng	Synthesis of saponins, sapogenins	cited in 218
Scorzonera hispanica	Synthesis of putrescine, spermidine	357,358
Stevia rebaudiana	Synthesis of stevioside	cited in 218

(iso-C_{31}) and n-C_{31} alkanes. However, normal callus cells contain lower molecular weight alkanes ranging in chain-length from C_{17} to C_{28}.[359] Vindoline was detected in cell suspensions of Catharanthus roseus derived from crown gall.[353] Misawa[218] cited examples showing that crown gall callus cells of Coptis japonica produced berberine, those of Panax ginseng accumulated both saponins and sapogenins; crown gall of N. tabacum accumulated ubiquinone-10,[355] and Stevia rebaudiana was cited as being able to produce stevioside. Crown gall callus cells of Bidens alba have been shown to produce several aromatic polyacetylenes characteristic of the intact plant. (Norton and Towers unpublished). Crown gall cells of Scorzonera hispanica[357,358] and N. tabacum [356] showed increased synthesis of polyamines.

These and other studies[360-365] show that plant cells transformed by Ti-plasmids are physiologically different from normal plant cells. For example, transformed cells have altered cytokinin metabolism; altered adenyl cyclase, cyclic AMP phosphodiesterase, ascorbic acid oxidase, and tyrosinase activities; altered membranes; altered cation concentrations; altered Feulgen DNA content and characteristic, novel amino acids called opines. However the molecular basis responsible for altered secondary metabolism in Ti-plasmid-stressed cultured cells remains to be identified.

Effects of Fungal-Induced Stress on Phytochemical Synthesis

It has been demonstrated repeatedly that fungal-induced stress of normal, intact plant tissues leads to the induction and accumulation of secondary phytochemicals in vivo called phytoalexins. Some of these are shown in Figure 6. The ability of cultured plant cells to produce certain stress-metabolites in response to molecules isolated from fungi appears to be a general phenomenon, and must be intimately correlated with physiological stress imposed on the cells.[28,366-376] Molecules which stimulate secondary metabolism leading to the induction of stress-metabolites are called elicitors and those derived from fungi may be referred to as "fungal elicitors". Recently, vacuum infiltration of leaves with water was shown to result in the accumulation of antifungal steroidal alkaloids in Solanum aviculare.[377]

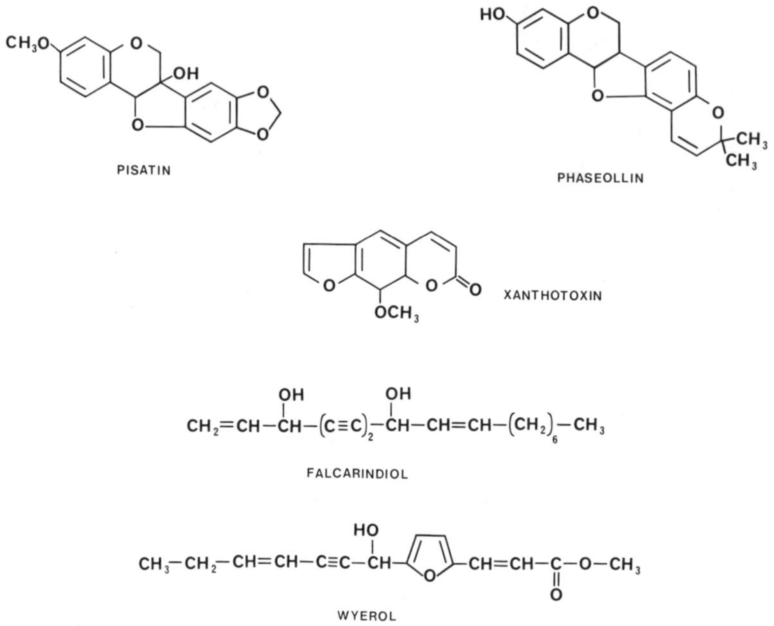

PISATIN

PHASEOLLIN

XANTHOTOXIN

$CH_2{=}CH{-}CH(OH){-}(C{\equiv}C)_2{-}CH(OH){-}CH{=}CH{-}(CH_2)_6{-}CH_3$

FALCARINDIOL

WYEROL

Fig. 6. Stress-metabolites produced by intact plants in vivo in response to fungal attack.

Purified glucan extracted from cell walls of Phytophthora megasperma var. sojae induced phenylpropanoid metabolism and increased the activity of PAL in cell suspensions of Glycine max. The cells also accumulated the antibiotic glyceollin in response to this fungal glucan. Similarly, the activity of PAL was stimulated in cell suspensions of Petroselinum hortense, and Acer pseudoplatanus treated with glucan.[370] Essentially the same effect was observed using cultured cells of Glycine max and glucan from P. megasperma f. sp. glycinea, and a quinoid red pigment related to glyceollin was isolated.[375]

The specific activity of PAL is usually low in cultured plant cells, but rises dramatically in response to "stress", for example, nitrate removal from the medium.[114] Glucan elicitors stimulate PAL activity independently of nitrate depletion. Furthermore, uptake of nitrate and cell growth are reduced or stopped in response to exogenously added glucan. Cells not treated with elicitor failed to synthesize glyceollin, or any related compounds, even after the onset of nitrate depletion. Isoflavonoid stress-metabolite synthesis, however, followed PAL increase in treated cells for up to 40 hours.[370] Elicitor from *Alternaria carthami* inhibits phosphate uptake by *Petroselinum hortense* cells which leads to an increase in vacuolar phosphate concentration and decreased cytoplasmic phosphate. The rapid decrease in cytoplasmic phosphate may be significant in regulation of phenylpropanoids.[378]

Enzymes of general phenylpropanoid metabolism were induced rapidly when cells of *P. hortense* contact carbohydrate elicitor from *Phytophthora megasperma* var. *sojae*,[372] *P. megasperma* f. sp. *glycinea* or *Alternaria carthami*.[379,380] The enzymes involved in flavonoid glycoside production were not detected under those conditions. However, the accumulation of furanocoumarins in this system in response to fungus-induced stress suggests the existence of a third group of enzymes involved in an elicitor-specific response.

Dixon et al.[367] studied elicitor-mediated enzyme induction in cultured cells of *Phaseolus vulgarus* using heat-released carbohydrate elicitors from cell walls of *Colletotrichum lindemuthianum*. They too found that these fungal products induced various enzymes of the flavonoid pathway, and led to accumulation of phaseollin.

A similar system operates in *Canavalia ensiformis* callus culture, and the isoflavonoid medicarpin and requisite enzymes are synthesized in response to viable spores of *Pithomyces chartarum*. The elicitor was not identified. Spores also elicited isoflavonoid O-methyltransferase.[373] Medicarpin was detected in alfalfa (*Medicago sativa*) callus tissue inoculated with zoospores of *Phytophthora megasperma*.[381]

Terpenoid synthesis can also be induced by elicitors. Eresk and Sziraki[371] showed that cell-free homogenates of *P. infestans* elicited synthesis of rishitin and phytuberin in callus tissue of *Solanum tuberosum*. Again, the active elicitor component was not identified. Helgeson et al.[382] were able to elicit capsidiol in callus tissues of *Nicotiana tabacum* in response to cell wall material and zoospores of *P. paristica* var. *nicotianae*. Budde and Helgeson[383] also showed that tobacco callus tissue challenged by zoospores of *P. parisitica* var. *nicotianae* synthesized four different stress-metabolites including rishitin and capsidiol.

Recently, 6-methoxymellein was induced in *Daucus carota* cell suspensions in response to infestation by *Chaetomium globosum*.[374]

Callus culture of *Ruta graveolens* when grown in the presence of fungi synthesized up to a ten-fold increase in acridone epoxide alkaloids.[384]

Tietjen and Matern[385] were the first to induce unspecified polyacetylenic stress-metabolites in cell suspensions of *Carthamus tinctorius* using cell wall material derived from *Alternaria carthami*. *Bidens alba* callus tissue has failed to produce any polyacetylenes (e.g. 1-phenylhepta-1,3,5-triyne) after 3 years of repeated subculturing on a variety of synthetic media with various hormonal regimes (Norton and Towers unpublished), in light or dark, even though these chemicals are prominent in the intact plant. Up to 430 μg phenylheptatriyne/gm fresh weight of leaves have been detected in intact leaves. However, callus cells of *B. alba* (cited as *B. pilosa*) produced up to 6.4 μg/gm callus of phenylheptatriyne after treatment with sterile, cell-free culture filtrates derived from the fungus *Pythium aphanidermatum*.[28] Culture filtrates from *Septoria astericola*, and *Rhizopus nigricans* also produced positive responses (DiCosmo and Towers unpublished).

Antibiotic stress-metabolites (phytoalexins) have been induced in nine plant species cultured in vitro by cell wall material and spores of various fungi belonging to two taxonomic classes. Some of these are illustrated in Figure 7. The active elicitor was a glucan in some cases. Phytoalexin production represents only a part of a major stimulation of secondary metabolism, a part that has been easily recognized because its products are antibiotic. The majority of studies have focused on the induced phenylpropanoid pathway and its enzymes. Less attention has been devoted to other metabolic and biochemical changes caused by fungal-induced stress in cultured plant cells.

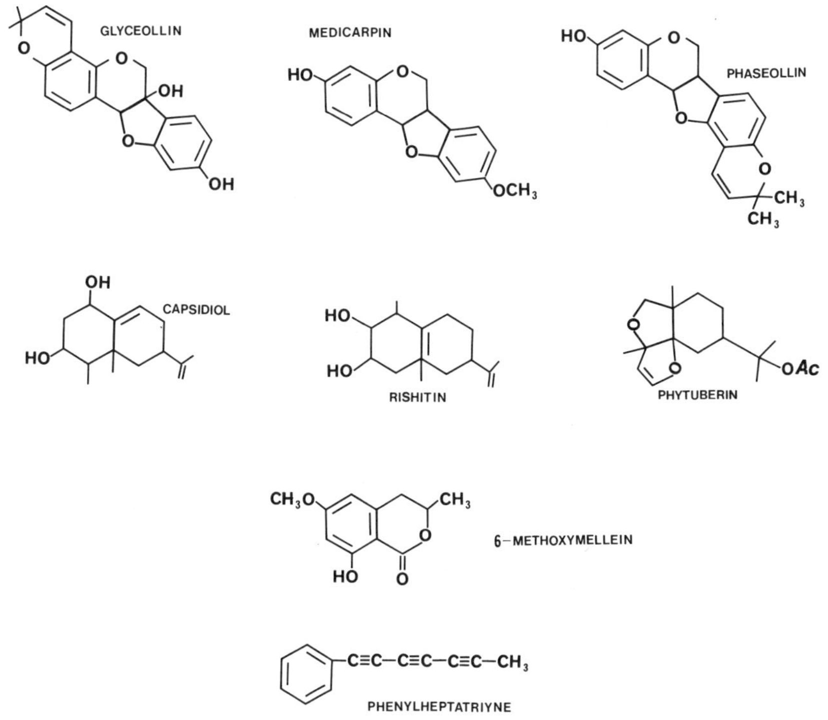

Fig. 7. Stress-metabolites produced by cultured plant cells in response to fungal contact.

Recently, two sesquiterpenoids, phytuberin and phytuberol, were induced in _N. tabacum_ callus cells inoculated with the bacterium _Pseudomonas solanacearum_.[386]

The use of tissue culture systems to study cell-cell interaction and the induction of secondary metabolism using various elicitors will become increasingly important in the coming years. One reason is that plant tissue cultures are potentially valuable sources of desirable pharmaceuticals and novel biological compounds. However, experience indicates that relatively few cultures produce secondary phytochemicals over extended periods in amounts comparable to those synthesized in intact plants. There are also the exciting possibilities that different fungal elicitor fractions activate different bioynthetic pathways, or the same elicitor may induce different responses in different plant species in culture.

CONCLUSIONS

We have presented considerable evidence which indicates that stress, i.e. nutritional, hormonal, and other environmental factors, including microbial insult, affect dramatically both the physiology and secondary metabolism of cultured plant cells. However, the phytochemical effects of certain types of stress are not predictable. One of the features of cultured cells is the activation of genes coding for compounds not usually produced. This response can occur through the stress-mediated induction of specific mRNA, which may result from altered nutrition, light regime, pH, temperature, anoxia or microbial contact. Alternatively, some compounds characteristic of the intact plant may not be synthesized.

Generally, cultured cells show a sigmoid growth curve characterized by an initial balanced metabolism of all essential nutrients with little accumulation of secondary products, followed by termination of this metabolic flow by exhaustion of nutrients or altered environment. It is probable that as growth rates decline, amino acids, nucleotides and precursors accumulate. The reduced primary metabolic flow imposes stress on the cells which results in declined growth and specific enzymatic perturbations. Evidence suggests that secondary product synthesis is greatest when the growth rates, carbohydrate, mineral and oxygen uptake decline. However, in some cases there are significant increases in growth concomitant to product synthesis. Other efficiently regulated changes that accompany the lowered growth rate, such as the accumulation of amino acids and other precursors, being more directly involved in stimulating secondary metabolism.

One of the goals of the study of secondary metabolism and altered phytochemical patterns of cultured cells is to be able to switch on specific metabolic sequences leading to rare or fine chemicals for medicine or industry. Although significant advances have been made, and interesting and novel compounds are isolated with increasing frequency, the regulation of secondary pathways is still relatively poorly understood.

The enzymes and partial regulation involved in the synthesis of lignin, flavones, and flavonoid glycosides and the biosynthetically related cinnamate esters are essentially known. This is certainly not true for the other classes of secondary phytochemicals.

One reason for this lacuna in our knowledge is that the majority of enzymes involved in specific phytochemical syntheses have not yet been characterized. Other reasons include the seemingly confused, confusing and unpredictable effects of various culture-induced stresses on phytochemical syntheses; this simply reflects unknown regulatory controls. Also, the interaction between primary metabolism and altered secondary metabolism needs careful scrutiny. These areas of molecular biology provide ample opportunity for novel approaches to our understanding of the control of secondary metabolism.

The ultimate goal of research involved with phytochemical production in cell cultures, is, of course, to predict the parameters necessary for formation of the various classes of secondary metabolites. Our survey has convinced us that this will eventually be possible and it will necessarily involve concerted efforts aimed at elucidating the mechanisms of regulatory control, and how to manipulate them.

ACKNOWLEDGMENTS

Undoubtedly, many interesting and pertinent studies were omitted from this survey, either through space limitation or oversight; we apologize both to our readers and especially to the authors of those reports.

Ms. Lindsay Brooke typed and retyped the original manuscript for which we are grateful.

This study was made possible by grants from the Natural Sciences and Engineering Research Council of Canada and an NSERC Postdoctoral Research Fellowship to F. DiCosmo.

REFERENCES

1. GRIME JP 1981 Plant strategies in shade. In Plants and the Daylight Spectrum (H Smith, ed) Academic Press New York pp. 159-186

2. DIXON RA, KW FULLER 1976 Effects of synthetic auxin levels on phaseollin production and phenylalanine ammonia lyase (PAL) activity in tissue cultures of Phaseolus vulgaris L. Physiol Plant Pathol 9: 299-312
3. GUSTINE DL 1981 Evidence for sulfydryl involvement in regulation of phytoalexin accumulation in Trifolium repens callus tissue cultures. Plant Physiol 68: 1323-1326
4. REILLEY JJ, WL KLARMAN 1980 Thymine dimer and glyceollin accumulation in U.V.-irradiated soybean suspension cultures. Environ Exp Bot 20: 131-134
5. GAFNI Y, I SHECHTER 1981 Diethylene glycol disulfide from castor bean cell suspension cultures. Phytochemistry 20: 2477-2479
6. DOLLER G, AW ALFERMAN, E REINHARD 1976 Production of indole alkaloids in tissue cultures of Catharanthus roseus. Planta Med 30: 14-20
7. TABATA M, N HIRAOKA 1976 Variation of alkaloid production in Nicotiana rustica callus cultures. Physiol Plant 38: 19-23
8. TANAKA H, Y MACHIDA, H TANAKA, N MUKAI, M MISAWA 1974 Accumulation of glutamine by suspension cultures of Symphytum officinale. Agric Biol Chem 38: 987-992
9. WATANABE K, S-I YANO, Y YAMADA 1982 The selection of cultured plant cell lines producing high levels of biotin. Phytochemistry 21: 513-516
10. ZENK MH, H EL-SHAGI, U SCHULTE 1975 Anthraquinone production by cell suspension cultures of Morinda citrifolia. Planta Med Suppl 79: 79-101
11. ZENK MH, H EL-SHAGI, H ARENS, H STOCKIGT, EW WEILER, B DEUS 1977 Formation of the indole alkaloids serpentine and ajmalicine in cell suspension cultures of Catharanthus roseus. In Plant Tissue Culture and its Bio-technological Application (W. Barz, W. Reinhard, M.H. Zenk, eds) Springer-Verlag, Berlin, New York, pp. 27-43
12. ELLIS BE 1980 Accumulation of hydroxyphenylethanol glucosides in cultured cells of Syringa vulgaris. Hoppe Seyler's Z. Physiol. Chem. 361: 242
13. BERLIN J, KH KNOBLOCH 1980 Biochemical characterization of a variant tobacco cell line accumulating high levels of cinnamoyl putrescine

derivatives. Hoppe Seyler's Z Physiol Chem 361: 219-220

14. BERGMAN L, H RENNENBERG 1978 Efflux und Production von Glutathion in Suspensiokulturen von Nicotiana tabacum. Z Pflanzenphysiol 88: 175-185
15. RADWAN SS, CK KOKATE 1980 Production of higher levels of trigonelline by cell cultures of Trigonella foenum graecum than by the differentiated plant. Planta 147: 340-344
16. STABA EJ, B KAUL 1971 Production of diosgenin by plant tissue culture technique. US Patent 3,628,287
17. FURUYA T, T YOSHIKAWA, H KIYOHARA 1983 Alkaloid production in cultured cells of Dioscoreophyllum cumminsii. Phytochemistry 22: 1671-1673
18. FUJITA Y, Y MAEDA, C SUGA, T MORIMOTO 1983 Production of shikonin derivatives by cell suspension cultures of Lithospermum erythrorhizon. III Comparison of shikonin derivatives of cultured cells and ko-shikon. Plant Cell Reports 2: 192-193
19. AKASU M, H ITOKAWA, M FUJITA 1976 Biscolaurine alkaloids in callus tissues of Stephania cepharantha. Phytochemistry 15: 471-473
20. BUTCHER DN, JD CONNOLLY 1971 An investigation of factors which influence the production of abnormal terpenoids by callus cultures of Andrographis paniculata Nees. J Exp Bot 22: 314-322
21. DELFEL NE, JA ROTHFUS 1977 Antitumor alkaloids in callus cultures of Cephalotaxus harringtonia. Phytochemistry 16: 1595-1598
22. STOCKIGT J, J TREIMER, MH ZENK 1976 Synthesis of ajmalicine and related indole alkaloids by cell free extracts of Catharanthus roseus cell suspensions. FEBS Lett 70: 267-270
23. NEUMANN D, E MULLER 1971 Beiträge zur Physiologie der Alkaloide. V. Alkaloidbildung in Kallus- und Suspensionskulturen von Nicotiana tabacum L. Biochem Physiol Pflanzen 162: 503-513
24. TABATA M, M TSUKUDA, H FUKUI 1982 Antimicrobial activity of quinone derivatives from Echium lycopsis callus cultures. Planta Med 44: 234-236
25. ARENS H, HO BORBE, B ULBRICH, J STOCKIGT 1982 Detection of pericine, a new CNS-active indole alkaloid from Picralima nitida cell suspension culture by opiate receptor binding studies. Planta Med 46: 210-214

26. BARZ W, E REINHARD, MH ZENK (eds) 1977 Plant Tissue Culture and its Bio-technological Application. Springer-Verlag, Berlin, New York, 419 pp
27. CAREW DP, RJ KRUEGER 1977 Catharanthus roseus tissue culture: the effects of medium modifications on growth and alkaloid production. J Nat Prod 4: 326-336
28. DiCOSMO F, RA NORTON, GHN TOWERS 1982 Fungal culture-filtrate elicits aromatic polyacetylenes in plant tissue culture. Naturwissenschaften 69: 550-551
29. FLECK J, A DURR, MC LETT, L HIRTH 1979 Changes in protein synthesis during the initial stage of life of tobacco protoplasts. Planta 145: 279-285
30. FLECK J, A DURR, C FRITSCH, MC LETT, L HIRTH 1980 Comparison of proteins synthesized in vivo and in vitro by mRNA from isolated protoplasts. Planta 148: 453-454
31. KURZ WGW, F CONSTABEL 1979 Plant cell cultures, a potential source of pharmaceuticals. Adv Appl Microbiol 25: 209-240
32. RADWAN, SS, F SPENER, HK MANGOLD, F STABA 1975 Lipids in plant tissue cultures. IV. The characteristic patterns of lipid classes in callus cultures and suspension cultures. Chem Phys Lipids 14: 72-80
33. REINERT J, YPS BAJAJ (eds) 1977 Applied and Fundamental Aspects of Plant Cell, Tissue and Organ Culture. Springer-Verlag, Berlin, New York, 716 pp
34. STECK W, BK BAILEY, JP SHYLUK, OL GAMBORG 1971 Coumarins and alkaloids from cell cultures of Ruta graveolens. Phytochemistry 10: 191-194
35. HALDER T, VN GADGIL 1983 Fatty acids of callus tissues of six species of Curcurbitaceae. Phytochemistry 22: 1965-1967
36. HIROTANI M, T FURUYA 1977 Restoration of cardenolide synthesis in redifferentiated shoots from callus cultures of Digitalis purpurea. Phytochemistry 16: 610-611
37. EBEL J, AR AYERS, P ALBERSHEIM 1971 Host-pathogen interactions XII. Response of suspension-cultured soybean cells to the elicitor isolated from Phytophthora megasperma var. sojae, a fungal pathogen of soybeans. Plant Physiol 57: 775-779
38. EBEL J, K HAHLBROCK 1977 Enzymes of flavone and flavonol-glycoside biosynthesis. Coordinated and

selective induction in cell-suspension cultures of Petroselinum hortense. Eur J Biochem 75: 201-209

39. HARGREAVES JA, C SELBY 1978 Phytoalexin formation in cell suspensions of Phaseolus vulgaris in response to an extract of bean hypocotyls. Phytochemistry 17: 1099-1102

40. DOKE N, K TOMIYA 1980 Effect of hyphal wall components from Phytophthora infestans on protoplasts of potato tuber tissues. Physiol Plant Pathol 16: 169-176

41. DIXON RA, PM DEY, DL MURPHY, IM WHITEHEAD 1981 Dose responses for Colletotrichum lindemuthianum elicitor-mediated enzyme induction in French bean cell suspension cultures. Planta 151: 272-280

42. HAHLBROCK K, CJ LAMB, C PURWIN, J EBEL, E FAUTZ, E SCHAFER 1981 Rapid responses of suspension-cultured parsley cells to the elicitor from Phytophthora megasperma var. sojae. Plant Physiol 67:768 773

43. KURZ WGW, F CONSTABEL 1979 Plant cell suspension cultures and their biosynthetic potential. In Microbial Technology Vol 1 (H.J. Peppler, D. Perlman, eds), Academic Press, New York, San Francisco, London, pp 389-416

44. BUTCHER DN 1977 Secondary products in tissue cultures. In Plant Cell, Tissue, and Organ Culture (J. Reinert, YPS Bajaj eds), Springer-Verlag, Berlin, pp.668-693

45. ZAPROMETOV MN 1978 Enzymology and regulation of the synthesis of polyphenols in cultured cells. In Frontiers of Plant Tissue Culture 1978 (TA Thorpe ed), International Association for Plant Tissue Culture, Calgary pp 335-343

46. ALFERMANN AW, E REINHARD eds 1978 Production of Natural Compounds by Cell Culture Methods. Gesellschaft fur Strahlen und Umweltforschung mbH, Munich. 361 pp

47. STABA EJ ed 1980 Plant Tissue Culture as a Source of Biochemicals. CRC Press, Boca Raton, Florida, 285 pp

48. SCHULER ML 1981 Production of secondary metabolites from plant tissue-culture - problems and prospects. Ann NY Acad Sci 369: 65-79

49. BARZ W, BE ELLIS 1981 Potential of plant cell cultures for pharmaceutical production. In Natural

Products as Medicinal Agents. (JL Beal, E Reinhard eds) Hippokrates-Verlag, Stuttgart, pp 471-507

50. BOHM H 1980 The formation of secondary metabolites in plant tissue and cell cultures. Int Rev Cytol Suppl 11B: 183-208
51. ELLIS BE Selection of chemically-variant plant cell lines for use in industry In Applications of Plant Cell and Tissue Culture to Agriculture and Industry (D Tomes, B Ellis, P Harney, K Kasha, RL Peterson eds) University of Guelph, Guelph, pp 63-80
52. KNOBLOCH KH, G BAST, J BERLIN 1982 Medium- and light-induced formation of serpentine and anthocyanins in cell suspension cultures of Catharanthus roseus. Phytochemistry 21: 591-594
53. TAL B, J GRESSEL, I GOLDBERG 1982 The effect of medium constituents on growth and diosgenin production by Dioscorea deltoidea cells grown in batch cultures. Planta Med 44: 111-115
54. DAVIES ME 1972 Polyphenol synthesis in cell suspension cultures of Paul's Scarlet Rose. Planta 104: 50-65
55. WESTCOTT RJ, GG HENSHAW 1976 Phenolic synthesis and phenylalanine ammonia-lyase activity in suspension cultures of Acer pseudoplatanus L. Planta 131: 67-73
56. CONSTABEL F 1968 Gerbstoffproduktion der Calluskulturen von Juniperus communis L. Planta 79: 58-64
57. MIZUKAMI H, M KONASHIMA, M TABATA 1977 Effect of nutritional factors on shikonin derivative formation in Lithospermum callus cultures. Phytochemistry 16: 1183-1186
58. MATSUMOTO T, K NISHIDA, M NOGUCHI, E TAMAKI 1973 Some factors affecting the anthocyanin formation by Populus cells in suspension culture. Agric Biol Chem 37: 561-567
59. AMORIN HB, DK DOUGALL, WR SHARP 1977 The effect of carbohydrate and nitrogen concentration on phenol synthesis in Paul's Scarlet Rose cells grown in tissue culture. Physiol Plant 39: 91-95
60. IKEDA T, T MATSUMOTO, M NOGUCHI 1976 Effects of nutritional factors on the formation of ubiquinone by tobacco plant cells in suspension culture. Agric Biol Chem 40: 1765-1770
61. GRAS M, J CRECHE, JC CHENIEUX, M RIDEAU 1982 Etude comparee des effects de al selection et des

facteurs de l'environnement sur l'accumulation alcoloidique des souches de Choisya ternata. Planta Med 46: 231-235

62. ABE S, Y OHTA 1983 Lunularic acid in cell suspension cultures of Marchantia polymorpha. Phytochemistry 22: 1917-1920
63. NAG TN, P KHANNA 1973 Effect of phenylalanine and glucose on growth and phyllemblin production in Emblica officinalis Gaertn tissue cultures. Indian J Pharm 35: 154-155
64. KADKADE PG 1982 Growth and podophyllotoxin production in callus tissue of Podophyllum peltatum. Plant Sci Lett 25: 107-115
65. HESS D, G LEIPOLDT, RD ILLG 1979 Investigations on the lactose induction of β-galactosidase activity in callus tissue cultures of Nemesia strumosa and Petunia hybrida. Pflanzenphysiol 94: 45-53
66. NETTLESHIP L, M SLAYTOR 1974 Adaption of Peganum harmala callus to alkaloid production. J Exp Bot 25: 114-123
67. TAKAYAMA S, M MISAWA, K KO, T MISATO 1977 Effect of cultural conditions on the growth of Agrostemma githago cells in suspension culture and the concomitant production of an anti-plant virus substance. Physiol Plant 41: 313-320
68. DOLLER G 1978 Influence of the medium on the production of serpentine by suspension cultures of Catharanthus roseus (L) G Don. In Production of Natural Compounds by Cell Culture Methods (AW Alfermann, E Reinhard, eds). Gesellschaft für Strahlen- und Umweltforschung mbH, Munich pp 109-116
69. KNOBLOCH K-H, J BERLIN 1980 Influence of medium composition on the formation of secondary compounds in cell suspension cultures of Catharanthus roseus (L) G. Don. Z. Naturforsch 35(c): 551-556
70. CAREW DP, RJ KREUGER 1976 Anthocyanidins of Catharanthus roseus callus cultures. Phytochemistry 15: 442
71. ZENK MH, H EL-SHAGI, B ULBRICH 1977 Production of rosmarinic acid by cell-suspension cultures of Coleus blumei. Naturwissenschaften 64: 585-586
72. BAUCH H-J, E LEISTNER 1978 Aromatic metabolism in cell suspension cultures of Galium mullugo. Planta Med 33: 105-123
73. KNOBLOCH K-H, G BEUTNAGEL, J BERLIN 1981 Influence of accumulated phosphate on culture growth and

formation of cinnamoyl putrescines in medium-induced cell suspension cultures of Nicotiana tabacum. Planta 153: 582-585

74. CHANDLER SF, JH DODDS 1983 The effect of phosphate, nitrogen and sucrose on the production of phenolics and solasodine in callus cultures of Solanum laciniatum. Plant Cell Reports 2: 205-208
75. RAVISHANKAR GA, AR MEHTA 1979 Control of ecdysterone biogenesis in tissue cultures of Trianthema portulacustrum. J Nat Prod 42: 152-158
76. OKAZAKI M, F HINO, K NAGASAWA, Y MIURA 1982 Effects of nutritional factors on formation of scopoletin and scopolin in tobacco tissue cultures. Agric Biol Chem 46: 601-607
77. EDLEMAN J, AD HANSON 1971 Sucrose suppression of chlorophyll synthesis in carrot callus cultures. Planta 98: 150-156
78. JASPARS EM, H VERDSTRA 1965 An α-amylase from tobacco crown-gall tissue cultures II. Measurement of the activity in media and tissues. Physiol Plant 18: 626-634
79. KARSTENS WHK, V DE MEESTER-MANGER CATS 1960 The cultivation of plant tissues in vitro with starch as a source of carbon. Acta Bot Neer 9: 263-274
80. MATSUMOTO T, K OKUNISHI, K NISHIDA, M NOGUCHI, E TAMAKI 1971 Studies on the culture conditions of higher plant cells in suspension culture. Part II. Effect of nutritional factors on the growth. Agric Biol Chem 35: 543-551
81. MITCHELL ED, BB JOHNSTON, T WHITTLE 1980 β-Galactosidase activity in cultured cotton cells: a comparison between cells grown on sucrose and lactose. In Vitro 16: 907-912
82. CHAUBET N, V ELIARD, A PAREILLEUX 1981 β-Galactosidases of suspension cultured Medicago sativa cells growing on lactose. Plant Sci Lett 22: 369-378
83. CHAUBET N, A PAREILLEUX 1982 Characterization of β-galactosidases of Medicago sativa suspension-cultured cells growing octose. Effect of the growth substrates on the activities. Z Pflanzenphysiol 106: 401-407
84. SACHER JA, KT GLASZIOU 1962 Regulation of invertase levels in sugar cane by auxin-carbohydrate mediated control system. Biochem Biophys Res Commun 8: 280-282

85. DEMAIN AL 1968 Regulatory mechanisms and the industrial production of microbial metabolites. Lloydia 31: 395-418
86. DREW SW, AL DEMAIN 1977 Effect of primary metabolites on secondary metabolism. Ann Rev Microbiol 31: 343-356
87. JINDAL HK, VS BANSAL, C KASINATHAN, S LARROYA, GK KHULLER 1983 Effect of carbon sources on the polar lipid fatty acids of Microsporum gypseum grown at different temperatures. Experientia 39: 151-153
88. BEHMER CJ, AL DEMAIN 1983 Further studies on carbon catabolite regulation of β-lactam antibiotic synthesis in Cephalosporium acremonium. Curr Microbiol 8: 107-114
89. AP REES T 1980 Integration of pathways of synthesis and degradation of hexose phosphates. In The Biochemistry of Plants, Vol 3 (J Preiss, ed), Academic Press, New York pp 1-42
90. DAVIES DD 1979 The central role of phosphoenolpyruvate in plant metabolism. Ann Rev Plant Physiol 30: 131-158
91. PREISS J, T KOSUGE 1976 Regulation of enzyme activity in metabolic pathways. In Plant Biochemistry (J Bonner, JE Varner eds), Academic Press, New York pp 277-336
92. PRYKE JA, T AP REES 1976 Activity of the pentose phosphate pathway during lignification. Planta 132: 279-284
93. PRYKE JA, T AP REES 1977 The pentose phosphate pathway as a source of NADPH for lignin biosynthesis. Phytochemistry 16: 557-560
94. MIFLIN BH 1973 Amino acid biosynthesis and its control in plants. In Biosynthesis and Its Control in Plants (BV Milborrow, ed) Academic Press, New York, pp 49-68
95. MIFLIN BJ 1976 Modification controls in time and space. In Regulation of Enzyme Synthesis and Activity in Higher Plants (H. Smith ed), Academic Press, New York pp 23-40
96. JESSUP W, MW FOWLER 1976 Interrelationships between carbohydrate metabolism and nitrogen assimilation in cultured plant cells. II. The effect of the nitrogen source and concentration on nutrient uptake and respiratory activity in cultured sycamore cells. Planta 132: 125-129

97. JESSUP W, MW FOWLER 1976 Interrelationships between carbohydrate metabolism and nitrogen assimilation in cultured plant cells. I. Effects of glutamate and nitrate as alternative nitrogen sources in cell growth. Planta 132: 119-123
98. SEENI S, A GNANAM 1982 Carbon assimilation in photoheterotrophic cells of peanut (*Arachis hypogaea* L.) grown in still nutrient medium. Plant Physiol 70: 823-826
99. DOUGALL DK 1977 Current problems in the regulatin of nitrogen metabolism in plant cell cultures. *In* Plant Cell Culture and its Bio-technological Application (W. Barz, W Reinhard, MH Zenk, eds), Springer-Verlag, Berlin, New York pp 76-94
100. BRADFORD JA, JS FLETCHER 1982 Influence of protein on NO_3^- reduction NH_4^+ accumulation, and amide synthesis in suspension cultures of Paul's Scarlet Rose. Plant Physiol 69: 63-66
101. JONES LH, JN BARRETT, PPS GOPAL 1973 Growth and nutrition of a suspension culture of *Pogostemon cablin* Benth. J Exp Bot 24: 145-158
102. ROSE D, SM MARTIN 1974 Parameters for growth measurement in suspension cultures of plants cells. Can J Bot 52: 903-912
103. ROSE D, SM MARTIN 1975 Effect of ammonium on growth of plant cells (*Ipomoea* sp.) in suspension cultures. Can J Bot 53: 1942-1949
104. MIFLIN BJ, PJ LEA 1976 The pathway of nitrogen assimilation in plants. Phytochemistry 15: 873-885
105. MIFLIN BJ, PJ LEA 1982 Ammonia assimilation and amino acid metabolism. *In* Nucleic Acids and Proteins in Plants Encyclopedia of Plant Physiology New Series, Vol. 14A (D Boulter, B Parthier eds). Springer-Verlag, Britain pp 5-64
106. ZIELKE HK, P FILNER 1971 Synthesis and turnover of nitrate reductase induced by nitrate in cultured tobacco cells. J Biol Chem 246: 1772-1779
107. CHROBOCZEK-KELKER H, P FILNER 1971 Regulation of nitrate and reductase and its relationship to the regulation of nitrate reductase in cultured tobacco cells. Biochim Biophys Acta 252: 69-82
108. FILNER P 1966 Regulation of nitrate reductase in cultured tobacco cells. Biochim Biophys Acta 118: 299-310
109. HEIMER YM, E RIKLIS 1979 On the mechanism of development of nitrate reductase activity in tobacco cells. Plant Sci Lett 16: 135-138

110. BEHREND J, RI MATELES 1975 Nitrogen metabolism in plant cell suspension cultures. Plant Physiol 56: 584-589
111. DOUGALL RK 1977 Current problems in the regulation of nitrogen metabolism in plant cell cultures. In Plant Tissue Culture and Its Bio-technological Application, Springer-Verlag, Berlin, pp 76-84
112. GAINES CG, GS BYNG, RJ WITAKER, RA JENSEN 1982 L-Tyrosine regulation and biosynthesis via arogenate dehydrogenase in suspension-cultured cells of Nicotiana silvestris Speg. et Comes. Planta 156: 233-240
113. JORDAN DB, JS FLETCHER 1979 The relationship between NO_2^- accumulation, nitrate reductase and nitrite reductase in suspension cultures of Paul's Scarlet Rose. Plant Sci Lett 17: 95-99
114. HAHLBROCK K 1974 Correlation between nitrate uptake, growth and changes in metabolic activities of cultured plant cells. In Tissue Culture and Plant Science (HE Street, ed) Academic Press London pp 363-378
115. PAHLICH E, C GERLITZ 1980 Deviations from Michaelis-Menten behaviour of plant glutamate dehydrogenase with ammonium as variable substrate. Phytochemistry 19: 11-13
116. OHTA S, O MATSUI, M YATAZAWA 1978 Culture conditions for nicotine production in tobacco tissue culture. Agric Biol Chem 42: 1245-1251
117. MEHTA AT, R SHAILAJA 1978 Front Plant Tissue Cult, IAPTC Congr 4th, Calgary Alberta, poster #1406
118. MISAWA M, K SAKATO, H TANAKA, M HAYASHI, M SAMEJIMA 1974 Production of physiologically active substances by plant cell suspension cultures. In Tissue Culture and Plant Science (HE Street, ed), Academic Press, London pp 405-432
119. MISAWA M, H TANAKA, O CHIYO, N MUKAI 1975 Production of a plasmin inhibitory substance by Scopalia japonica suspension cultures. Biotechnol Bioeng 17: 305-334
120. RENNENBERG HL, L BERGMAN 1979 Einfluss von Ammonium und Sulfat auf die Glutathion-Produktion in Suspensionkulturen von Nicotiana tabacum. Z Pflanzenphysiol 92: 133-142
121. SASSE F, U HECKENBERG, J BERLIN 1982 Accumulation of β-carboline alkaloids and serotonin by cell cultures of Peganum harmala L. Plant Physiol. 69: 400-404

122. WINK M, T HARTMAN, L WITTE 1980 Biotransformation of cadaverine and potential biogenetic intermediates of lupanine biosynthesis by plant cell suspension cultures. Planta Med 40: 31-39
123. STECK W, OL GAMBORG, BK BAILEY 1973 Increased yields of alkaloids through precursor biotransformation in cell suspension Ruga graveolens. Lloydia 36: 93-95
124. TABATA M, H YAMAMOTO, N HIRAOKA, M KONOSHIMA 1972 Organization and alkaloid production in tissue cultures of Scopolia parviflora. Phytochemistry 11: 949-955
125. HIRAOKA N, M TABATA, M KONOSHIMA 1973 Formation of acetyltropine in Datura callus cultures. Phytochemistry 12: 795-799
126. RAZZAQUE A, BE ELLIS 1977 Rosmarinic acid production in Coleus cell cultures. Planta 137: 287-291
127. SAIRAM TV, P KHANNA 1971 Effect of tyrosine production of alkaloids in Datura tatula tissue cultures. Lloydia 34: 170-174
128. KAMIMURA S, M AKUTSU, M NISHIKAWA 1976 Formation of thebaine in suspension cultures Papaver bracteatum. Agric Biol Chem 40: 913-919
129. KUANG-CHICH C, L CHENG 1981 Callus cultures of the three well-known Chinese herbs and their medicinal contents. In Plant Tissue Culture. (H Hu ed) Pitman Publishing Ltd London pp 469-479
130. See reference 124.
131. KRUEGER RJ, DP CAREW 1978 Catharanthus roseus tissue culture: the effects of precursors on growth and alkaloid production. Lloydia 41: 327-331
132. DEUS B 1978 Zellkulturen von Catharanthus roseus. In Production of Natural Compounds by Cell Culture Methods (AW Alfermann, E Reinhard, eds). Gesellschaft für Strahlen- und Umweltforschung mbH, Munich, pp 118-123
133. BERLIN J, L WITTE 1982 Metabolism of phenylalanine and cinnamic acid in tobacco cell lines with high and low yields of cinnamoyl putrescines. J Nat Prod 45: 88-93
134. VELIKY IA 1972 Synthesis of carboline alkaloids by plant cell cultures. Phytochemistry 11: 1405-1406
135. VELIKY IA, KM BARBER 1975 Biotransformation of tryptophan by Phaseolus vulgaris suspension culture. Lloydia 38: 125-130

136. KHANNA P, R KHANNA, M SHARMA 1978 Production of free ascorbic acid and effect of exogenous ascorbic acid and tyrosine on production of major opium alkaloids from in vitro tissue cultures of Papaver somniferum Linn. Indian J Exp Biol 16: 110-112
137. HUS A-F 1981 Effect of protein synthesis inhibitors on cell cultures of Papaver somniferum. J Nat Prod 44: 408-414
138. NETTLESHIP L, S SLAYTOR 1974 Limitations of feeding experiments in studying alkaloid biosynthesis in Peganum harmala callus cultures. Phytochemistry 13: 735-742
139. KAUL B, SJ STOHS, EJ STABA 1969 Dioscorea tissue cultures. III. Influence of various factors on diosgenin production by Dioscorea deltoidea callus and suspension cultures. Lloydia 32: 347-359
140. MARGNA U 1977 Control at the level of substrate supply - an alternative in the regulation of phenylpropanoid accumulation in plant cells. Phytochemistry 16: 419-426
141. WIDHOLM JM 1972 Tryptophan biosynthesis in Nicotiana tabacum and Daucus carota cell cultures: site of action of inhibitory tryptophan analogs. Biochim Biophys Acta 261: 44-51
142. BELSER WL, JB MURPHY, DP DELMER, SE MILLS End product control of tryptophan biosynthesis in extracts and intact cells of higher plant Nicotiana tabacum var. Wisconsin 38. Biochim Biophys Acta 237: 1-10
143. FLOSS HG, JE ROBBERS, PF HEINSTEIN 1974 Regulatory control mechanisms in alkaloid biosynthesis. Recent Adv Phytochem 8: 141-178
144. BUCHOLZ D, B REUPKE, H BICKEL, G SCHULTZ 1979 Reconstitution of amino acid synthesis by combining spinach chloroplasts with other leaf organelles. Phytochemistry 18: 1109-1111
145. BICKEL H, L PALME, G SCHULTZ 1978 Incorporation of shikimate and other precursors into aromatic amino acids and prenylquinones of isolated spinach chloroplasts. Phytochemistry 17: 119-124
146. AHARONOWITZ Y 1980 Nitrogen metabolite regulation of antibiotic biosynthesis. Ann Rev Microbiol 34: 209-233
147. SWUTZER RK 1977 The inactivation of microbial enzymes in vivo. Ann Rev Microbiol 31: 135-157

148. LORIMER GH, HS GEWITZ, W VOELKER, LP SOLOMONSON, B VENNESLAND 1974 The presence of bound cyanide in the naturally inactivated form of nitrate reductase of Chlorella vulgaris. J Biol Chem 249: 6074-6079
149. EDWARDS GE, SC HUBER 1981 The C_4 pathway. In The Biochemistry of Plants, a Comprehensive Treatise, Vol 8 (MD Hatch, NK boardman, eds), Academic Press, New York pp 237-281
150. WISKICH JT 1980 Control of the Krebs cycle. In The Biochemistry of Plants, a Comprehensive Treatise, Vol 2 (DD Davies, ed) Academic Press, New York pp 243-278
151. TURNER JF, DH TURNER 1980 The regulation of glycolysis and the pentose phosphate pathway. In The Biochemistry of Plants, a Comprehensive Treatise, Vol 2 (DD Davies, ed) Academic Press, New York pp 279-316
152. ROSS CE 1981 Biosynthesis of Nucleotides. In The Biochemistry of Plants, a Comprehensive Treatise, Vol 6 (A Marcus, ed) Academic Press, New York pp 169-205
153. MUDD JB 1980 Phospholipid biosynthesis. In The Biochemistry of Plants, a Comprehensive Treatise, Vol 4 (PK Stumpf, ed) Academic Press, New York pp 249-282
154. KNOBLOCH K-H, J BERLIN 1981 Phosphate mediated regulation of cinnamoyl putrescine biosynthesis in cell suspension cultures of Nicotiana tabacum. Planta Med 42: 167-172
155. HINZ H, MH ZENK 1981 Production of protoberine alkaloids by cell suspension cultures of Berberis species. Naturwissenschaften 68: 620-621
156. DOBBERSTEIN RH, EJ STABA 1969 Ipomoea, Rivea and Argyreia tissue cultures: influence of various chemical factors on indole alkaloid production and growth. Lloydia 32: 141-152
157. KURPINSKI VM, JE ROBBERS, HG FLOSS 1976 Physiological study of ergot: induction of alkaloid synthesis by tryptophan at the enzymatic level. J Bacteriol 125: 158-165
158. ERGE D, M MAIER, D GROEGER 1973 Untersuchungen über enzymatische Umwandlung von Chanoclavin -I. Biochem Physiol Pflanzen 164: 234-247
159. WEINBERG ED 1974 Secondary metabolism: control by temperature and inorganic phosphate. Dev Ind Microbiol 15: 70-81

160. TAL B, JS ROKEM, I GOLDBERG 1983 Factors affecting growth and product formation in plant cells grown in continuous culture. Plant Cell Reports 2: 219-222
161. BENJAMIN BD, MR HEBLE, MS CHADHA 1978 Alkaloid synthesis in tissue cultures and regenerated plants by *Tylophora indica* Merr. (Asclepidaceae). Z Pflanzenphysiol 92: 77-84
162. BEEVERS L 1982 Post-transitional modifications. *In* Nucleic Acids and Proteins in Plants. Encyclopaedia of Plant Physiology New Series Vol 14 A, (D Boulter, B Parthier, eds). Springer-Verlag, Berlin, pp 136-168
163. DEVLIN DM 1975 Plant Physiology. D Van Nostrand Co. New York 600 pp
164. RAINS DW 1976 Mineral metabolism. *In* Plant Biochemistry (J Bonner, JE Varner, eds). Academic Press, New York, pp 561-597
165. WEBSTER GC 1953 Enzymatic synthesis of gamma-glutamyl-cysteine in higher plants. Plant Physiol 28: 728-730
166. CROTEAU R, F KARP 1976 Enzymatic synthesis of camphor from neryl pyrophosphate by a soluble preparation from sage (*Salvia officinalis*). Biochem Biophys Res Commun 72: 440-447
167. FROST RG, CA WEST 1977 Properties of kaurene synthetase from *Marah macrocarpa*. Plant Physiol 59: 22-29
168. RAPPAPORT L, D ADAMS 1978 Gibberellins: a synthesis, compartmentation and physiological process. Phil Trans R Soc London B 284: 521-539
169. SCHIFF JA, RC HODSON 1973 The metabolism of sulfate. Ann Rev Plant Physiol 24: 381-414
170. ANDERSON JW 1980 Assimilation of inorganic sulphate in cysteine. *In* The Biochemistry of Plants, Vol. 5 (BJ Miflin, ed) Academic Press, New York pp 203-223
171. GIOVANELLI J, SH MUDD, AH DATKO 1980 Sulfur amino acids in plants. *In* The Biochemistry of Plants, Vol. 5 (BJ Miflin, ed) Academic Press, New York, pp 453-505
172. SCHWENN JD, U SCHRIEK, H KILTZ 1983 Dissimilation of methionine in cell suspension cultures from *Catharanthus roseus*. Planta 158: 540-549
173. CANTONI GL 1977 S-Adenosylmethionine: present status and future perspectives. *In* The

Biochemistry of Adenosylmethionine, (F. Salvatore, E Borek, V Zappia, HG Williams-Ashman, F Schlenk, eds), Columbia Univ Press, New York pp 557-577

174. CASTLE M, G BLONDIN, WR NES 1963 Evidence for the origin of the ethyl group of β-sitosterol. J Am Chem Soc 85: 3306-3308
175. CHEN Y, E ZAHAVI, P BARAK, N UMIEL 1980 Effects of salinity stresses on tobacco I. The growth of *Nicotiana tabacum* callus cultures under seawater, NaCl, and mannitol stresses. Z Pflanzenphysiol 98: 141-153
176. LIPETZ J 1962 Calcium and the control of lignification in tissue cultures. Amer J Bot 49:460-464
177. KLAPHECK S, W GROSSE, L BERGMANN 1982 Effect of sulfur deficiency on protein synthesis and amino acid accumulation in cell suspension cultures of *Nicotiana tabacum*. Z Pflanzenphysiol 108: 235-245
178. DIX PH, RS PEARCE 1981 Proline accumulation in NaCl resistant and sensitive cell lines of *Nicotiana sylvestris*. Z Pflanzenphysiol 102: 243-248
179. HEYSER JW, MW NABORS 1981 Osmotic adjustment of cultured tobacco cells (*Nicotiana tabacum* var. Samsum) grown on sodium chloride. Plant Physiol 67: 720-727
180. GARCIA FG, JW EINSET 1983 Ethylene and ethane production in 2,4-D treated and salt treated tobacco tissue cultures. Ann Bot 51: 287-295
181. MACKENZIE IA, HE STREET 1970 Studies of the growth in culture of plant cells. VIII. Production of ethylene by suspension cultures of *Acer pseudoplatanus* L. J Exp Bot 21: 824-834
182. SHAH RR, KV SUBBAIAH, AR MEHTA 1976 Hormonal effect on polyphenol accumulation in *Cassia* tissues cultured in vitro. Can J Bot 54: 1240-1245
183. McKENZIE E, L NETTLESHIP, M SLAYTOR 1975 New natural products from *Peganum harmala*. Phytochemistry 14: 273-275
184. LAM TH, HE STREET 1977 The effects of selected aryloxyalkanecarboxylic acids on the growth and levels of soluble phenols in cultured cells of *Rosa damascena*. Z Pflanzenphysiol 84: 121-128
185. See reference 180
186. KURKDJIAN A, Y MATHIEU, J GUERN 1982 Evidence for an action of 2,4-dichlorophenoxyacetic acid on the

vacuolar pH of Acer pseudoplatanus cells in suspension culture. Plant Sci Lett 27: 77-86
187. TABATA M, N HIRAOKA, M IKENOUE, Y SANO, M KONOSHIMA 1975 The production of anthaquinones in callus cultures of Cassia tora. Lloydia 38: 131-134
188. MULDER-KRIEGER T, R VERPOORTE, YP DE GRAAF, M VANDER KREEK, A BAERHEIM-SVENDSEN 1982 The effects of plant growth regulators and culture conditions on the growth and the alkaloid content of cultures of Cinchona pubsecens. Planta Med 46: 15-18
189. IKUTA A, K SYONO, T FURUYA 1975 Alkaloids in plants regenerated from Coptis callus cultures. Phytochemistry 14: 1209-1210
190. OZEKI Y, A KOMAMINE 1981 Induction of anthocyanin synthesis in relation to embryogenesis in a carrot suspension culture: correlation of metabolic differentiation with morphological differentiation. Physiol Plant 53: 570-577
191. ROMEIKE A 1975 Versuche zur Veresterung von Tropin und Tropasäure an Pflanzenzellkulturen. Biochem Physiol Pflanzen 168: 87-92
192. HIRAOKA N, M TABATA 1983 Acetylation of tropane derivatives by Datura innoxia cell cultures. Phytochemistry 22: 409-412
193. HARBORNE JB, J ARDITTI, E BALL 1970 (Abstract) The anthocyanins of callus culture from the stem of Dimorphotheca auriculata (Cape marigold, Compositae). Am J Bot 57: 763
194. MARSHALL JG, EJ STABA 1976 Hormonal effects of diosgenin biosynthesis and growth in Dioscorea deltoidea tissue cultures. Phytochemistry 15: 53-55
194a. HEINSTEIN P, H EL-SHAGI 1981 Formation of gossypol by Gossypium hirsutum L. cell suspension cultures. J Nat Prod 44:1-6
195. FUKUI H, M TSUKADA, H MIZUKAMI, M TABATA 1983 Formation of stereoisomeric mixtures of naphthoquinones derivatives in Echium lycopsis callus cultures. Phytochemistry 22: 453-456
196. STRICKLAND RG, N SUNDERLAND 1972 Production of anthocynins, flavonols, and chlorogenic acids by cultured callus tissues of Haplopappus gracilis. Ann Bot 36: 443-457
197. CONSTABEL F, JP SHYLUK, OL GAMBORG 1971 The effect of hormones on anthocyanin accumulation in cell

cultures of Haplopappus gracilis. Planta 96: 306-316

198. IBRAHIM RK, ML THAKUR, B PERMANAND 1971 Formation of anthocyanins in callus tissue cultures. Lloydia 34: 175-182
199. FURUYA T, H KOJIMA, K SYONO 1971 Regulation of nicotine biosynthesis by auxins in tobacco callus tissues. Phytochemistry 10: 1529-1532
200. SHIIO I, S OHTA 1973 Nicotine production by tobacco callus tissues and effect of plant growth regulators. Agric Biol Chem 37: 1857-1964
201. OKAZAKI M, F HINO, K KOMINAMI, Y MIURA 1982 Effects of plant hormones on formation of scopoletin and scopolin in tobacco tissue cultures. Agric Biol Chem 46: 609-614
202. ZITO SW, EJ STABA 1982 Thebaine from root cultures of Papaver bracteatum. Planta Med 45: 53-54
203. HODGES CC, H RAPOPORT 1982 Morphinan alkaloids in callus cultures of Papaver somniferum. J Nat Prod 45: 482-485
204. SMITH SL, GW SLYWKA, J KRUEGER 1981 Anthocyanins of Strobilanthes dyeriana and their production in callus culture. J Nat Prod 44: 609-610
205. VAN DEN BERG AJJ, RP LABADIE 1981 The production of acetate derived hydroxyanthraquinones, -dianthrones, -naphthalenes and L-benzenes in tissue cultures from Rumex alpinus. Planta Med 41: 169-173
206. PHILLIPS R, GG HENSHAW 1977 The regulation of synthesis of phenolics in stationary phase cell cultures of Acer pseudoplatanus L. J Exp Bot 28: 785-794
207. KING PJ 1976 Studies on the growth culture of plant cells. XX. Utilization of 2,4-dichlorophenoxyacetic acid by steady-state cell cultures of Acer pseudoplatanus. J Exp Bot 27: 1053-1072
208. AL-ABTA S, IJ GALPIN, HA COLIN 1979 Flavour compounds in tissue cultures of celery. Plant Sci Lett 16: 129-134
209. KAUL B, EJ STABA 1968 Dioscorea tissue cultures. I. Biosynthesis and isolation of diosgenin from Dioscorea deltoidea callus suspension cells. Lloydia 31: 171-179
210. NISHI A, I TSURITANI 1983 Effect of auxin on the metabolism of mevalonic acid in suspension-cultured carrot cells. Phytochemistry 22: 399-401

211. SHIMIZU K, T KIKUCHI, N SUGANO, A NISHI 1979 Carotenoid and steroid synthesis by carrot cells in suspension. Physiol Plant 46: 127-132
212. BARZ W, H HERZBECK, W HUSEMANN, G SCHNEIDERS, HK MANGOLD 1980 Alkaloids and lipids of heterotrophic, photomixotrophic and photoautotrophic cell suspension cultures of Peganum harmala. Planta Med 40: 137-148
213. RUCKER W, K JENTZSCH, W WICHTL 1981 Organdifferenzierung und Glykosidbildung bei in vitro kultivierten Blattexplantaten von Digitalis purpurea L.; Einfluss verschiedener Wuchsstoffe, Nahrlösungen und Lichtverhältnisse. Z Pflanzenphysiol 102: 207-220
214. BIESBOER DD, PG MAHLBERG 1979 The effect of medium modification on selected precursors on sterol production by short-term callus cultures of Euphorbia tirucalli. J Nat Prod 42: 648-657
215. TABATA M, H MIZUKAMI, N HIRAOKA, M KONOSHIMA 1974 Pigment formation in callus cultures of Lithospermum erythrorhizon. Phytochemistry 13: 927-932
216. SALEH MM 1982 (Abstract). Effect of environment on the growth and glycosides of callus tissue of Digitalis lanata. Planta Med 45: 135
217. STEARNS EM, WT MORTON 1975 Effects of growth regulators on fatty acids of soybean suspension cultures. Phytochemistry 14: 619-622
218. MISAWA M 1977 Production of natural substances by plant cell cultures described in Japanese patents. In Plant Tissue Culture and Its Bio-technological Application (W. Barz, E. Reinhard, MH Zenk eds.). Springer-Verlag, Berlin, pp 17-26
219. MILLER CO 1969 Control of deoxyisoflavone synthesis in soybean tissue. Planta 87: 26-35
220. MULLER W-U, E LEISTNER 1978 Aglycones and glucosides of oxygenated naphthalenes and glycosyl transferase from Juglans. Phytochemistry 17: 1739-1742
221. IBRAHIM RK, D EDGAR 1976 Phenolic synthesis in Perilla cell suspension cultures. Phytochemistry 15: 129-131
222. See reference 205
223. SCHMITZ M, U SEITZ 1972 Hemmung der Anthocyansynthese durch Gibberellinsäure A_3 bei

Kalluskulturen von Daucus carota. Z Pflanzenphysiol 68: 259-265

224. BRAUN G, U SEITZ 1975 Verlauf der Akkumulation von Kaffe-, Ferula- und Chlorogensäure in Beziehung zur Cyanodinakkumulation bei 2 Zellinien von Daucus carota. Biochem Physiol Pflanzen 168: 93-100
225. LUI JHC, EJ STABA 1981 Effects of age and growth regulators on serially propogated Digitalis lanata leaf and root cultures. Plant Med 41: 90-95
226. YAMADA Y, T KUBOI 1976 Significance of caffeic acid-O-methyltransferase in lignification of cultured tobacco cells. Phytochemistry 15: 395-396
227. KUBOI T, Y YAMADA 1976 Caffeic acid-O-methyltransferase in a suspension cell aggregates of tobacco. Phytochemistry 15: 397-400
228. COTTLE W, PE KOLATTUKUDY 1982 Abscissic acid stimulation of suberization. Inducation of enzymes and deposition of polymeric components and associated waxes in tissue cultures of potato tuber. Plant Physiol 70: 775-780
229. SOLIDAY CL, BB DEAN, PE KOLATTUKUDY 1978 Suberization: inhibition by washing and stimulation of abscissic acid in potato disks and tissue culture. Plant Physiol 61: 170-174
230. STOHS SJ, CS TAMG, H ROSENBERG 1977 Influence of cyclic AMP on growth and steroid formation in Corchorus olitorius L. suspension cultures. Lloydia 40: 370-373
231. MATHEW T, IC DAVE, BK GAUR 1978 Effect of gibberellic acid, indole-3-acetic acid and L-ascorbic acid on the regulation of RNA metabolism in de-etoliated barley shoot. Z Pflanzenphysiol 90: 391-396
232. NETTLESHIP L, M SLAYTOR 1971 Ruine: a glycosidic β-carboline from Peganum harmala. Phytochemistry 10: 231-234
233. ALLEN JRF, BR HOLMSTEDT 1980 The simple β-carboline alkaloids. Phytochemistry 19: 1573-1582
234. IKUTA A, H ITOKAWA 1982 Berberine and other protoberine alkaloids in callus tissue of Thalictrum minus. Phytochemistry 21: 1419-1421
235. GREGOR HD 1974 Einfluss von Gibberellinsäure A_3 auf die PAL-Aktivität und die Synthese von Phenylpropanderivaten in Zellkulturen von Haplopappus gracilis. Protoplasma 80: 273-277

236. SUGANO N, K HAYASHI 1967 Dynamic interrelation of cellular ingredients relevant to the biosynthesis of anthocyanin during tissue culture of carrot aggregen. Bot Mag (Tokyo) 80: 440-449

237. DAVIES ME 1972 Effects of auxin on polyphenol accumulation and the development of phenylalanine ammonia lyase activity in dark suspension cultures of Paul's Scarlet Rose. Planta 104: 66-77

238. WALLER GR, S MANGIAFICO, RC FOSTER, RH LAWRENCE 1981 Sterols of Delphinium ajacis production and metabolic relationships in whole plants and callus tissue. Planta Med 42: 344-355

239. RUCKER W, K JENTZSCH, M WICHTL 1981 Organdifferenzierung und Glykosidbildung bei in vitro kultivierten Blattexplantaten von Digitalis purpurea L.; Enfluss verschiedener Wuchsstoffe, Nahrlösungen und Lichtverhältnisse. Z Pflanzenphysiol 102: 207-220

240. BALAGUE C, A LATCHE, J FALLOT, J-C PECH 1983 Some physiological changes occurring during the senescence of auxin-deprived pear cells in culture. Plant Physiol 69: 1339-1343

241. ENTSCH B, CW PARKER, DS LETHAM 1983 An enzyme from lupin seeds forming alanine derivatives of cytokinins. Phytochemistry 22: 373-381

242. SEIBERT M, PG KADKADE 1980 Environmental factors: A. Light. In Plant Tissue Culture as a Source of Biochemicals (E.J. Staba ed.). CRC Press, Florida, pp 123-141

243. SENGER H 1982 The effect of blue light on plants and microorganisms. Phytochemistry 35: 911-920

244. GRISEBACH H, K HAHLBROCK 1974 Enzymology and regulation of flavonoid and lignin biosynthesis in plants and cell suspension cultures. Recent Adv Phytochem 8: 21-52

245. LACKMAN I 1971 Wirkungsspektren der Anthocyansynthese in Gewebekulturen und Keimlingen von Haplopappus gracilis. Planta 98: 258-269

246. STRICKLAND RG, N SUNDERLAND 1972 Photocontrol of growth, and of anthocyanin and chlorogenic acid production in cultured callus tissues of Haplopappus gracilis. Ann Bot 36: 671-685

247. GREGOR HD, J REINERT Induktion der Phenylalanin Ammoniumlyase in Gewebekulturen von Haplopappus gracilis. Protoplasma 74: 307-319

248. IBRAHIM RK, ML THAKUR, B PERMANAND 1971 Formation of anthocyanins in callus tissue cultures. Lloydia 34: 175-182
249. TAKEDA R, K KATOH 1981 Growth and sesquiterpenoid production by Calypogeia granulata Inoue cells in suspension culture. Planta 151: 525-530
250. WELLMAN E 1971 Phytochrome-mediated flavone glycoside synthesis in cell suspension cultures of Petroselinum hortense after preirradation with ultraviolet light. Planta 101: 283-286
251. DUELL-PFAFF N, E WELLMANN 1982 Involvement of phytochrome and a blue light photoreceptor in UV-B induced flavonoid synthesis in parsley (Petroselinum hortense Hoffm) cell suspension cultures. Planta 156: 213-217
252. HAHLBROCK K, J EBEL, R ORTMAN, A SUTTER, E WELLMAN, H GRISEBACH 1971 Regulation of enzyme activities related to the biosynthesis of flavone glycosides in cell suspension cultures of parsley (Petroselinum hortense). Biochem Biophys Acta 244: 7-15
253. HAHLBROCK K, K-H KNOBLOCH, F KREUZALER, JRM POTTS, E WELLMANN 1976 Coordinated induction and subsequent activity changes of two groups of metabolically interrelated enzymes. Light-induced synthesis of flavonoid glycosides in cell suspension cultures of Petroselinum hortense. Eur J Biochem 61: 199-216
254. EBEL J, K HAHLBROCK 1977 Enzymes of flavone and flavonol-glycoside biosynthesis. Coordinated and selective induction in cell-suspension cultures of Petroselinum hortense. Eur J Biochem 75: 201-209
255. MURPHY TM, GB MATSON, SL MORRISON 1983 Ultraviolet-stimulated $KHCO_3$ efflux from rose cells. Regulation of cytoplasmic pH. Plant Physiol 73: 20-24
256. CORDUAN G, E REINHARD 1972 Synthesis of volatile oils in tissue cultures of Ruta graveolens. Phytochemistry 11: 917-922
257. JENTE R 1971 Polyacetylenverbindungen in Gewebekulturen von Centaurea ruthenica Lam. Tetrahedron 27: 4077-4083
258. ICHIHARA K, M NODA 1977 Distribution and metabolism of polyacetylenes in safflower. Biochim Biophys Acta 487: 249-260

259. RADIN DN, HM BEHL, P PROKSCH, E RODRIGUEZ 1982 Rubber and other hydrocarbons produced in tissue cultures of guayule (Parthenium argentatum). Plant Sci Lett 26: 301-310
260. ROLLER U 1978 Selection of plants and plant tissue cultures of Catharanthus roseus with high content of serpentine and ajmalicine. In Production of Natural Compound by Cell Culture Methods (AW Alfermann, E Reinhard, eds.). Gesellschaft fur Strahlen-und Umweltforschung mbH. Munich. pp 95-104
261. OHTA S, M YATAZAWA 1978 Effect of light on nicotine production in tobacco tissue culture. Agric Biol Chem 42: 873-877
262. HAHLBROCK K 1972 Isolation of apigenin from illuminated cell suspension cultures of soybean, Glycine max. Phytochemistry 11: 165-167
263. BRUNET G, RK IBRAHIM 1973 Tissue culture of citrus peel and its potential for flavonoid synthesis. Z Pflanzenphysiol 69: 152-162
264. GREGOR HD, J REINERT 1972 Induktion der Phenylalanine Ammonia-Lyase in Gewebekulturen von Haplopappus gracilis. Protoplasma 74: 307-319
265. FRITSCH H, K HAHLBROCK, H GRISEBACH 1971 Biosynthese von Cyanidin in Zellsuspension-kulturen von Haplopappus gracilis. Z Naturforsch 266: 581-585
266. JONES RW, RW SHEARD 1977 Effects of blue and red light on nitrate reductase level in leaves of maize and pea seedlings. Plant Sci Lett 8: 305-311
267. APARICIO PJ, JM ROLDAN, F CALERO 1976 Blue light photoreaction of nitrate reductase from green algae and higher plants. Biochem Biophys Res Comm 70: 1071-1077
268. ROLDAN JM, F CALERO, PJ APARICIO 1978 Photoreactivation of spinach nitrate reductase: role of flavins. Z Pflanzenphysiol 90: 467-474
269. See reference 168.
270. MADYASTHA KM, JE RIDGWAY, JG DWYER, CJ COCIA 1977 Subcellular localization of a cytochrome P450-dependent monooxygenase in vescicles of the higher plant Catharanthus roseus. J Cell Biol 72: 302-313
271. GRUNWALD C 1978 Function of sterols. Phil Trans R Soc Lond B 284: 541-558

272. WELLBURN AR 1982 Bioenergetic and ultrastructural changes associated with chloroplast development. Int Rev Cytol 80: 133-191
273. HARDING RW, W SHROPSHIRE 1980 Photocontrol of carotenoid biosynthesis. Ann Rev Plant Physiol 31: 217-238
274. THOMAS RL, JJ JEN 1975 Phytochrome-mediated carotenoids, biosynthesis in ripening tomatoes. Plant Physiol 56: 452-453
275. SCHOPFER P 1977 Phytochrome control of enzymes. Ann Rev Plant Physiol 28: 233-252
276. ANDERSON LE, TCL NG, KEY PARK 1974 Interaction of pea leaf chloroplastic and cytoplasmic glucose-6-phosphate dehydrogenase by light and dithiothreitol. Plant Physiol 53: 835-839
277. REED AJ, DT CANVIN, JH SHERRARD, RH HAGEMAN 1983 Assimilation of [^{15}N] nitrate and [^{15}N] nitrate in leaves of five plant species under light and dark conditions. Plant Physiol 71: 291-294
278. BUCHANAN BB 1980 Role of light in the regulation of chloroplast enzymes. Ann Rev Physiol 31: 341-374
279. SMITH H, EE BILLET, AB GILES 1977 The photocontrol of gene expression in higher plants. _In_ Regulation of Enzyme Synthesis and Activity in Higher Plants (H. Smith ed.). Academic Press, New York, pp 93-127
280. HAREL E 1978 Chlorophyll biosynthesis and its control. _In_ Progress in Phytochemistry, Vol. 5 (L Reinhold, JB Harborne, T Swain eds.). Pergamon Press Toronto. pp 127-180
281. HAHLBROCK K, H GRISEBACH 1979 Enzymic controls in the biosynthesis of lignin and flavonoids. Ann Rev Plant Physiol 30: 105-130
282. FORREST GE 1969 Studies of polyphenol metabolism of tissue cultures derived from the tea plant (_Camellia sinensis_ L.). Biochem J 113: 765-772
283. RADWAN SS, HK MANGOLD, H HUSEMANN, W BARZ 1979 Lipids in plant tissue cultures. VII. Heterotrophic and mixotrophic cell cultures of _Chenopodium rubrum_. Chem Phys Lipids 24: 79-84
284. THOMAS DR, AK STOBART 1970 Lipids of tissue cultures of _Kalanchoe crenata_. J Exp Bot 21: 274-285
285. RADWAN SS, S GROSSE-OETRINGHAUS, HK MANGOLD 1978 Lipids in plant tissue cultures. VI. Effect of temperature on the lipids of _Brassica napus_ and _Tropaeolum majus_ cultures. Chem Phys Lipids 22: 177-184

286. MANGOLD HK 1977 The common and unusual lipids of plant cell cultures. In Plant Tissue Culture and Its Biotechnological Application (W. Barz, E. Reinhard, MH Zenk eds.). Springer-Verlag, Berlin, pp 55-65
287. GOWER M, A SANSONETTI, P MAZLIAK 1983 Lipid composition of tobacco cells cultivated at various temperatures. Phytochemistry 22: 855-859
288. COURTOIS D, J GUERN 1980 Temperature response of Catharanthus roseus cells cultivated in liquid medium. Plant Sci Lett 17: 473-482
289. BAQAR MR, TH LEE 1978 The interaction of CPTA and high temperature on carotenoid synthesis in tomato fruit. Z Pflanzenphysiol 88: 431-435
290. CHEN HH, PH LI 1980 Biochemical changes in tuber bearing Solanum species in relation to frost hardiness during cold acclimation. Plant Physiol 66: 414-421
291. LI PH, CJ WEISER 1966 The relation of cold resistance to the status of phosphorus and certain metabolites in redosier dogwood (Cornus stolonifera Michx.). Plant Cell Physiol 7: 475-484
292. POMEROY MK, D SIMINOVITCH, F WIGHTMAN 1970 Seasonal biochemical changes in the living bark and needles of red pine (Pinus resinosa) in relation to adaptation to freezing. Can J Bot 48: 953-967
293. HWEI-HWANG C, PH LI, ML BRENNER 1983 Involvement of abscissic acid in potato cold acclimation. Plant Physiol 71: 362-365
294. KACPERSKA-PALACZ A 1978 Mechanism of cold acclimation in herbaceous plants. In Plant Cold Hardiness and Freezing Stress (PH Li, A Sakai, eds.). Academic Press, New York, pp 139-152
295. RIKIN A, M WALKMAN, AE RICHMOND, A DOVRAT 1975 Hormonal regulation of morphogenesis and cold-resistance: I. Modifications by abscissic acid and by gibberllic acid in alfalfa Medicago sativa L. seedlings. J Exp Bot 26: 175-183
296. WALDMAN M, A RIKIN, A DOVRAT, AE RICHMOND 1975 Hormonal morphogenesis and cold resistance. II. Effect of cold acclimation and exogenous abscissic acid on gibberellic acid and abscissic acid activities in alfalfa (Medicago sativa L.) seedlings. J Exp Bot 26: 853-859
297. DALE J, WF CAMPBELL 1981 Response of tomato plants to stress temperatures: increase in abscissic acid concentrations. Plant Physiol 67: 26-29

298. NAKAMOTO H, GE EDWARDS 1983 Influence of oxygen and temperature on the dark inactivation of pyruvate, orthophosphate dikinase and NADP-malate dehydrogenase in maize. Plant Physiol 71: 568-573
299. KEY JL, CY LIN, YM CHEN 1981 Heat shock proteins of higher plants. Proc Natl Acad Sci USA 78: 3526-3530
300. DOUGALL DK, S LABRAKE, GH WHITTEN 1983 Growth and anthocyanin accumulation rates of carrot suspension cultures grown with excess nutrients after semicontinuous culture with different limiting nutrients at several dilution rates, pHs, and temperatures. Biotechnol Bioeng 25: 581-594
301. DOUGALL DK, S LABRAKE, GH WHITTEN 1983 The effects of limiting nutrients, dilution rate, culture, and temperature on the yield constant and anthocyanin accumulation of carrot cells grown in semicontinuous chemostat cultures. Biotechnol Bioeng 25: 569-579
302. DOBBERSTEIN RH, EJ STABA 1966 Chlorophyll production in Japanese mint suspension cultures. Lloydia 29: 50-57
303. YAMADA Y, Y HARA, H KATAGI, M SENDA 1980 Protoplast fusion. Effect of low temperature on the membrane fluidity of cultured cells. Plant Physiol 65: 1099-1102
304. WAGNER F, H VOGELMANN 1977 Cultivation of plant tissue cultures in bioreactors and formation of secondary metabolites. _In_ Plant Tissue Culture and Its Biotechnological Application (W. Barz, E. Reinhard, MH Zenk eds.). Springer-Verlag, Berlin pp 245-252
305. VELIKY IA 1977 Effect of pH on tryptophol formation by cultured _Ipomoea_ sp. plant cells. Lloydia 40: 482-486
306. KOUL S, A AHUJA, S GREWAL 1983 Growth and alkaloid production in suspension cultures of _Hyoscyamus muticus_ as influenced by various cultural parameters. Planta Med 47: 11-16
307. RADWAN SS, HK MANGOLD 1975 Lipids in plant tissue cultures. V. Effect of environmental conditions on the lipids of _Glycine soja_ and _Brassica napus_ cultures. Chem Phys Lipids 14: 87-91
308. HOOK DD, RMM CRAWFORD (eds.) 1978 Plant Life in Anaerobic Environments, Ann Arbor Sci Publ, Ann Arbor, MI 564 pp

309. APP AA, AN MEISS 1958 Effect of aeration on rice alcohol dehydrogenase. Arch Biochem Biophys 77: 181-190
310. HAGEMAN RH, D FLEHER 1960 The effect of anaerobic environment on the activity of alcohol dehydrogenase and other enzymes of corn seedlings. Arch Biochem Biophys 87: 203-209
311. CRAWFORD RMM, M McMANMON 1968 Inductive response of alcohol and malic dehydrogenase in relation to flooding tolerance in roots. J Exp Bot 19: 435-441
312. WEBER N, HK MANGOLD 1982 Metabolism of long-chain alcohols in suspension cultures of soya and rape. Planta 155: 225-230
313. RADWAN SS, HK MANGOLD, F SPENER 1974 Lipids in plant tissue cultures. III. Very long-chain fatty acids in the lipids of callus cultures and suspension cultures. Chem Phys Lipids 13: 103-107
314. WICKREMASINGHE RL, T SWAIN, JL GOLDSTEIN 1963 Accumulation of amino acids in plant cell tissue cultures. Nature (London) 199: 1302-1303
315. STABA EJ 1980 Secondary metabolism and biotransformation. In Plant Tissue Culture as a Source of Biochemicals (EJ Staba ed.). CRC Press, Boca Raton, FL pp 59-97
316. CRAWFORD RMM 1978 Metabolic adaptations to anoxia. In Plant Life in Anaerobic Environments (DD Hook, RMM Crawford, eds.). Ann Arbor Sci Publ, Ann Arbor, MI pp 119-136
317. CHIRKOVA TV 1978 Some regulatory mechanisms of plant adaptation to temporal anaerobiosis. In Plant Life in Anaerobic Environments (DD Hook, RMM Crawford eds.). Ann Arbor Sci Publ, Ann Arbor, MI pp 137-154
318. LEBLOVA S 1978 Pyruvate conversions in higher plants during anaerobiosis. In Plant Life in Anaerobic Environments (DD Hook, RMM Crawford eds.). Ann Arbor Sci Publ, Ann Arbor, MI pp 119-136
319. COSSINS EA 1978 Ethanol metabolism in plants. In Plant Life in Anaerobic Environments (DD Hook, RMM Crawford eds.). Ann Arbor Sci Publ, Ann Arbor, MI pp 169-202
320. CAMERON DS, EA COSSINS 1967 Studies of intermediary metabolism in germinating pea cotyledons. The pathway of ethanol metabolism and the role of the tricarboxylic acid cycle. Biochem J 105: 323-331

321. WAGER HG 1983 The labelling of glucose-6-phosphate in pea slices supplied with ^{14}C glucose, and the assessment of the respiratory pathways in air, in 10% CO_2 and in 2% O_2 in N. J Exp Bot 34: 211-220
322. MALMSTROM BG 1982 Enzymology of oxygen. Ann Rev Biochem 51: 21-59
323. BONNET R 1981 Oxygen activation and tetrapyroles. In Essays in Biochemistry, (PN Campbell, RD Marshall eds.). Academic Press, New York, pp 1-51
324. MILLER CO 1972 Modification of the cytokinin promotion of deoxyisoflavone synthesis in soybean tissue. Plant Physiol 49: 310-313
325. See reference 9
326. HAHLBROCK K, H RAGG 1975 Light-induced changes of enzyme activities in parsley cell suspension cultures. Arch Biochem Biophys 166: 41-46
327. de KLERK-KIEBERT YM, TJA KNEPPERS, LHW VAN DER PLAS 1982 Influence of chloramphenicol on growth and respiration of soybean (Glycine max L.) suspension cultures. Physiol Plant 55: 98-102
328. NOE W, C LANGEBARTELS, HU SEITZ 1980 Anthocyanin accumulation and PAL activity in suspension culture of Daucus carota. Planta 149: 283-287
329. NOE W, HU SEITZ 1982 Induction of de novo synthesis of phenylalanine ammonialyase by L-α-aminooxy-β-phenylpropionic acid in suspension cultures of Daucus carota. Planta 154: 454-458
330. BERLIN J, KG KOKOSCHKE, K-H KNOBLOCH 1981 Selection of tobacco cell lines with high yields of cinnamoyl putrescines. Plant Med 42: 173-180
331. BERLIN J, L WITTLE, J HAMMER, KG KUKOSCHKE, A ZIMMER, D PAPE 1982 Metabolism of p-fluorophenylalanine in p-fluorophenylalanine sensitive and resistant tobacco cell cultures. Planta 155: 244-250
332. BERLIN J, K-H KNOBLOCH, G HOGLE, L WITTE 1982 Biochemical characterization of two tobacco cell lines with diffeent levels of cinnamoyl putrescines. J Nat Prod 45: 83-87
333. LEA PJ, RD NORRIS 1976 The use of amino acid analogues in studies on plant metabolism. Phytochemistry 15: 585-595
334. WIDHOLM JM 1980 Selection of plant cell lines which accumulate certain compounds. In Plant Tissue Culture as a Source of Biochemicals (EJ Staba ed.). CRC Press, Boca Raton, FL pp 99-112

335. BERLIN J 1980 p-Fluorophenylalanine resistant cell lines of tobacco. Z Pflanzenphysiol 97: 317-324
336. PALMER JE, JM WIDHOLM 1975 Characterization of carrot and tobacco cell cultures resistant to p-fluorophenylalanine. Plant Physiol 56: 233-238
337. GATHERCOLE RWE, HE STREET 1976 Isolation, stability and biochemistry of p-fluorophenylalanine resistant cell line of Acer pseudoplatanus L. New Phytol 77: 29-41
338. GATHERCOLE RWO, HE STREET 1978 A p-fluorophenylalanine resistant cell line of sycamore with increased contents of phanylalanine, tyrosine and phenolics. Z Pflanzenphysiol 89: 283-287
339. BERLIN J, JM WIDHOLM 1977 Correlation between phenylalanine ammonia lyase activity and phenolic biosynthesis in p-fluorophenylalanine-sensitive and resistant tobacco and carrot tissue cultures. Plant Physiol 59: 550-553
340. BERLIN J, JM WIDHOLM 1978 Metabolism of phenylalanine and tyrosine in tobacco cell lines resistant and sensitive to p-fluorophenylalanine. Phytochemistry 17: 65-68
341. See Reference 334
342. WIDHOLM JM 1972 Anthranilate synthetase from 5-methytryptophan-susceptible and resistant cultured Daucus carota cells. Biochim Biophys Acta 279: 48-57
343. WIDHOLM JM 1972 Cultured Nicotiana tabacum cells with an altered anthranilate synthase which is less sensitive to feedback inhibition. Biochim Biophys Acta 261: 52-58
343a.WOOD, RKS (ed.) 1982 Active Defense Mechanisms in Plants. Plenum Press, New York, 381 pp
344. BAILEY JA, JW MANSFIELD (eds.) 1982 Phytoalexins. Blackie and Son Ltd., Glasgow, 334 pp
345. BELL AA, ME MACE 1981 Biochemistry and Physiology of Resistance. In Fungal Wilt Diseases of Plants. (ME Mace, AA Bell, CH Beckman, eds.). Academic Press, New York, pp 431-486
346. CRUICKSHANK IAM 1977 A review of the role of phytoalexins in disease resistance mechanisms. In Natural Products and the Protection of Plants (GB Marini-Bettolo, ed.), Elsevier, Amsterdam, pp 509-561

347. STOESSL A 1980 Phytoalexins - a biogenetic perspective. Phytopath Z 99: 251-272
348. FRIEND J 1977 Biochemistry of plant pathogens. In Plant Biochemistry II, (DN Northcote, ed.), University Park Press, Baltimore, MD pp 141-182
349. FRIEND J 1981 Plant phenolics, lignification and plant disease. Progr Phytochem 7: 197-261
350. WEST CA 1981 Fungal elicitors of the phytoalexin response in higher plants. Naturwissenschaften 68: 447-457
351. DIXON RA, DS BENDALL 1978 Changes in phenolic compounds associated with phaseollin production in cell suspension cultures of *Phaseolus vulgaris*. Physiol Plant Pathol 13: 283-294
352. DIXON RA, DS BENDALL 1978 Changes in the levels of enzymes of phenylpropanoid and isoflavonoid synthesis during phaseollin production in cell suspension cultures of *Phaseolus vulgaris*. Physiol Plant Pathol 13: 295-306
353. BODER GB, M GORMAN, IS JOHNSON, PJ SIMPSON 1964 Tissue culture studies of *Catharanthus roseus* crown gall. Lloydia 27: 328-333
354. BROWN SA, M TENNISWOOD 1974 Aberrant coumarin metabolism in crown gall tissues of tobacco. Can J Bot 52: 1091-1094
355. IKEDA T, T MATSUMOTO, M NOGUCHI 1976 Formation of ubiquinone by tobacco plant cells in suspension culture. Phytochemistry 15: 568-569
356. AUDISIO S, N BAGNI, DS FRACASSINI 1976 Polyamines during the growth in vitro of *Nicotiana glauca* R. Grah. habituated tissue. Z Pflanzenphysiol 77: 146-151
357. SPERANZA A, N BAGNI 1977 Putrescine biosynthesis in *Agrobacterium tumefaciens* and in normal and crown gall tissues of *Scorzonera hispanica* L. Z Pflanzenphysiol 81: 226-233
358. BAGNI N, DS FRACASSINI, E CORSINI 1972 Tumors of *Scorzonera hispanica*: their content in polyamines. Z Pflanzenphysiol 67: 19-23
359. WEETE JD, S VENKETESWARAN, JL LASETER 1971 Two populations of aliphatic hydrocarbons of teratoma and habituated tissue culture of tobacco. Phytochemistry 10: 939-943
360. WOOD HN, AC BRAUN 1961 Studies on the regulation of certain essential biosynthetic systems in normal

and crowngall tumor cells. Proc Nat Acad Sci USA 47: 1907-1913

361. ROBERTS WP 1982 The molecular basis of crown gall induction. Int Rev Cytol 80: 63-92
362. KAHL G, JS SCHELL eds 1982 Molecular Biology of Plant Tumors. Academic Press, New York, 615 pp
363. SPURR HW, AC HILDEBRANDT, AJ RIKER 1967 Ascorbic acid oxidase and tyrosinase activities in relation to crown-gall development. Phytopathology 52: 1079-1086
364. MEINS F 1982 Habituation of cultured plant cells. In Molecular Biology of Plant Tumors (G. Kahl, JS Schell, eds.). Academic Press, New York, pp 3-31
365. TEMPE J, P GUYON, D TEPFER, A PETIT 1979 The role of opines in the ecology of the T_i plasmids of Medical, Environmental and Commercial Importance (KN Timmis, A Puhler, eds.), Elsevier/North Holland Biomedical Press, Amsterdam, pp 353-372
366. DIXON RA, PM DEY, MA LAWTON, CJ LAMB 1983 Phytoalexin induction in French bean. Intercellular transmission of elicitation in cell suspension cultures and hypocotyl sections of Phaseollus vulgaris. Plant Physiol 71: 251-256
367. DIXON RA, PM DEY, DL MURPHY, IM WHITEHEAD 1981 Dose responses for Colletotrichum lindemuthianum elicitor-mediated enzyme induction in French bean cell suspension cultures. Planta 151: 272-280
368. DIXON RA, KW FULLER 1977 Characterization of components from culture filtrates of Botrytis cinerea which stimulate phaseollin biosynthesis in Phaseolus vulgaris cell suspension cultures. Physiol Plant Pathol 11: 287-296
369. DIXON RA, KW FULLER 1978 Effects of growth substances on non-induced and Botrytis cinerea culture filtrate-induced phaseollin production in Phaseollus vulgaris cell suspension cultures. Physiol Plant Pathol 12: 279-288
370. EBEL J, AR AYERS, PA ALBERSHEIM 1976 Host-pathogen interactions. XII. Responses of suspension-cultured soybean cells to the elicitor isolated from Phytophthora megasperma var. sojae, a fungal pathogen of soybeans. Plant Physiol 57: 775-779
371. ERESK T, I SZIRAKI 1980 Production of sesquiterpene phytoalexins in tissue culture callus of potato tubers. Phytopath Z 97: 364-368

372. HAHLBROCK K, CJ LAMB, C PURWIN, J EBEL, E FAUTZ, E SCHAFER 1981 Rapid response of suspension-cultured parsley cells to the elicitor from Phytophthora megasperma var. sojae. Plant Physiol 67: 768-773
373. GUSTINE DL, RT SHERWOOD 1978 Regulation of phytoalexin synthesis in jackbean callus cultures. Stimulation of phanylalanine ammonia-lyase and O-methyltransferase. Plant Physiol 61: 226-230
374. KUROSAKI F, A NISHI 1983 Isolation and antimicrobial activity of the phytoalexin 6-methoxymellein from cultured carrot cells. Phytochemistry 22: 669-672
375. ZAHRINGER U, E SCALLER, H GRISEBACH 1981 Induction of phytoalexin synthesis in soybean. Structure and reactions of naturally occurring and enzymatically prepared prenylated pterocarpans from elicitor-treated cotyledons and cell cultures of soybean. Z Naturforsch 36(c): 234-241
376. HAHLBROCK K, F KREUZALER, H RAGG, E FAUTZ, DN KUHN 1982 Regulation of flavonoid and phytoalexin accumulation through mRNA and enzyme induction in cultured plant cells. Hoppe-Seylers Z Physiol Chem 363: 121-122
377. ROWAN DD, PE MAC DONALD, RA SKIPP 1983 Antifungal stress metabolites from Solanum aviculare. Phytochemistry 22: 2102-2104
378. STRASSER H, KG TIETJEN, K HIMMELSPACH, U MATERN 1983 Rapid effect of an elicitor on uptake and intracellular distribution of phosphate in cultured parsley cells. Plant Cell Reports 2: 140-143
379. TIETJEN KG, D HUNKLER, U MATERN 1983 Differential response of cultured parsley cells to elicitors from two non-pathogenic strains of fungi. 1. Identification of induced products as coumarin derivatives. Eur J Biochem 131: 401-407
380. TIETJEN KG, U MATERN 1983 Differential response of cultured parsley cells to elicitors from two non-pathogenic strains of fungi. 2. Effects on enzyme activities. Eur J Biochem 131: 409-413
381. MILLER SA, LB GRAVES, DP MAXWELL 1981 (Abstract) Hyphal development and phytoalexin accumulation in alfalfa tissue culture - Phytophtohora megasperma system. Phytopathol 71: 895
382. HELGESON JP, AD BUDDE, GT HABERLACH 1978 (Abstract) Capsidiol-phytoalexin produced by tobacco callus tissues. Plant Physiol Suppl 61: 58

383. BUDDE AD, JP HELGESON 1981 (Abstract). Phytoalexins in tobacco callus tissues challenged by zoospores of *Phytophthora parasitica* var. *nicotianae*. Phytopathol 71: 206
384. EILERT U, B ENGEL, E REINHARD, B WOLTERS 1983 Acridone epoxides in cell cultures of *Ruta* species. Phytochemistry 22: 14-15
385. TEITJEN K, U MATERN 1981 (Abstract) Mode of action of *Alternaria carthami* toxin in Safflower (*Carthamus tinctorius* L.). Phytochem Soc N Amer Newsletter, July, p. 12
386. FUJIMORI TH, H TANAKA, K KATO 1983 Stress compounds in tobacco callus infiltrated by *Pseudomonas solanacearum*. Phytochemistry 22: 1038

Chapter Six

TEMPERATURE STRESS AND MEMBRANE LIPID MODIFICATION

M. N. CHRISTIANSEN

Plant Physiology Institute
U.S. Department of Agriculture
Science and Education
Agricultural Research Service
Beltsville, Maryland 20705

INTRODUCTION

Changes in the lipid composition of organisms in response to temperature alteration were documented very early by Ivanov.[1] The universality of temperature alteration of lipid unsaturation is supported by similar reports quite well documented in other plant species, land and marine animals, and micro-organisms. It is generally recognized, for example, that the iodine number of seed

---

Abbreviations: Lipids -- PC, phosphatidylcholine; PS, phosphatidylserine; PI, phosphatidylinositol; PE, phosphatidylethanolamine; PA, phosphatidic acid; PG, phosphatidylglycerol; PX, unidentified phospholipid; SL, sulfolipid; DG, digalactosyl diglyceride; MG, monogalactosyl diglyceride; NL, neutral lipids. Fatty acids are denoted by two numbers, representing the number of carbon atoms in the hydrocarbon chain and the number of double bonds respectively.

lipid varies within species with change in production latitude. In animals, the unsaturation of body triglycerides varies with their location within the body or extremities. For example, lipid unsaturation of body fat of the cow is lower than fat of the leg.

The observation that temperature extremes induce variation in lipid characteristics has stimulated a considerable volume of research relating lipid quantity and quality to chilling and freezing resistance of plants or plant parts. Temperature extremes limit the geographic distribution and productivity of crop plants. Subjection of plants to extremes of temperature causes marked physiological dysfunction.

Levitt[2] categorizes plants in the following temperature response classes:

1. Chilling sensitive - species sensitive to temperatures above 0°C.

2. Tender - those species sensitive to light frost at or near 0°C.

3. Slightly hardy - species capable of developing some resistance to frost as a consequence of membrane lipid unsaturation. Limited to survival above -5°C.

4. Moderately hardy - in addition to membrane lipid modification as in group three, these species develop a higher cell sap concentration during hardening. Survival limited to -5 to -10°C.

5. Very hardy - possess highly unsaturated membrane lipids and thus have osmotically functional membrane at low temperature; high osmotic cell sap; cryoprotectant systems and resistance to dehydration at -10 to - 20°C.

6. Extremely hardy - resistance to loss of membrane mediated water control via lipid peroxidation. Ability to maintain free water in super-cooled liquid state; accumulation of special phospholipids and membrane proteins.

From the description of these classes, it becomes evident that membranes and particularly the lipid constituents, are considered important in determining the survival and functionality of plant cells at low temperature.

MEMBRANE COMPOSITION

A number of early observations prompted researchers to implicate membrane composition with cell functionality at temperature extremes. Sachs observed cessation of protoplast streaming at or near 10°C in chilling sensitive species.[3] Such functions as water or mineral uptake by roots cease at temperatures near 0°C.[4] Cellular exudation from leaf discs or from root tips is greatly increased by chilling or freezing.[5,6] Greater loss of potassium from yam roots,[7] and of carbohydrates and amino acids from cotton root tips[6] at chilling temperatures is reported. Cell organelles likewise show a loss of function as a consequence of cold, and the responses can be related to membrane lipid composition. Generally plants that exhibited tolerance to low temperature had mitochondria that carried on active respiration at low temperature.[8] Mitochondrial lipids of these species were generally highly unsaturated. Conversely, chilling sensitive species such as cotton had more saturated mitochondrial lipids. The subject was recently reviewed by Raison.[9] It has been theorized that a change in molecular ordering of the membrane lipids is the major alteration in low temperature response of cold sensitive plant species. The alteration in lipid matrix alters the conformation and activity of membrane bound enzymes. Metabolic dysfunction and concurrent impact on plant function are believed to be secondary to the initial structural change in membranes. Yamaki and Uritani suggested that prolonged cold leads to changes in protein-phospholipid bonding as a consequence of phase transition of membrane phospholipid.[10] In mango, chilling induced a marked discontinuity of Arrhenius plots of mitochondrial succinate oxidase activity concurrent with a decrease in the palmitoleic/palmitic ratio.[11] Miller et al. studied changes in structure and function of wheat mitochondrial membranes from cold treatment.[12] Plants grown at 2°C differed from those grown at 24°C in unsaturated fatty acid levels. Total unsaturation and particularly levels of linolenic acid increased; substrate oxidation decreased along with efficiency of energy coupling. Lipid unsaturation was suggested as the controlling factor in temperature effect on respiration.

Ashworth et al. noted different responses in wheat.[13] Cold hardening at 10°C caused no alteration of the phospholipid fractions (Table 1), whereas, lower temperatures altered the total fatty acid composition (Table 2), inducing the production of greater quantities of total linolenic acid at 10°C. Changes were also reflected in the fatty acid complement of phospholipids (Tables 3 and 4). Although linolenic acid levels increased with low temperature, no change occurred in the activation energy of respiration (Table 5). These data disagree with results of Miller et al.[12], but agree with those data reported by Pomeroy and Andrews[14] from studies conducted with wheat and rye. Based on our studies[13] we concluded that cold hardening-induced increases in linolenic acid did not affect the rate of respiration at low temperature or activation energy of respiration. The effect on other membrane mediated processes such as water or nutrient uptake may be considerable.

The major site of fatty acid unsaturation in chilled

Table 1. Phospholipid composition of wheat seedling roots

Phospholipids were separated into five fractions using high performance liquid chromatograph. The fractions were identified as phosphatidyl ethanolamine (PE), phosphatidyl glycerol (PG), phosphatidyl inositol (PI), phosphatidic acid (PA), phosphatidyl serine (PS), and phosphatidyl choline (PC). Fractions were collected and the phosphorus content determined by Bartlett's modification of the Fiske-SubbaRow method.

Temperature	Growth Temperature	Phospholipid fractions				
		PE	PG	PI	PA-PS	PC
	°C	% lipid phosphorus				
Control	25	34	5	7	4	51
	20	37	5	4	6	48
	15	31	8	7	3	52
	10	38	6	7	4	45
BASF 13-338 (100 μM)	25	35	6	5	4	50
	20	33	6	5	2	54
	15	34	5	7	4	50
	10	37	5	4	3	51

Table 2. The effect of temperature and BASF 13-338 on the fatty acid composition of wheat seedling roots

Seedlings were grown for either 5 days at 25 and 20°C, 8 days at 15°C, or 12 days at 10°C.

Treatment	Growth Temperature	Fatty acid				
		16	18	18:1	18:2	18:3
	°C	% by weight of fatty acids				
Control	25	23	1	4	50	21
	20	24	1	3	44	28
	15	22	1	4	36	38
	10	22	2	5	32	39
BASF 13-338 (100 µM)	25	23	1	4	63	9
	20	22	0	3	62	12
	15	20	1	4	64	11
	10	20	0	3	61	16

cotton seedling root tips was in the microsomes, with little alteration of mitochondrial or nuclear lipids,[15] so possibly mitochondria are the wrong site to study. The evidence obtained from mitochondrial respiration and membrane leakage studies is paralleled by similar research with chloroplasts from tomato and bean.[16]

Chloroplast evidence provided by Kaniuga and Michalski[17] showed that the Hill reaction was inhibited by cold via a release of free fatty acids (particularly linolenic) from chloroplasts. Peoples and Koch reported a reduction in CO_2 exchange rate in alfalfa as a consequence of one 12-hour chilling at 5°C.[18] Peoples et al. later observed that chilling-induced reduction of photosynthesis was inversely correlated with the fatty acid double bond index of chloroplast membranes.[19]

Not all the reports agree that chilling sensitivity is related to chloroplast membrane dysfunction. Nolan and Smillie compared Hill activity of chloroplasts from chilling sensitive (mung bean and corn) and resistant (barley and pea) plants and noted a change in activation

Table 3. Fatty acid composition of phosphatidylcholine from roots of wheat seedlings

Phospholipids were separated using high performance liquid chromatography. The phosphatidylcholine fraction was collected and the fatty acid composition determined.

Treatment	Growth Temperature	Fatty acid				
		16	18	18:1	18:2	18:3
	°C	% by weight				
Control	25	27	2	5	49	16
	20	22	3	10	44	21
	15	26	5	9	34	26
	10	24	2	6	32	36
BASF 13-338 (100 μM)	25	25	2	5	63	5
	20	23	3	7	61	5
	15	21	4	9	58	7
	10	19	2	5	63	11

energy requirement for all species that was different and distinct for each species. These changes were not related to chilling sensitivity.[20]

CHANGES DURING TEMPERATURE STRESS

Considerable research has attempted to elucidate the importance of membrane lipid phase changes that are associated with various plant functions including respiration, photosynthesis, membrane bound enzyme activity such as ATPase, water and solute uptake, or other osmotic relations. The following has been recorded:

1. Membrane phase changes as measured by electron spin resonance occur in mitochondria, chloroplasts, glyoxysomes and protoplastids.[21]

2. Phase transition occurs over a temperature zone but not at a precise temperature point.

3. Phase change is reversible.

4. Phase change temperature increases with increasing fatty acid chain length and decreases with increased unsaturation.

5. Membrane surface area decreases during a change from liquid crystalline to gel state.

Affirmation of the relation between unsaturation of fatty acids and chilling resistance was provided by St. John and Christiansen.[22] We showed that cotton seedlings could be cold hardened by gradually lowering temperature to 8°C. At lowered temperature, linolenic acid of polar lipids increased along with cold resistance (Tables 6 and 7). A specific chemical inhibitor of linolenic acid synthesis (BASF 13-338) was used to block low temperature induced-increase of linolenate which also blocked increase in cold resistance of the plant.

In later work, St. John et al.[23] showed that freezing resistance in wheat, barley and rye could be blocked by BASF 13-338. The unsaturation of membrane lipids of leaves was altered in direct relation to freeze susceptibility and survival (Figure 1, Table 8).

Table 4. Fatty acid composition of phosphatidyl ethanolamine from roots of wheat seedlings

Phospholipids were separated using high performance liquid chromatography. The phosphatidylethanolamine fraction was collected and the fatty acid composition determined.

Treatment	Growth Temperature	Fatty acid				
		16	18	18:1	18:2	18:3
	°C	% by weight				
Control	25	32	3	4	47	14
	20	25	3	6	45	21
	15	32	4	7	32	24
	10	27	6	7	25	34
BASF 13-338 (100 μM)	25	27	5	7	56	4
	20	27	5	7	56	4
	15	26	4	8	55	7
	10	24	3	5	56	11

Table 5. Respiration rate of wheat root tips at different temperatures

Wheat seedlings were grown at either 25 or 10°C to an equivalent size. Plants were grown in distilled water (control) or in 100 μM BASF 13-338. Values are the mean of three readings on four replicates.

	25°C Wheat		10°C Wheat	
Temperature	Control	BASF 13-338	Control	BASF 13-338
°C	μl O_2 consumed/mg dry wt·h			
30	9.5	10.3	10.0	9.6
27	8.5	7.8	9.0	9.1
25	8.3	8.3	7.1	7.1
22	5.8	6.1	7.0	7.4
20	5.9	5.6	7.1	6.6
18	4.8	5.3	5.6	5.3
17	4.6	4.9	4.8	4.6
15	3.7	3.8	4.2	4.2
12	2.6	3.0	2.7	2.9
10	2.2	2.2	2.6	2.6
8	1.6	1.8	2.5	2.6
7	1.5	1.6	1.8	1.7
6			1.5	1.7
5	1.1	1.3	1.7	1.6
4			1.2	1.3

An explanation for the accumulation of polyunsaturated fatty acids associated with freezing resistance as herein discussed was given by Harris and James.[24] Based on [^{14}C]acetate studies with bean, flax and castor bean seed, they concluded that O_2 content of the cell was the controlling factor in fatty acid desaturation. Mazliak later showed 20% oxygen to be the maximum response level for desaturation of oleate.[25] He concluded that temperature and O_2 control jointly desaturase activity. Nozawa et al. suggested that membrane fluidity itself is a regulating factor in controlling the membrane-bound fatty acid desaturases.[26]

Although somewhat out of place in this discussion, Downton and Hawker found that soluble and starch grain-bound starch synthase from chilling sensitive corn, avocado and sweet potato showed discontinuities on Arrhenius plots at 12°C.[27] The same enzyme activity from chilling resistant potato showed no deflection down to 3° C. Although starch synthase is not membrane bound, it is associated with the lipid lysolecithin. Thus, their studies suggest that non-membrane lipids associated with enzyme systems can influence activity as a function of temperature.

Phospholipids

There is a paucity of data on polar head changes in plants at low temperature. As temperatures become lower there is a marked increase in PC and PE in poplar tree bark with no alteration of other phospholipids.[28]

In rape plants, Smolenska and Kuiper reported marked changes in phospholipids (Tables 9 and 10) as a consequence of low temperature.[29] They noted increases in leaf PC and PE and a decrease in phosphatidylglycerol PG at 5°C; and reverse trends in roots. However, the total

Table 6. Fatty acid composition of polar lipids of 1-cm root tips of cotton seedling radicles

The data are the average of three experiments.

Treatment	Growth Temperature	Fatty acid				
		C 16	C 18	C 18:1	C 18:2	C 18:3
	°C	% by weight				
Control (water)	30	32.6	3.7	7.5	36.3	13.6
	25	35.3	2.8	5.7	33.4	22.9
	20	32.9	3.3	5.4	31.7	26.8
	15	31.6	4.8	4.2	31.9	27.5
BASF 13-338 (10 μM)	30	34.7	2.8	7.6	44.0	10.9
	25	34.4	2.7	5.7	43.1	14.0
	20	34.5	3.7	6.4	41.7	13.7
	15	30.9	4.1	4.6	43.7	17.1

Table 7. C 18:2/C 18:3 Fatty acid ratios of cotton seedling root tip polar lipids

Growth temperature	Treatment	
	Control	Sandoz 9785 (10 μM)
°C		
30	2.67	4.04
25	1.46	3.08
20	1.18	3.07
15	1.16	2.56

phospholipid level of roots was 45% less in plants grown at 25°C/20°C (day/night). Linolenic acid increased in PC, PE, PG and phosphatidic acid at 5°C; similar increases were noted in neutral lipid and the galactolipids of 5°C treatments as compared to 25°C/20°C. Smolenska and Kuiper carefully indicate that linolenic acid may play a role in freezing tolerance in that it is important for plant functions at low temperature.

Sterols

There is little information on the effect of temperature on sterol production. Davis and Finkner observed minimal effect of temperature on total sterol content of wheat plants.[30] Likewise, sterols were unaffected by temperature in potato tubers.[31] de la Roche, however, reported a significant decrease in the ratio of campesterol to sitosterol from low temperature exposure of wheat embryos.[32] Willemot found little change in total sterol profile or in the lipid P to sterol ratio.[33] Thus there is disagreement concerning temperature effect on this lipid class and little to suggest a role for sterols in cold hardening of plants.

Waxes

The quantity and quality of cuticular waxes can have marked effect on plants' resistance to desiccation and temperature extremes. Radiant energy reflection or absorption and water retention are influenced by wax

Table 8. Effect of pre-emergence treatment with BASF 13-338 on membrane fatty acid composition of shoot tissue from small grains

	BASF 13-338	Fatty acid composition by weight* (%)				
		16:0	18:0	18:1	18:2	18:3
	kg/ha					
Arthur wheat	0.0	10.1 a	0.8 a	1.8 a	11.7 a	75.6 a
	2.8	8.8 b	0.8 a	2.4 a	35.4 b	52.8 b
	5.6	8.4 b	0.6 a	2.1 a	50.0 c	38.9 c
	11.2	8.7 b	0.8 a	2.2 a	61.4 d	27.1 d
Potomac wheat	0.0	10.0 a	1.1 a	1.9 a	11.7 a	75.4 a
	2.8	9.1 a	1.0 a	2.6 a	35.4 b	52.4 b
	5.6	8.8 a	1.0 a	2.3 a	44.0 c	44.0 c
	11.2	9.4 a	1.2 a	2.4 a	54.5 d	32.5 d
Monroe barley	0.0	11.6 a	1.3 a	1.7 a	10.0 a	75.4 a
	2.8	10.0 b	1.0 a	1.4 a	20.8 b	66.9 b
	5.6	10.4 ab	1.2 a	1.7 a	31.9 c	54.9 c
	11.2	10.4 ab	1.3 a	1.5 a	40.2 d	46.6 d
Abruzzi rye	0.0	11.3 a	1.0 a	2.4 a	14.2 a	71.3 a
	2.8	10.2 ab	0.8 a	2.1 a	18.4 b	65.7 b
	5.6	9.8 bc	0.9 a	2.4 a	32.4 c	54.5 c
	11.2	8.8 c	0.8 a	1.7 a	30.8 c	57.6 c

* Values for individual fatty acids within a cereal followed by common letters are not significantly different at the 5% level with Duncan's Multiple range test.

coatings of tissue. Baker showed that increases in temperature caused increased cuticular wax (Table 11).[34] The morphology of the cuticular surface was also altered; at 15°C wax on leaves of Brussels sprouts was in the form of hollow tubes projecting from the leaf surface, while at 21°C the wax appeared as a mixture of tubules and dendrites on the leaf surface. At 35°C a mesh work of large dendrites cover the surface in a flat plane. An elevated growth temperature decreases the content of

Table 9. The effect of a 2-week cold-treatment on growth, frost tolerance, and lipid level of the winter rape plants

Parameter	Tissue	Start	25/20° C control	5°C
Fresh weight, g	Leaves	2.4	10.9	4.4
	Roots	1.0	5.3	2.2
Dry weight, g	Leaves	0.24	1.70	0.49
	Roots	0.05	0.34	0.14
Frost tolerance, T_{k50} °C	Leaves	-11.1	-9.7	-13.3
	Roots	-8.3	-8.1	-8.3
Lipid level, mg/g fresh weight	Leaves	14.0 ± 1.2	11.8 ± 1.6	16.0 ± 2.7
	Roots	2.9 ± 0.2	2.4 ± 0.1	2.9 ± 0.3
Lipid level, mg/g dry weight	Leaves	140.0 ± 12.4	75.6 ± 10.9	144.0 ± 25.9
	Roots	60.3 ± 2.9	31.3 ± 1.2	47.2 ± 3.8
Fatty acid level, mg/g dry weight	Leaves	45.2	29.9	46.9
	Roots	14.0	7.8	12.4
Linolenic acid level, mg/g dry weight	Leaves	19.2	12.6	20.7
	Roots	5.5	3.0	6.1

Table 10. Lipid composition of the leaves and roots of winter rape plants expressed as μg fatty acid/g dry weight and as percentage of saponifiable lipids and of phospholipids (within parentheses)

	Leaves						Roots					
	Start		Control, 25/20°C		5°C		Start		Control, 25/20°C		5°C	
	mg/g	%	mg/g	%	mg/g	%	mg/g	%	mg/g	%	mg/g	%
NL	9.80	21.6	5.20	17.4	11.5	24.6	2.00	14.3	0.93	11.9	2.10	16.8
MG	10.80	23.8	8.20	27.4	11.3	24.0	0.31	2.2	0.20	2.6	0.32	2.6
DG	6.40	14.2	4.05	13.6	6.25	13.3	0.43	3.1	0.35	4.4	0.52	4.2
SL	2.20	4.9	2.10	7.1	2.10	4.4	-	-	-	-	-	-
PX	1.35	3.0(8.2)	1.30	4.4(12.6)	1.75	3.8(11.2)	0.46	3.3(4.1)	0.41	5.2(6.5)	0.75	6.1(8.0)
PG	4.20	9.3(26.3)	2.20	7.4(21.3)	2.35	5.0(14.8)	0.52	3.7(4.6)	0.25	3.2(4.0)	0.55	4.4(5.8)
PA	1.00	2.2(6.1)	0.65	2.2(6.4)	0.85	1.8(5.4)	0.36	2.5(3.1)	0.28	3.3(4.0)	0.32	2.6(3.4)
PE	2.90	6.5(18.2)	1.90	6.3(18.1)	3.80	8.1(24.0)	4.55	32.5(40.4)	2.70	34.5(42.6)	3.65	29.3(38.4)
PI	1.55	3.4(9.6)	1.15	3.8(10.9)	1.65	3.5(10.5)	0.89	6.3(7.9)	0.43	5.5(6.8)	0.91	7.4(9.6)
PS	-	- (-)	-	- (-)	-	- (-)	0.19	1.4(.17)	0.14	1.8(2.2)	0.18	1.4(1.8)
PC	5.05	11.2(31.5)	3.15	10.6(30.7)	5.35	11.4(34.1)	4.30	30.7(38.2)	2.15	27.5(33.9)	3.15	25.2(33.0)
Phospholipids	16.05	35.4(100)	10.35	34.6(100)	15.75	33.6(100)	11.27	80.5(100)	6.36	81.1(100)	9.51	76.4(100)

aldehydes and increases that of alkanes. The alteration of waxes by environmental extremes is an interesting and possibly fruitful research area with potential to benefit

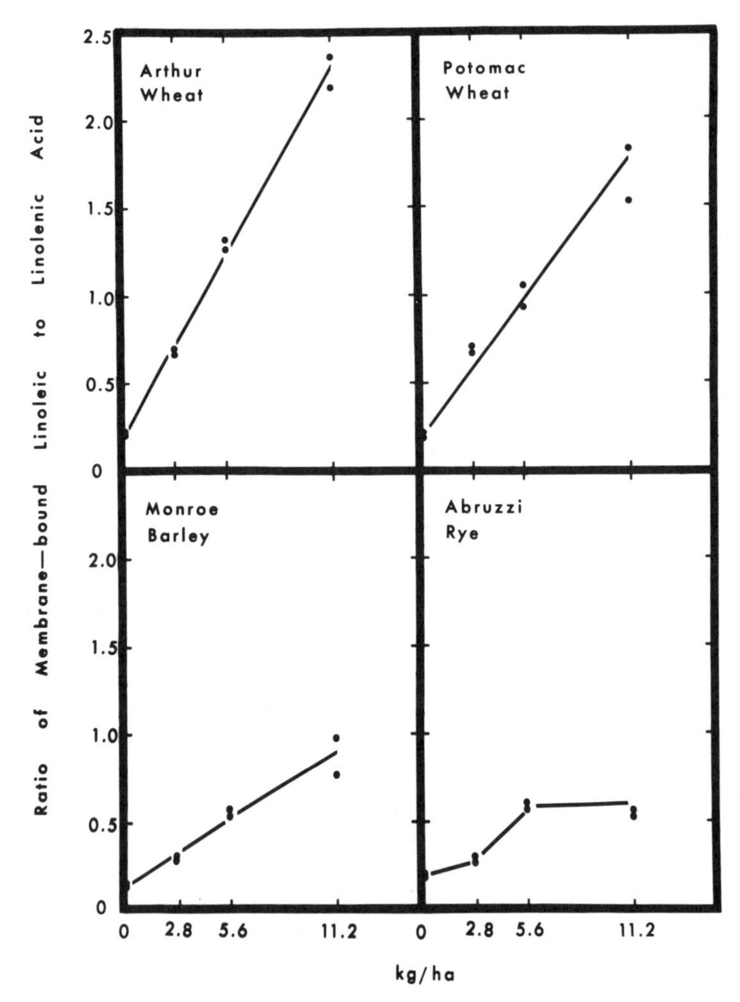

Fig. 1. Effect of BASF 13-338 application rate (abscissa) on the ration of membrane linoleic (18:2) to linolenic acid (18:3) of field cultured wheat, barley, and rye. Duplicate points for each rate represent duplicate assays.

adaptation of plants to stressful environments.

SUMMARY

Although temperature regulation of membrane centered biochemical activity remains unresolved, the involvement of membrane lipid unsaturation in mediation of water and solute movement during cold hardening seems somewhat better explained by available circumstantial evidence, e.g., chilling injury in plants can in many cases be experimentally prevented by chilling in high humidity, indicating that balance of water uptake by roots and water loss by leaves is uncontrolled under chilling conditions. Plants can be chill hardened at marginal temperatures which results in marked increase in linolenic acid in root polar lipid; if linolenic acid increases are chemically blocked, chill hardening is also prevented.

Freeze hardening is a stepwise process; at marginally low temperatures (2-10°C) membrane lipids become more unsaturated thereby ensuring osmotic functionality at low temperature; as temperatures drop below 0°C a membrane controlled loss of water to ice crystals in the intercellular spaces occurs with a concurrent increase in cell osmotic concentration. The key to cold survival is membrane control of cell dehydration. The same end result of cold resistance can be obtained by drought stressing plants to increase cell osmoticum. Thus, temperature relations in plants can be partially explained on the simple basis of membrane control of water uptake and loss from cells.

Numerous authors have shown that cultivars within a species vary widely in cold tolerance but show the same increases in fatty acid unsaturation.[34] Lack of differences in fatty acid unsaturation does not necessarily prove a lack of involvement. Unsaturation may be an initial prerequisite to development of cold resistance and less hardy cultivars may lack other attributes that contribute to survival at low temperature. The hypothesis that, "over a temperature range membrane form and function is dependent upon its lipid constituents", is subject to many interpretations and to much debate. The relative newness of the concept and the faults of techniques used to characterize the chemistry of membranes and to measure

Table 11. Variations in the constituent classes of the leaf waxes of Brussels sprouts with environmental conditions

	Radiant energy rate									
	80 Wm^{-2}					38 Wm^{-2}				
	Temperature (°C)/Relative humidity (%)									
	15/70	21/40	21/70	21/98	35/70	15/70	21/40	21/70	21/98	35/70
Normal										
Alkane	51	47	40	45	36	51	46	41	38	34
Alkyl ester	1	2	tr	2	3	1	2	1	2	2
Ketone	25	22	35	16	22	26	21	24	16	19
Aldehyde	tr	6	7	10	14	2	5	11	13	14
Secondary alcohol	15	14	12	15	13	14	14	14	12	11
Primary alcohol	3	5	4	7	7	4	6	6	10	12
Fatty acid	3	3	1	3	4	1	4	2	7	6

the response of membrane function combine to present an uncertain picture.

The use of Arrhenius plots to present and interpret data is questionable. Electron spin resonance, differential thermal analysis, differential scanning calorimetric or fluorescence probes have limitations. Lipid fractionation presents problems. Certainly it is invalid to relate bulk lipid chemistry to function as a basis for membrane involvement in temperature response. Rather, specific sites, whether they be mitochondria, microsomes, chloroplasts, or plasma membrane need to be investigated to establish the role of specific kinds of lipid in temperature accommodation of plants. It is a challenging problem and the possibilities are many.

REFERENCES

1. IVANOV SM 1929 Dependence of the chemical composition of oil-containing plants on the climate. Chem Zbl 1928: 1971 (Chem Abstr 22: 4149, 1929)
2. LEVITT J 1980 Responses of plants to environmental stress. Vol 1 Freezing and High Temperature Stresses. Academic Press, New York

3. SACHS J 1964 Uber die Obere Temperatur-Grenze der Vegetation Flora 22: 5-12
4. KRAMER PJ 1949 Plant and soil water relationships. 1st Ed McGraw-Hill, New York, p 227
5. GUINN G 1971 Chilling injury in cotton seedlings: Changes in permeability of cotyledons. Crop Sci 11: 101-102
6. CHRISTIANSEN MN, HR CARNS, DJ SLYTER 1970 Stimulation of solute loss from radicles of Gossypium hirsutum L. by chilling, anaerobiosis, and low pH. Plant Physiol 46: 53-56
7. LIEBERMAN J, CC CRAFTS, WV AVDIA, MS WILCOX 1958 Biochemical studies of chilling injury in sweet potatoes. Plant Physiol 33: 307-311
8. LYONS JM 1973 Chilling injury in plants. Annu Rev Plant Physiol 24: 445
9. RAISON J 1980 Membrane lipids: structure and function. In PK Stumpf, ed, The Biochemistry of Plants Vol 4 Lipids: Structure and Function. Academic Press, New York pp 57-83
10. YAMAKI S, I URITANI 1974 Mechanism of chilling injury in sweet potato. XII Temperature dependency of succinoxidase activity and lipid-protein interaction in mitochondria from healthy or chilling-stored tissue. Plant Cell Physiol 15: 669-676
11. KANE O, P MARCELLIN, P MAZLIAK 1978 Incidence of ripening and chilling injury on the oxidative activities and fatty acid compositions of the mitochondria from mango fruits. Plant Physiol 61: 634-63812
12. MILLER RW, I de la ROCHE, MK POMEROY 1974 Structural and functional response of wheat mitochondrial membranes to growth at low temperature. Plant Physiol 53: 426-431
13. ASHWORTH EN, MN CHRISTIANSEN, JB ST JOHN, GW PATTERSON 1981 Effect of temperature and BASF 13-338 on the lipid composition and respiration of wheat roots. Plant Physiol 67: 711-715
14. POMEROY MK, CJ ANDREWS 1975 Effect of temperature on respiration of mitochondria and shoot segments of cold-hardened and nonhardened wheat and rye seedlings. Plant Physiol 56: 703-706
15. BARTKOWSKI EJ 1979 Chill-induced changes in organelle membrane fatty acids. In JM Lyons, D Graham and JK Raison eds, Low Temperature Stress in

Crop Plants, The Role of the Membrane. Academic Press, New York. pp 431-435

16. SHNEYOUR A, JK RAISON, RM SMILLIE 1973 The effect of temperature on the rate of photosynthetic electron transfer in chloroplasts of chilling sensitive and chilling resistant plants. Biochem Biophys Acta 292: 152-161
17. KANIUGA Z, W MICHALSKI 1978 Photosynthetic apparatus in chilling-sensitive plants. II Changes in fatty acid composition and photoperoxidation in chloroplasts following cold storage and illumination of leaves in relation to Hill reaction activity. Planta 140: 129-134
18. PEOPLES TR, DW KOCH 1978 Physiological response of three alfalfa cultivars to one chilling night. Crop Sci 18: 255-259
19. PEOPLES TR, DW KOCH, SC SMITH 1978 Relationships between chloroplast membrane fatty acid composition and photosynthetic response to chilling temperature in four alfalfa cultivars. Plant Physiol 61: 472-481
20. NOLAN WG, RM SMILLIE 1977 Temperature induced changes in Hill activity of chloroplasts isolated from chilling-sensitive and chilling-resistant plants. Plant Physiol 59: 1141-1145
21. WADE NL, WR BREIDENBACH, JM LYONS, AC KEITH 1974 Temperature-induced phase changes in the membranes of glyoxysomes, mitochondria and pro-plastids from germinating castor bean endosperm. Plant Physiol 54: 320-323
22. ST JOHN JB, MN CHRISTIANSEN 1976 Inhibition of linolenic acid synthesis and modification of chilling resistance in cotton seedlings. Plant Physiol 57: 257-259
23. ST JOHN JB, MN CHRISTIANSEN, EN ASHWORTH, WA GENTNER 1979 Effect of BASF 13-338, a substituted pyridazinone on linolenic acid levels and winter hardiness of cereals. Crop Sci 19: 65-69
24. HARRIS P, AT JAMES 1969 Effect of low temperature on fatty acid biosynthesis in seeds. Biochim Biophys Acta 187: 13-18
25. MAZLIAK P 1979 Temperature regulation of plant fatty acyl desaturases. In JM Lyons, D Graham and JK Raison, eds, Low Temperature Stress in Crop Plants, The Role of the Membrane. Academic Press, New York pp 391-404
26. NOZAWA Y, R KASAI 1978 Mechanism of thermal

adaptation of membrane lipids in Tetrahymena pyriformis NT-1. Possible evidence for temperature mediated inductation of palmitoyl-CoA desaturase. Biochim Biophys Acta 529: 54-66

27. DOWNTON WJS, JS HAWKER 1975 Evidence for lipid-enzyme interaction in starch synthesis in chilling sensitive plants. Phytochemistry 14: 1259-1263
28. YOSHIDA S 1974 Studies on lipid changes associated with frost hardiness in cortex in woody plants. Contrib Inst Low Temp Sci, Ser B 18: 1-41
29. SMOLENSKA G, PJC KUIPER 1977 Effect of low temperature upon lipid and fatty acid composition of roots and leaves of winter rape plants. Physiol Planta 41: 29-35
30. DAVIS DL, VC FINKNER 1973 Influence of temperature on sterol biosynthesis in Triticum aestivum. Plant Physiol 52: 324-326.
31. GALLIARD T, BERKELEY HD, MATTHEW JA 1975 Sci Food Agric 26: 1163-1170 (cited by Willemot Ref. 33)
32. de la ROCHE IA 1979 In LS Underwood, ed, Comparative Mechanisms of Cold Adaptation in the Arctic. Symp Am Inst Biol Sci, Academic Press, New York
33. WILLEMOT C 1979 Chemical modification of lipids during frost hardening of herbaceous species. In JM Lyons, D Graham, JK Raison, eds, Low Temperature Stress in Crop Plants. The Role of the Membrane. Academic Press, New York pp 411-430
34. BAKER EA 1974 The influence of environment on leaf wax development in Brassica oleracea var. gemmifera. New Phytol 73: 955-966
35. de la ROCHE IA, MK POMEROY, CJ ANDREWS 1975 Changes in fatty acid composition in wheat cultivars of contrasting hardiness. Cryobiology 12: 506-512

Chapter Seven

MORPHOLOGY, CHEMISTRY, AND GENETICS OF *GOSSYPIUM* ADAPTATIONS TO PESTS

ALOIS A. BELL

National Cotton Pathology Research Laboratory
Agricultural Research Service
United States Department of Agriculture
College Station, Texas 77841

INTRODUCTION

Gossypium is one of eight genera in the plant tribe Gossypieae in the family Malvaceae.[1] The tribe Gossypieae is distinguished from related tribes by the production of spherical lysigenous glands located below the pallisade cells of cotyledons and leaves and throughout the bark of old roots and stems. The glands contain pigments which distinguish them from surrounding cells. Most of the pigments and other compounds in the pigment glands are toxic terpenoids and flavonoids. Thus, the glands apparently have evolved because of the added protection that they give against insects, rodents, and other herbivores.

Various classifications have been proposed for *Gossypium* species. A recent one distinguishing 33 species is shown in Table 1.[2] The species are listed in genome groups; species in genomes A through G are diploids (n = 13); and those in the AD genome are amphidiploids (n = 26). Most genome groupings were determined from cytogenetic data, especially chromosome pairing during meiosis in interspecific hybrids (Table 2).[2-5] Some tentative genome assignments were made on the basis of assumed relations (e.g., Australia originally was thought to have a single genome group). The amphidiploid species (AD genome) apparently arose from hybridization of A and D genome diploid species followed by chromosome doubling. The A chromosomes in amphidiploids have their greatest similarity to those found in G. herbaceum, while the D chromosomes are most closely related to those in G. raimondii.[4,5]

Generally, each diploid genome group is geographically isolated from other diploid genome groups.[2,5] The only wild ancestor of the A genome, G. herbaceum L. var. africanum (Watt) Hutch. and Ghose, grows in southeastern Africa. The A genome species G. herbaceum and G.

Table 1. Species of Gossypium listed according to genome groups.[2]

A Genome	D Genome
G. arboreum L.	G. aridum (Rose & Standl.) Skov.
G. herbaceum L.	G. armourianum Kearn.
	G. davidsonii Kell.
B Genome	G. gossypioides (Ulbr.) Standl.
G. anomalum Wawr. & Peyr.	G. harknessii Brandg.
G. capitis-viridis Mauer	G. klotzschianum Anderss.
G. triphyllum (Harv. & Sond.) Hochr.	G. laxum Phillips
	G. lobatum Gentry
	G. raimondii Ulbr.
C Genome	G. thurberi Tod.
G. australe F. Muell.	G. trilobum (D.C.) Skov.
G. costulatum Tod.[a]	G. turneri Fryx.[a]
G. cunninghamii Tod.[a]	
G. nelsonii Fryx.[a]	
G. pilosum Fryx.[a]	E. Genome
G. pulchellum (C.A. Gordn.) Fryx.[a]	G. areysianum (Defl.) Hutch.
G. populifolium (Benth.) Tod.	G. incanum (Schwartz) Hillc.
G. robinsonii F. Muell	G. somalense (Gurke) Hutch.
G. sturtianum J.H. Willis	G. stocksii Mast. in Hook.
F Genome	G Genome
G. longicalyx Hutch. & Lee	G. bickii Prokh.
AD Genome	
G. barbadense L.	G. lanceolatum Tod.
G. darwinii Watt	G. mustelinum Watt
G. hirsutum L.	G. tomentosum Seem.

[a]Genome identification is uncertain, pending genetic studies.

Table 2. Average univalent frequencies for intergenomic hybrids.[3,4]

Genome	A	B	C	D	E	AD
A	< 1	3	10	12	17	13
B	-	< 1	11	18	22	27
C	-	-	< 1	11	24-25	27
D	-	-	-	< 1[a]	23-25	13[a]
E	-	-	-	-	< 1	38
F	-	-	-	-	-	22
AD	-	-	-	-	-	< 1

[a]Except for hybrids with *G. gossypioides* which show 1 to 2 univalents with other D genome species and 21 with AD species.

arboreum, however, have been cultivated throughout Asia and parts of Africa for hundreds of years. Species of the B genome occur in northwestern, southwestern, and central Africa; those of the E genome occur in northeastern Africa and Arabia; and *G. longicalyx* (F genome) grows in east central Africa. The C genome species are spread over northern and central Australia, and *G. bickii* (G genome) is found in north central Australia. All but two D genome species occur in locations isolated from one another in the western half of Mexico; *G. klotzschianum* and *G. raimondii* occur only in the Galapagos islands and Peru, respectively. *G. thurberi* occurs in the southwestern United States as well as northern Mexico.

The amphidiploid (AD) genome group occurs only in the western hemisphere. Four species are highly restricted: *G. tomentosum* in the Hawaiian Islands, *G. darwinii* in the Galapagos islands, *G. mustelinum* in northeastern Brazil, and *G. lanceolatum* in west central Mexico. *G. hirsutum* occurs over a wide range from southern Florida and southwestern United States over most of the Caribbean islands, Mexico, northern Central America, and parts of Brazil. *G. barbadense* also occurs over a wide range including the Caribbean islands, Central America, and northern and central South America.

Most cotton fiber in the world now is produced from *G. hirsutum* (Upland cotton) or *G. barbadense* (Egyptian, Tanguis, or Pima cotton). Appreciable amounts of *G. arboreum* and *G. herbaceum* (Asiatic cotton), however, are still grown on small farms in many Asiatic countries, usually without pesticides. The latter species have been especially valuable as sources of pest resistance.

Various approaches besides cytogenetics have been used to study relationships among and within Gossypium species. These include phenetic analysis of morphological characters,[6] electrophoretic analysis of enzymes and other proteins in seed,[7-11] chromatographic analysis of flavonoids in flower petals[12] and chromatographic analysis of terpenoids in pigment glands.[13] In all cases, the studies generally support the genome groupings based on cytogenetics. Variations among species from different genome groups are generally greater than those among species within a group, and the fewest variations are found within species. These studies also indicate that several distinct subgroups of species may be distinguished in the D and C genome groups.

Because of considerable morphological variation in cultivated species, various subdivisions of these species have been proposed. For example, G. herbaceum var. africanum is used to distinguish the wild strain of G. herbaceum from the cultivated strains. Various other varietal or subspecies designations have been proposed for strains of cultivated species, but these are not used uniformly. Collections of G. hirsutum from different geographical areas have been designated as races 'latifolium,' morilli,' 'palmeri,' 'richmondii,' 'marie-galante' and 'yucatanense'.[3,14] The Texas Race Collection[14] includes more than 1000 collections belonging to these various subdivisions. While there is little genetic or biochemical justification for most subdivisions, some have proven to be more useful than others as sources of host plant resistance.

Because of high production losses (ca 40%) caused by pests and high costs of pest control, extensive searches for pest resistance in the Gossypium species have been conducted over the past 20 years. Most searches for resistances have centered on cultivated species. However, in several instances useful characters have been obtained from noncultivated species. The remainder of this review will consider the character adaptations in Gossypium species that suppress pests, the genetic and physiological control of these characters, and approaches for transferring these characters into cultivated cottons.

GOSSYPIUM ADAPTATIONS TO PESTS

Certain physiological and morphological characters that effect host plant resistance are present before the plant is attacked by pests and thus are constitutive defenses. In contrast, other characters develop only in tissues challenged by the pests, and thus are active defenses. The constitutive defenses generally are most important for resistance to insects, arachnids, and other herbivores, whereas active defenses are essential for resistance to most microbial pathogens.

Constitutive Defenses

Known Defenses. Several characters, and the pests that they are known to suppress, are listed in Table 3. With few exceptions these characters originated from mutants or exotic strains of cultivated amphidiploids *G. hirsutum* and *G. barbadense*. A smooth leaf character also was transferred from *G. armourianum*, and reduced antheridium also occurs in male-sterile lines containing *G. hirsutum* chromosomes in *G. harknessii* cytoplasm. Nectariless originated from *G. tomentosum*. Pilose and okra leaf generally have been obtained from mutants or exotic strains of *G. hirsutum*, but also occur naturally in *G. tomentosum* and *G. lanceolatum*, respectively.

The usefulness of constitutive defense characters depends to a great extent on any adverse effects associated with the character. Pilose, for example, gives outstanding resistance to many insects, but results in short fiber and low yield, and greatly increases fine trash in harvested cotton fiber. Pilose also increases damage from the cotton bollworm and tobacco budworm. Consequently, the character has not been used in commercial varieties. Red color, pubescence, smooth leaf, frego bract, and high pigment gland density (high terpenoid) also are associated with increased damage from a few nontarget insects, and red color adversely effects agronomic performance. Okra leaf has no adverse affects on insects, but is associated with determinancy and reduced yields in some geographical areas. The pubescent, glabrous, and okra characters have been used in commercial cultivars in geographical areas where their benefits outweight their limitations.

No adverse affects have been associated with the nectariless character, and this character is now used in several commercial varieties to suppress insects. The nectariless character also suppresses populations of certain beneficial insects, such as the red fire ant, but does not appreciably reduce overall predation of detrimental cotton insects.[19] Nectariless and okra leaf characters often enhance the value of other characters such as smooth leaf and frego bract when they are used in

Table 3. Constitutive characters in *Gossypium* that suppress pests.[a]

Resistance character	Pest or disease suppressed by character[b]
Plant color:	
Red plant	aph, bw
Red stem	aph, bw
Leaf conformation:	
Okra leaf	pbw, wfl, BR
Pubescent	cfh, jas, lyb, wfl
Pilose	arw, bw, cal, cfh, clp, dlm, jas, lyb, pbw, sbw, thr, wfl
Smooth leaf	cbw, cfh, clp, pbw, tbw, wfl
Flower conformation:	
Frego bract	bw, cbw, BR
Yellow anthers	tbw
Reduced antheridium	bw, BB
Seed conformation:	
Hard seeded	SD
Glands:	
Nectariless	cal, cbw, cfh, clh, clp, clw, lyb, pbw, tbw, BR
Pigment glands	bw, cbw, cfh, clh, clw, lyb, pbw, sbw, tbw
Secondary products:	
Terpenoid aldehydes	aph, bw, cbw, cfh, clh, lyb, pbw, sbw, tbw
Condensed tannins	cbw, clp, pbw, sbw, smi, tbw
Anthocyanins	sbw, tbw

[a]Adapted from references 2, 5, 15, 16, 17, 18

[b]Abbreviations: aph = aphids; arw = army worms; bw = boll weevil; cal = cabbage looper; cbw = cotton bollworm; cfh = cotton fleahopper; clp = cotton leaf perforator; clh = cotton leaf hopper; clw = cotton leaf worm; dlm = dipterous leaf minors; jas = jassids; lyb = Lygus bugs; pbw = pink bollworm; sbw = spotted bollworm; smi = spider mite; tbw = tobacco budworm; thr = thrips; wfl = whitefly; BR = boll rots; BB = bacterial blight; SD = seed deterioration by fungi.

combination.[20] More details on the effects of known resistance characters on insects are available in recent reviews.[15,16]

Potential Defenses. Frequently, nonpreference for oviposition or antibiosis to insects have been demonstrated in exotic strains of cultivated *Gossypium* species and in wild species, but causal factors have not been identified.[21-26] Some of the causal factors may be the same as those already identified. Other characters that might contribute to resistance are the glandular trichomes (capitate hairs) and various secondary products, such as the volatile terpenes, flavonoid glycosides, and cyclopropenoid fatty acids. Qualitative variations in terpenoid aldehydes and tannins might also effect resistance.

Glandular trichomes have been shown to act as resistance factors in various plants.[27] Secretions from these glands include various secondary compounds that attract, repel, or poison insects. Young glandular trichomes on cotton leaves show positive histochemical tests for catechins. Glandular trichomes are preferentially extracted when whole leaves are dipped in diethyl ether for 10 to 20 seconds, and catechin, gallocatechin, and oligomers of these compounds are found in these extracts. Thus, catechins and probably tannin derivatives appear to occur in young trichomes. Older trichomes often do not show a catechin reaction, but yield isoquercitrin as a major product when extracted with methanol. No other compounds are reported from glandular trichomes of cotton. Considerable variation in the density of glandular trichomes has been demonstrated among cultivars and species.[28,29]

Numerous volatile terpenes have been isolated and identified from Upland cotton, but their cellular localization is unknown.[18] Low concentrations of (+)-limonene, (+)-α-pinene, β-caryophyllene oxide, (-)-β-caryophyllene, or (+)-β-bisabolol are attractive to boll weevils.[30] Optimum combinations of these compounds are more attractive than the crude volatile oil from cotton. The possible impor- tance of volatile terpenes as resistance factors to insects has not been evaluated, but two experiments indicate their probable importance First, volatile chemicals and extracts from other odoriferous plants suppress oviposition by cotton leaf hoppers, when these substances are applied to cotton leaves.[31] Second, the tobacco budworm has been shown to accumulate β-carophyllene, α-humulene, γ-bisabolene, β-caryophyllene oxide, and β-bisabolol from consumed cotton tissue, making this pest more attractive to the beneficial parasitoid wasp, Campoletis sonorensis.[32] These observations indicate that volatile terpenes may both repel pests and act as attractants to beneficial insects attacking pests. Both of these possibilities need further study especially with volatile oils obtained from different Gossypium species. Volatile terpenes in essential oils of G. barbadense vary in quality and quantity from those in G. hirsutum,[17,33] but the significance of these variations for resistance to pests is not known.

Flavonoids have shown mixed effects on insects. At low concentrations isoquercitrin (quercetin-3-glucoside) and rutin (quercetin-3-rhamnoglucoside) stimulate feeding by larvae of cotton bollworm and tobacco budworm,[34] and quercetin-7-glucoside and quercetin-3'-glucoside stimulate feeding of boll weevil.[35] At concentrations above 0.2% in artificial diets, rutin and isoquercitrin inhibit larval growth and especially pupation in cotton bollworm, tobacco budworm, and pink bollworm.[18,36-38] Levels of 2 to 4% flavonoids in dry flower petals have been reported.[17] The toxicity of isoquercitrin (a monoglycoside) to cotton insects has been consistently greater than that of rutin (a diglycoside), indicating that flavonols with single sugar moieties may be more toxic than those with two sugar moieties.[37] Also, flavonols with 3',4'-dihydroxy groups are much more toxic than those with 4'-hydroxy groups. Thus, both qualitative and quantitative variations in flavonoids could contribute to pest resistance. Considerable qualitative variations have been found among flavonoids of the Gossypium species;[12] no data are available on quantitative variations.

Methyl esters of the cyclopropenoid fatty acid, malvalic acid, caused 50% inhibition of larval growth of pink bollworm, cotton bollworm, and tobacco budworm when included in artificial diets at concentrations of 0.28, 0.64 and 0.66% respectively.[39] A concentration of 0.5% interfered with pupation and with desaturation of fatty acids in the cotton boll worm. Concentrations of cyclopropenoid acids in crude oils from seed vary considerably among cultivated species. Mean concentrations of 0.67, 0.83, 1.14, and 1.40% were found in *G. barbadense*, *G. hirsutum*, *G. herbaceum*, and *G. arboreum*, respectively.[40] Concentrations for different cultivars ranged from 0.69 to 1.02% in *G. hirsutum* and from 1.17 to 1.51% in *G. arboreum*. Since seed is 20 to 25% oil, concentrations of cyclopropenoid acids in *G. arboreum* are sufficiently high to contribute resistance to insects that might feed on seed. Sterculic and malvalic acid make up 5 to 8% of the fatty acids in oils of flower buds. However, this amounts to only 0.02 to 0.10% of the total dry weight because of the low oil content of the buds.[36]

Numerous studies have shown relationships between pest resistance and total concentrations of terpenoid aldehydes and tannins in cotton cultivars or tissues.[41-55] However, structural variations in these antibiotics, which may be equally important characters for increasing resistance, have received only limited attention.

Surveys of terpenoid aldehydes in pigment glands of *Gossypium* species revealed several variations in these compounds, especially in green tissues (Fig. 1).[13,55] These variations include: 1) the degree of methylation of the phenolic hydroxyl located *meta* to the aldehyde, 2) the presence or absence of *para*-naphthoquinones formed from hemigossypol or its methyl ether, 3) quantitative differences in heliocides (H_1, H_4, B_1, B_4) derived from ocimene, 4) the presence or absence of heliocides (H_2, H_3, B_2, B_3) derived from myrcene, and 5) the presence or absence of raimondal (Fig. 2).[56] The enzymes required for quinone formation and raimondal synthesis occur only in green tissues. Consequently, quinones, heliocides, and raimondal are not found in seeds, roots, or internal glands of stem bark, even if they are the predominant compounds in leaves, bracts, or carpel walls.

Species in the A genome group lack the ability to methylate terpenoids, and the D genome species, except *G. gossypioides*, lack the ability to form quinones. Consequently, evolution of AD amphidiploids allowed the production of hemigossypolone methyl ether and heliocides B_1 and B_4 (predominant compounds in *G. barbadense*, *G. darwinii*, and *G. tomentosum*). These antibiotics do not occur in either the A or D diploid species. This added synthetic capability in the amphidiploid may have

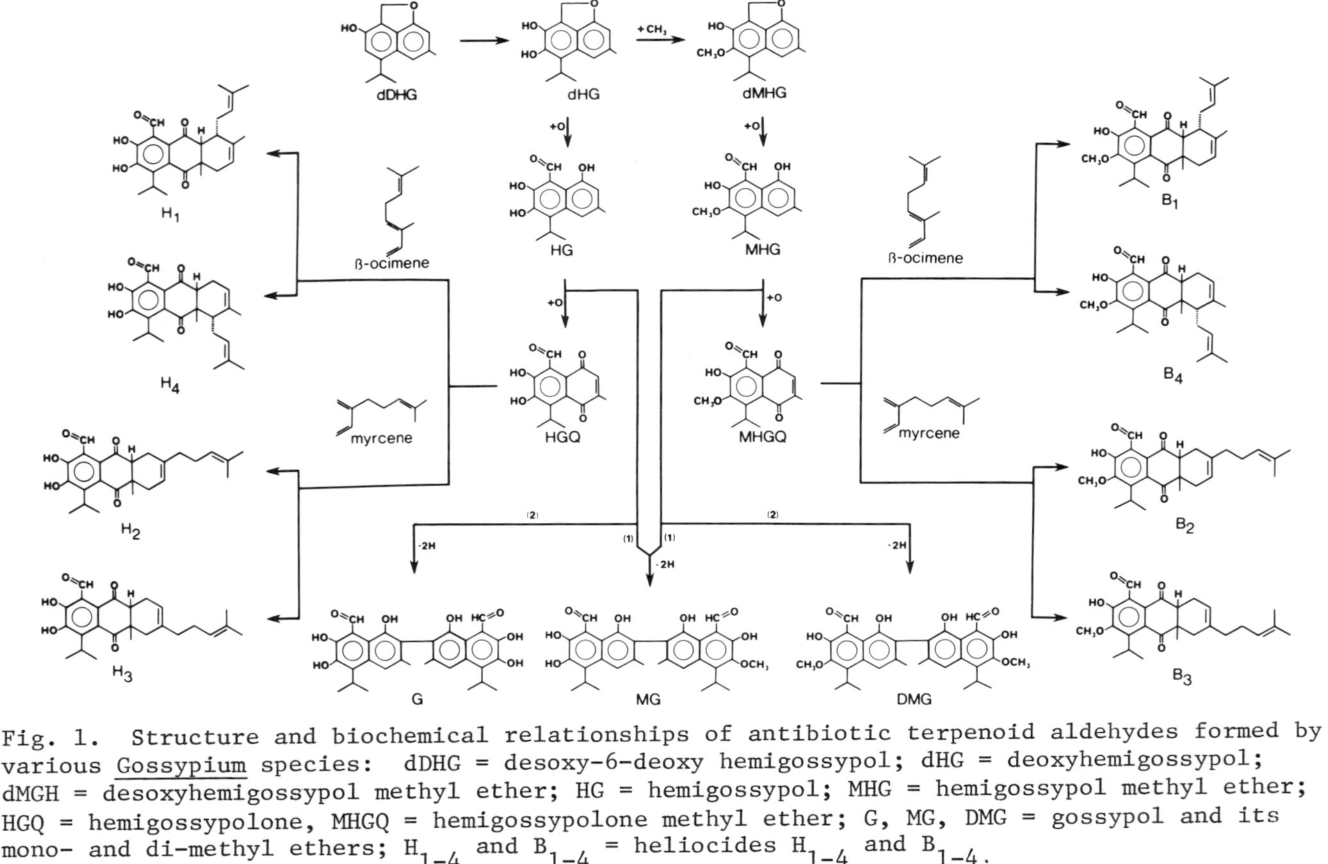

Fig. 1. Structure and biochemical relationships of antibiotic terpenoid aldehydes formed by various Gossypium species: dDHG = desoxy-6-deoxy hemigossypol; dHG = deoxyhemigossypol; dMGH = desoxyhemigossypol methyl ether; HG = hemigossypol; MHG = hemigossypol methyl ether; HGQ = hemigossypolone, MHGQ = hemigossypolone methyl ether; G, MG, DMG = gossypol and its mono- and di-methyl ethers; H_{1-4} and B_{1-4} = heliocides H_{1-4} and B_{1-4}.

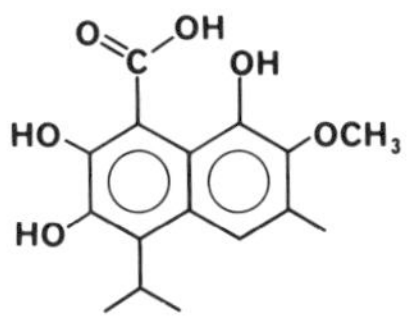

Raimondal

Fig. 2. Structure of raimondal.

increased pest resistance by diversifying the natural antibiotics in cotton.

The relative advantages or disadvantages of different terpenoid aldehydes in pigment glands are still not clear. An initial study indicated that heliocides formed from ocimene may be more toxic to tobacco budworm than heliocides formed from myrcene.[57] However, other studies have found little difference in the toxicity of the various heliocides and gossypol.[36,58] In most cases, these compounds have been added individually to artificial diets or bioassay media, rather than in the combinations that are genetically possible. Mixed terpenoids are much more toxic to nematodes than gossypol alone,[59] and it is possible that natural mixtures may also be more effective against other pests.

Qualitative variations in the condensed tannins of cotton also appear to be a source of pest resistance. Tannins extracted with boiling water from *G. barbadense* leaves have greater antisporulant activity against the fungal pathogen *Verticillium dahliae* than do comparable tannins from *G. hirsutum*.[60] Likewise, tannins from spider mite-resistant *G. hirsutum* strains (Texas race collection numbers 254, 1055, 1123, and 1124) precipitate hemoglobin from human blood more efficiently than tannins from mite-susceptible cottons.[48] Only a small percentage (<5%) of tannin molecules from any cultivar stimulate muscle contraction in the hind gut of the South American cockroach, and some cultivars lack these active tannins (G.A. Greenblatt, R.L. Ziprin, R.D. Stipanovic, and A.A. Bell, unpublished data). The specific reasons for these differences in various activities are not known.

Tannin molecules in cotton probably vary in (+)-catechin, (−)-epicatechin, (+)-gallocatechin, and

(-)-epigallocatechin content, _cis_ and _trans_ isomers of the C3-C8 linkages, polymer size, and possibly methylation.[61,62] The effects of such variations on pest resistance are unstudied. Gallocatechin dimers precipitate hemoglobin from human blood more efficiently than catechin dimers.[62] Higher gallocatechin/catechin ratios were found in tannins from a _G. arboreum_ and a _G. barbadense_ cultivar than in those from a _G. hirsutum_ cultivar.[61] The highly astringent tannin from Texas race collection number 1055 also has a high gallocatechin/catechin ratio.[48] Thus, the monomeric composition of tannin may be one important determinant of biological activity.

Active Defenses

Active Defenses Against Pests. Plant tissues invaded by microorganisms or attacked by insects respond with active defenses. These active defenses include: synthesis of new antibiotics (phytoalexins); thickening, lignification, and suberization of cell walls; plugging of xylem vessels by tyloses and gels; synthesis of polyphenols that oxidize to tannins; synthesis or activation of hydrolytic enzymes that act on microbial cell walls; synthesis of inhibitors of pest enzymes, such as protease inhibitors; and hypertrophy and hyperplasia of cells.[63,64] Cells undergoing active defense reactions to bacteria or fungi in leaves often become necrotic within 12 to 36 hours after the responses begin. Consequently, active defense is often referred to as a hypersensitive or incompatible reaction. I prefer to call this response "necrogenic resistance," recognizing that it is a normal resistant reaction of plants to most, if not all, microbial pests.

The active defense reactions most studied in cotton are the synthesis of terpenoid phytoalexins, the synthesis of tannins, and the formation of tyloses and gels in xylem vessels invaded by fungal wilt pathogens. The predominant phytoalexins formed in germinating seed,[65] young roots,[66] hypocotyls,[47] xylem vessels,[60] cambial tissues,[60] and boll endocarp[60] are desoxyhemigossypol, hemigossypol, and their methyl ethers (Fig. 1). These compounds are occasionally accompanied by much smaller amounts of gossypol, hemigossypolone, and their methyl ethers. In cotyledons, leaves and other green tissue the phytoalexins that have been isolated are 2,7-dihydroxycadalene, lacinilene C, and their methyl ethers (Fig. 3).[67,68] The phytoalexins with the greatest antibiotic activity against Verticillium dahliae are the desoxyterpenoids (M.E. Mace, R.D. Stipanovic, and A.A. Bell, unpublished data). Detailed structural analyses of tannins formed during active defenses have not been made.

The effectiveness of active defense reactions depends primarily on the quickness of their initiation after the pest contacts the host. More rapid synthesis of terpenoid phytoalexins in resistant than in susceptible cotton cultivars has been demonstrated following infection by Xanthomonas campestris pv. malvacearum (bacterial blight),[59,65] Fusarium oxysporum f.sp. malvacearum (Fusarium wilt),[69] Verticillium dahliae (Verticillium wilt),[59,70] and Meloidogyne incognita (Root knot nematode).[66] More rapid formation of tyloses[70,71] and tannins[59,72] in resistant than in susceptible cotton cultivars also has been shown with wilt diseases. Necrogenic resistance probably also occurs against southwestern cotton rust, tropical rust, aerolate mildew, anthracnose, reniform nematode, and

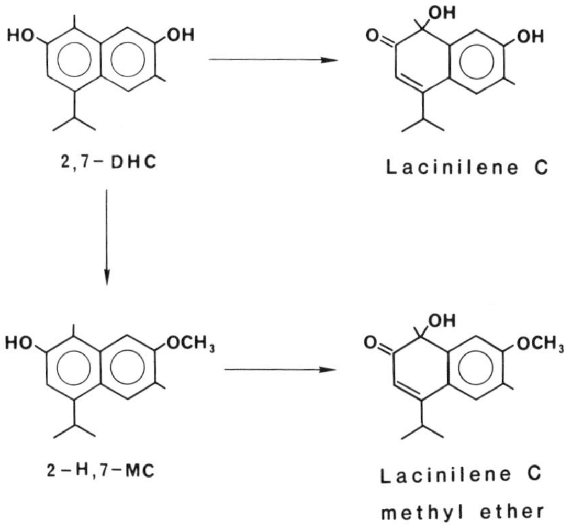

Fig. 3 Structures and biochemical relationships of 2,7-dihydroxycadalene, 2-hydroxy-7-methoxycadalene, lacinilene C, and lacinilene C methyl ether.

various virus diseases of cotton, since cell necrosis is normally associated with resistant responses.[73-76] In susceptible reactions with microbial pathogens, the necrogenic responses either fail to develop or develop 24 to 96 hours later than in resistant reactions.

Some of the sources of necrogenic resistance are shown in Table 4. Seventeen different genes controlling necrogenic resistance to bacterial blight have been found in various *Gossypium* species; most have been transferred to Upland or Egyptian cotton.[77-79] Thus, the genetic transfer of active defense reactions from wild to cultivated cottons is feasible.

The biochemistry of "pest recognition" leading to necrogenic resistance in cotton is only partially known. Wilt resistant cultivars of *G. barbadense* give necrogenic responses to dead as well as live conidia of *Verticillium dahliae*, whereas susceptible *G. hirsutum* cultivars give no response to dead cells and a delayed response to live cells.[60] A glycoprotein isolated from the cell wall of *V. dahliae* also initiates synthesis of terpenoid phytoalexins.[80] Likewise, cultivars resistant to *Xanthomonas campestris*, but not susceptible cultivars, give necrogenic responses to extracellular polysaccharides from the bacterium.[60] The cotton pathogen *Fusarium oxysporum* f.sp. *vasinfectum* has a cell wall antigen that cross-reacts with antiserum to cotton roots, indicating common antigens in the host and its pathogen.[81] Similar antigens were not found in nonpathogenic fungi. Collectively, these observations indicate that resistant cotton plants somehow recognize cell walls or capsular materials from microorganisms, and that this reaction may be suppressed by common antigens in pathogens and susceptible cotton plants.

Carbohydrate oligomers released from fungal cell walls by plant enzymes act as potent elicitors of necrogenic resistance in soybean.[82,83] Carbohydrate oligomers released from plant cell walls by the action of fungal pectinases or the plant's own enzymes also may act as elicitors.[63,84] Thus, the active resistance of exotic strains or wild species of cotton to microbial pests might be due to differences in the enzymes or the structural components of cell walls. Exact mechanisms of recognition, however, remain to be determined.

Table 4. Sources of necrogenic resistance to various cotton diseases and nematodes

Disease or nematode	Source of necrogenic resistance[a]
Bacterial blight	*G. arboreum*, *G. anomalum*, *G. barbadense* exotic strains, mutants
Fusarium wilt	*G. arboreum*, *G. herbaceum*, *G. barbadense*, exotic strains
Verticillium wilt	*G. barbadense*, exotic strains
Southwestern cotton rust	*G. arboreum*, *G. anomalum*
Tropical rust	*G. barbadense*
Anthracnose	*G. arboreum*
Thielaviopsis root rot	*G. arboreum*
Aerolate mildew	blight resistant strains
Blue disease virus	*G. arboreum*
Leaf curl and mosaic viruses	exotic strains
Anthocyanosis virus	blight resistant cultivars
Root knot nematode	*G. darwinii*, exotic strains
Reniform nematode	*G. longicalyx*, *G. anomalum*, *G. arboreum*

[a] Adapted from references 74, 75, 77-79; exotic strains or mutants are from *G. hirsutum*.

Active Defenses in Genetic Lethals. Crosses between certain *Gossypium* species give normal appearing seed, but after germination the hybrid plant develops typical disease symptoms, and it dies or is markedly stunted. These genetic lethal reactions may begin at any point from early embryogenesis until after early boll development depending on the strains and species involved. Upland cotton (*G. hirsutum*) normally gives rise to genetic lethals when crossed with *G. davidsonii* (or *G. klotzschianum*),[85] *G. gossypioides*,[86] or *G. arboreum* var. *sanguineum*.[87] The interspecific hybrids from *G. davidsonii* become necrotic during late embryo development or soon after seedlings emerge; those from *G. gossypioides* die after 6 to 18 true leaves are formed; and those from *G. arboreum* var. *sanguineum* die just before or after initial flowering.

The lethal reaction can be prevented by transferring compatibility genes from a third compatible species into *G. hirsutum*. For example, two compatibility genes have been transferred from an exotic strain of *G. barbadense* into *G. hirsutum* to allow hybridization with *G. davidsonii* and *G. klotzschianum*.[85,88] I have recently transferred similar compatibility genes from *G. barbadense* and Texas race collection number 1055 that allow hybridization of the recipient *G. hirsutum* strains with *G. gossypioides* and *G. arboreum* var. *sanguineum*, respectively. These various compatibility genes now allow complete use of all *Gossypium* species for cotton improvement by removing crossing barriers with Upland cotton.

The genetic lethal syndrome in *G. hirsutum* x *G. gossypioides* hybrids has a striking resemblance to that of plants dying from fungal wilt diseases, including symptoms such as epinasty, wilting, yellowing and defoliation of leaves, vascular browning, and hypertrophy and hyperplasia of the cambium. Close examination of these plants shows responses typical of those associated with necrogenic resistance, including enhanced phytoalexin and tannin synthesis.[86] However, these responses apparently are so extensive that they are detrimental in plants dying of the genetic lethal response. Concentrations of toxic terpenoids and tannins accumulated in plants killed by genetic lethal reactions were similar to those in plants killed by wilt fungi.[61] I have recently examined the other genetic lethal systems, and in all cases found extensive spontaneous induction of terpenoid aldehyde and tannin synthesis (active defense reactions) occurring concurrently with the lethal response. Thus, genetic lethal reactions in *Gossypium* appear to parallel autoimmune diseases found in humans and other animals. Further, these observations suggest that many plant disease symptoms may result from immune reactions of the plant.

Because active defense responses are associated with both necrogenic resistance and interspecific genetic incompatibility, common biochemical or genetic mechanisms may control both phenomena. To test this possibility I evaluated 26 *Verticillium*-resistant plant selections for compatibility to *G. gossypioides*. These plants were selected after three generations of selfing of second backcross progeny from an original cross of resistant *G. barbadense* x susceptible *G. hirsutum*. Assuming random distribution of genes approximately 3% of the plants should have been homozygous and 6% heterozygous for G. gossypioides compatibility from G. barbadense. In fact, 25% were homozygous and 60% were heterozygous. The heterozygous plants were both selfed and backcrossed to G. hirsutum, and progeny were again selected for resistance to Verticillium wilt. Again, nearly all resistant selections were either heterozygous or homozygous for compatibility to G. gossypioides. These results indicate that necrogenic resistance to Verticillium wilt and compatibility to G. gossypioides are associated and might even be controlled by the same gene or closely linked genes from G. barbadense.

Multiple Species Introgression: A Source of Active Defense. If common genes (and biochemical mechanisms) control both interspecific and host-pest recognition in a Gossypium species, it should be possible to generate new patterns of pest recognition (and thus new active defenses) by creating multiple species plants. The recombination of recognition systems in such plants should create novel systems that previously did not exist. To test this hypothesis several three-species and multiple-species plants were created (see Methods 4 and 6 under Transfer of Resistance from Wild to Cultivated Cottons) and crossed with cultivars of Upland cotton. Progeny from these crosses were then tested for resistance to bacterial blight, Verticillium wilt, and root-knot nematode. Reactions of progeny from one triple species plant to Verticillium wilt and root-knot nematode are shown in Tables 5 and 6. Twelve other different three- and multiple-species plants also gave transgressive segregation for pest resistance; both plants more resistant and more susceptible than the original species were usually obtained. The highest level of resistance to Verticillium wilt found in the experiment shown in Table 5 has been backcrossed twice into G. hirsutum, while maintaining the resistance level. Thus, multiple species plants may be useful to create levels of necrogenic resistance greater than now available in individual species.

REGULATION OF RESISTANCE CHARACTERS

Genetic Regulation

The major genes controlling known constitutive defenses have been identified and often located on

Table 5. Frequencies of plants in cultivars and in progeny from three-species males crossed with these cultivars showing different levels of Verticillium wilt resistance.

Verticillium Index[a]	Cultivar or hybrid progeny[b]				
	SBSI	P464	C-E	P464 x AR-RA	C-E x AR-RA
0-0.5	0	0	0	0	0
0.6-1.5	2	0	0	0	2
1.6-3.4	9	1	0	25	14
3.5-4.4	4	2	1	17	5
4.5-5.0	1	13	15	38	18

[a]Verticillium index: 0 = no symptoms, 3 = 50% defoliation, and 5 = dead. Plants with grades below 3.5 exhibit moderate to high levels of resistance in the field.

[b]Abbreviations: SBSI = the resistant standard 'Seabrook Sea Island 12B2', P464 = cultivar 'Paymaster 464', CE = cultivar 'TAMCOT CAMD-E', and AR-RA = the three-species hybrid *G. hirsutum* x (*G. arboreum* x *G. raimondii*).

specific chromosomes.[89] Several characters in amphidiploids are controlled by a single gene in only the A or D genome complement of chromosomes. For example, the red plant character is controlled by the single $\underline{R}_1$ gene on chromosome 16 of the D genome; the $\underline{R}_2$ gene on homoeologous chromosomes 7 of the A genome conditions petal spot. Other characters controlled by a single dominant gene are okra leaf, dense trichomes, and smooth leaf. Frego bract is controlled by a single recessive gene. Characters that vary mostly in degree are often controlled by allelic genes, i.e. different states of the same gene that give different effects. Pubescence, pilose, and smooth leaf are controlled by the alleles $\underline{H}_1$, $\underline{H}_2$ and $\underline{Sm}_2$ respectively, on chromosome 6 of the A genome. These alleles give different densities and lengths of trichomes. The gene $\underline{Sm}_1$ also contributes smoothness but occurs on homeologous chromosome 25 of the D genome. It sometimes, but not always, is used in combination with the $\underline{Sm}_2$ allele.

Several characters are controlled by two genes located on separate but homoeologous chromosomes, i.e.

separate chromosomes from the A and D genome that each carry similar genes and linkages. The separate genes controlling nectariless (ne_1, ne_2), pigment glands (Gl_2, Gl_3) and genetic lethality with *G. davidsonii* (Le_1, Le_2) occur on chromosome 12 of the A genome or chromosome 26 of the D genome.[90,91] Genes (P_1, P_2) on separate chromosomes also control yellow and orange pollen color. Reduced antheridium is the only resistance character known to be controlled by nuclear-cytoplasmic interaction.[92]

High terpenoid aldehyde concentrations in flower buds are achieved by substituting the Gl_3' or Gl_3^r alleles for the normal Gl_3 allele.[93,94] The special alleles probably were obtained from the exotic Socorro Island wild strain of *G. hirsutum*. Various undefined minor genes also affect terpenoid concentrations.[95-98] Increases in terpenoids apparently involve increases in both pigment gland volume and density. Tannin levels apparently are controlled by several additive genes that have not been identified.[96-98]

Only limited information is available on the genetic control of variations in terpenoid structure. The segregation of F_1 and F_2 progeny from *G. hirsutum* x *G. barbadense* cross indicated methylation in *G. hirsutum* is controlled by both a structural gene and a dominant regulator gene that repressed methylation in all tissues except seedling radicles.[99] The production of ocimene (evidenced by heliocides H_1, H_4, B_1, and B_4) and myrcene (evidenced by heliocides H_2, H_3, B_2, and B_3) appear to be under the control of separate dominant genes. More recently, I have studied methylation in various interspecific hybrids, involving combinations of *G. australe*, *G. barbadense*, *G. hirsutum*, *G. longicalyx*, *G. robinsonii*, *G. sturtianum*, and *G. tomentosum*. The regulator gene in *G. hirsutum* is effective in limiting methylation in all hybrids except those involving *G. sturtianum*. Methylation was dominant in all tetraploids, hexaploids, and pentaploids containing a complete set of *G. sturtianum* chromosomes. Thus, *G. sturtianum* apparently carries a distinct allele or gene for methylation that is not regulated by the *G. hirsutum* genes. I also have studied the formation of the terpenoid raimondal in various interspecific crosses. This character is always dominant and apparently is due to a single gene.

The genetic regulation of necrogenic resistance has been studied in only a few disease interactions. Seventeen different dominant genes have been reported to confer necrogenic resistance to certain races of *Xanthomonas malvacearum*.[77,79] However, there is not adequate information to determine whether some of these might be different alleles of the same genes or homoeologues on separate chromosomes. Resistance to Fusarium and Verticillium wilt in *G. barbadense* is apparently controlled by a pair of dominant genes, while that in strains of *G. hirsutum* is controlled by a single gene.[78] Disease severity classes normally are not distinct with wilt diseases. Consequently, inheritance patters are difficult to interpret. Necrogenic resistance to most other diseases is also partially dominant but detailed genetic studies have not been performed.[74]

Physiological Regulation

The expression of resistance characters varies with the age of the plant and with the environment. The constitutive defenses are often least expressed in the first true leaves and then increase in intensity with each leaf formed until maximum expression is first shown between leaves 8 to 12. As full fruit load is reached, the expression of the character often declines. These patterns have been shown for red color,[3] pubescence,[100] pigment gland density,[101] and tannin content.[45,48] Similar patterns probably exist for other resistance characters.

Because the speed of necorgenic resistance is the important determinant of its effectiveness, expression of active defenses is sensitive to environment.[102] Any factor that slows the host response relative to the speed of infection by the pest will decrease the level of resistance. Consequently, active defense mechanisms may not be apparent under cold or hot temperatures, high moisture levels, or any other condition that decreases the speed of development of active defenses in cotton tissues.[102,103]

It is important to recognize the physiological changes in resistance characters in order to measure associated genetic differences in resistance most accurately. For example, comparisons of constitutive

Table 6. Frequencies of plants in cultivars and in progeny from three-species males crossed with cultivars supporting different levels of root-knot nematode reproduction.[a]

Eggs	Cultivar or hybrid progeny[b]			
gram root	M-8	A623	P464	P464 x AR-RA
0-500	0	10	0	2
500-1500	0	0	0	8
1500-3500	0	0	2	15
3500-7500	2	0	5	23
7500-11500	2	0	8	9
11500-13500	2	0	3	0
Over 13500	4	0	2	3

[a]Studies performed in cooperation with J.A. Veech.

[b]Abbreviations: M-8 = the susceptible standard 'M-8', A623 = the resistant standard 'Auburn 623', P464 = cultivar 'Paymaster 464', and AR-RA = the triple species hybrid *G. hirsutum* x (*G. arboreum* x *G. raimondii*).

resistance characters in the eighth true leaf may be far more meaningful than in the first or second true leaves. Likewise, temperatures and other environmental conditions should be controlled to optimize expression of active defense characters.

TRANSFER OF RESISTANCE FROM WILD TO CULTIVATED COTTONS

Resistance characters in wild amphidiploid species or strains normally can be transferred readily into the cultivated amphidiploids by standard techniques of hybridization and selection. Selections based on progeny row analyses generally give better results than individual plant selections when more than one gene controls a character. This and other traditional methods of breeding for pest resistance have been reviewed.[104,105]

The transfer of resistance from diploid species into cultivated amphidiploid species presents greater difficulties, because chromosome doubling must be introduced at some point before the desired genes can be transferred. Also, endosperm failure and genetic lethal reactions often prohibit the formation of hybrids among species. The recent refinement of ovule culture techniques for cotton, including *in vitro* fertilization, have overcome endosperm failure.[106] Likewise, the introduction of compatibility genes from a third mutually compatible species has allowed hybridization of otherwise incompatible species.

The following methods are used for the transfer of resistance characters from diploid to amphidiploid species.[2,3,5]

1. Formation of two-species hexaploids. The diploid species is crossed onto an amphidiploid female to give sterile triploid plants. Buds of the triploid plants are treated with colchicine (usually 0.5-1.0% in water or lanolin) to give fertile hexaploid branches or whole plants. Subsequently, the hexaploid, pentaploid, and aneuploid generations are backcrossed as female parents to the cultivated amphidiploid, until finally the desired character is obtained along with the original number (2n = 52) of chromosomes. This technique was used to transfer characters such as the $Sm_1\sigma\}$ allele from *G. armourianum*, the Le^{dav} allele from *G. davidsonii*, and bacterial blight resistance genes (B_4 and B_6) from *G. arboreum* into the cultivated amphidiploid species. I have also used it to tranfer raimondal from *G. raimondii* and terpenoid methylation from *G. sturtianum* to *G. hirsutum*.

2. Formation of two-species tetraploids from synthetic autotetraploids. The diploid species is treated with colchicine to double chromosome numbers, yielding an autotetraploid. This is then crossed with the amphidiploid, and after several backcrosses the desired gene is incorporated into the cultivated amphidiploid species. This technique was used to transfer bacterial blight resistance from *G. herbaceum* to cultivated amphidiploids.[77] Although this technique sounds simple, I have found that the formation of autotetraploids is extremely difficult, if not impossible, for most diploid species, including most strains of *G. arboreum*.

3. Formation of two-species tetraploids from hexaploids. The two-species hexaploid obtained by method 1 is backcrossed to the diploid species to yield the tetraploid. This is then backcrossed to the cultivated amphidiploid until the desired character is incorporated. I used *G. sturtianum*-*G. hirsutum* hexaploids as either male or female in crosses with *G. sturtianum*. Consequently,

this technique allowed transfer of *G. hirsutum* chromosomes into *G. sturtianum* cytoplasm to give a new source of male sterility. The frequency of successful crosses, however, is greatest when both the hexaploid and the derived two-species tetraploid were used as females. This approach was used to transfer flavonoid and terpenoid characters from *G. sturtianum* into *G. hirsutum*.

4. Formation of three-species tetraploids from two-species artifical tetraploids. Species from two different diploid genomes are crossed to yield sterile diploid plants. These are treated with colchicine to yield fertile tetraploid branches or plants. Crossing with the cultivated amphidiploid then yields a hybrid plant with single sets of chromosomes from each of the three species. This plant is then backcrossed to the cultivated species for several generations to transfer the desired character. Three-species hybrid plants have been used to transfer resistance to bacterial blight, southwestern cotton rust, and blue disease virus from diploid to cultivated amphidiploid species.[73,74,77]

In the past, three-species hybrid plants have been formed from crosses of A genome species, especially *G. arboreum*, with D genome species. However, I have recently used this technique with crosses between cultivars of *G. arboreum* and *G. anomalum* (B genome). Synthetic AB tetraploids, unlike most synthetic AD tetraploids, are reasonably self-fertile, female fertile, and day-neutral. Consequently, they may facilitate the transfer of chromosomes and cytoplasms from different strains of Asiatic cottons to Upland and Egyptian cottons.

5. Formation of three-species tetraploids from hexaploids. Crosses between many diploid species result in endosperm failure, and thus it is not possible to form synthetic tetraploids by the usual means. The endosperm failure in diploid hybrids may be overcome by ovule culture.[106] However, an alternative approach is to cross one diploid species into a hexaploid formed between the other diploid and the amphidiploid as described in Method 1. Endosperm failure is not frequent in this situation because diploids are always crossed onto female plants with higher ploidy levels. The synthetic tetraploid obtained by this method is then used as in method 4. Method 5 has been used to incorporate *G. harknessii* and *G.*

armourianum into triple species plants with G. arboreum and G. hirsutum.[3,5]

6. Formation of a multiple-species tetraploids from three-species artificial tetraploids. In this technique two diploid species within a genome group or in the closely related A and B genome groups are crossed, and the hybrid is crossed with a third diploid species in another genome group to give a sterile three-species diploid. This is treated with colchicine to form a fertile three-species tetraploid that is then crossed with an amphidiploid species or an F_1 hybrid from two amphidiploid species to give a four- or five-species hybrid plant, respectively. I have used this technique extensively starting with three-species tetraploids formed from crosses of G. anomalum, G. arboreum, G. capitis-viridis, or G. herbaceum with (G. thurberi x G. raimondii). In other crosses the starting F_1 was (G. raimondii x G. gossypioides). In general, the A and B genome species cross more readily with F_1 hybrids of two D genome species than with either pure species. Consequently, multiple-species tetraploids are formed more readily than three-species tetraploids. Recently, I used this same technique to cross all D genome species onto G. arboreum x G. anomalum diploid hybrids. However, I have not yet tried to develop multiple-species hybrid plants from these three-species diploid hybrids.

CONCLUSIONS

Increasingly, host plant resistance characters are being used to control pests in cotton production. Prior to 1940 more than 5% of the crop was lost annually to anthracnose, bacterial blight, and Fusarium wilt. Incorporation of various characters to improve active defenses against these diseases in commercial cultivars reduced losses to less than 1% from 1975-80.[74] Similar outstanding gains in controlling bacterial blight with host plant resistance have been made in various African countries where the disease once destroyed 10% or more of the crop. Likewise, large losses to Verticillium wilt in the United States in the late 1960's have been partially overcome by the use of improved resistance, although higher levels of resistance are still needed. Cultivars carrying the nectariless character have given appreciable insect control, and acreages of these cultivars have increased rapidly since their release in the late 1970's.[107,108] Breeding lines with improved resistance to nematodes are approaching the point where they will be used in commercial cultivars, and should decrease losses from disease complexes as well as from nematodes.[109,110]

One of the most successful breeding programs for using host plant resistance characters in cotton is the MAR (Multiple Adversity Resistance) System at Texas A&M University.[111] This program has placed special emphasis on resistance to seed deterioration, seedling diseases, and bacterial blight. In addition, all known characters conferring resistance to pests are used in the basic germplasm pools, and characters suppressing more than one pest are given special emphasis. Cultivars produced by the MAR System are now the predominant cottons grown on many production areas in Texas, especially where pests are the major limitations to production.

In the future, wild species will probably be the greatest source of new characters for pest resistance. Various studies of flavonoids, terpenoids, seed proteins, and morphological characters have shown greater variation among genome groups than among species within a genome group; the least variation occurs within species. The same patterns would be expected for resistance characters affecting pests. The development of improved ovule culture techniques and the discovery of transferable compatibility genes that prevent genetic lethal reactions with *G. arboreum* var. *sanguineum*, *G davidsonii*, and *G. gossypioides* now make it possible to transfer genes from any *Gossypium* species into Upland cottons. Multiple-species hybrid plants originally developed to study genetic relationships among species are also useful tools to transfer specific morphological and biochemical characters from diploid species to cultivated amphidiploid species. In addition, the multiple hybrid species plants may give rise to progeny with higher levels of resistance to pests than is found in any of the component species.

REFERENCES

1. FRYXELL PA 1968 A redefinition of the tribe Gossypieae Bot Gaz 129: 296-308
2. Technical Committee Regional Research Project S-77 1981 Preservation and utilization of germplasm in

cotton 1968-1980 Southern Cooperative Ser Bull No 256
3. Technical Committee Regional Research Project S-1 1956 Genetics and cytology of cotton 1948-1955. Southern Cooperative Ser Bull No 47
4. PHILLIPS LL 1966 The cytology and phylogenetics of the diploid species Gossypium. Am J Bot 53: 328-335
5. Technical Committee Regional Research Project S-1 1968 Genetics and cytology of cotton 1956-1967. Southern Cooperative Ser Bull No 139
6. FRYXELL PA 1971 Phenetic analysis and phylogeny of the diploid species of Gossypium L. (Malvaceae). Evolution 25: 554-562
7. CHERRY JP, FRH KATTERMAN, JE ENDRIZZI 1970 Comparative studies of seed proteins of species Gossypium by gel electrophoresis. Evolution 24: 431-447
8. CHERRY JP, FRH KATTERMAN, JE ENDRIZZI 1972 Seed esterases, leucine aminopeptidases and catalases of species of genus Gossypium. Theor Appl Genet 42: 218-226
9. JOHNSON BL, MM THEIN 1970 Assessment of evolutionary affinities in Gossypium by protein electrophoresis. Am J Bot 57: 1081-1092
10. JOHNSON BL 1975 Gossypium palmeri and a polyphyletic origin of the New World cottons. Bull Torrey Bot Club 102: 340-349
11. HANCOCK JF 1982 Alcohol dehydrogenase isozymes in Gossypium hirsutum and its putative diploid progenitors: The biochemical consequences of enzyme multiplicity. Plant Syst Evol 140: 141-149
12. PARKS CR, WL EZELL, DE WILLIAMS, DL DREYER 1975 VII The application of flavonoid distribution to taxonomic problems in the genus Gossypium. Bull Torrey Bot Club 102: 350-361
13. BELL AA, RD STIPANOVIC, DH O'BRIEN, PA FRYXELL 1978 Sesquiterpenoid aldehyde quinones and derivatives in pigment glands of Gossypium. Phytochemistry 17: 1297-1305
14. Technical Committee Regional Research Project S-77 1974 The regional collection of Gossypium germplasm. USDA Publ ARS-H-2
15. MAXWELL FG 1980 Advances in breeding for resistance to cotton insects. Beltwide Cotton Prod Res Conf, St Louis, MO pp 141-147

16. SCHUSTER MJ 1980 Insect resistance in cotton. In Biology and Breeding for Resistance to Arthropods and Pathogens in Agricultural Plants. Texas Agr Exp Sta and Univ Calif pp 101-112
17. BELL AA 1984 Physiology of secondary products in cotton. In JM Stewart, JR Mauney, eds, Cotton Physiology--A Treatise. In press
18. HEDIN PA, JN JENKINS, DH COLLUM, WH WHITE, WL PARROTT 1983 Multiple factors in cotton contributing to resistance to the tobacco budworm, *Heliothis virescens* F. In PA Hedin, ed, Plant Resistance to Insects. Am Chem Soc Symp Ser 208 pp 347-365
19. AGNEW CW, WL STERLING, DA DEAN 1982 Influence of cotton nectar on red imported fire ants and other predators. Environ Entomol 11: 629-634
20. MAXWELL FG 1982 Current status of breeding for resistance to insects. J Nematol 14: 14-23
21. JENKINS JN, WL PARROTT, JC McCARTY, LN LATSON 1977 Evaluation of cotton, *Gossypium hirsutum* L., lines for resistance to the tarnished plant bud, *Lygus lineolaris*. Tech Bull 89, USDA and Mississippi State Univ
22. JENKINS JN, WL PARROTT, JC McCARTY, AT EARNHEART 1978 Evaluation of primitive races of *Gossypium hirsutum* L. for resistance to boll weevil. USDA Tech Bull 91
23. LAMBERT L, JN JENKINS, WL PARROTT, JC MCCARTY 1980 Evaluation of foreign and domestic cotton cultivars and strains for boll weevil resistance. Crop Sci 20: 804-806
24. McCARTY JC JR, JN JENKINS, WL PARROTT 1982 Partial suppression of boll weevil oviposition by a primitive cotton. Crop Sci 22: 490-492
25. MULLINS W, EP PIETERS 1982 Effect of resistant and susceptible cotton strains on larval size, developmental time, and survival of the tobacco budworm. Environ Entomol 11: 363-366
26. WILSON FD, BW GEORGE, RL WILSON 1981 Screening cotton for resistance to pink bollworm. USDA Agr Rev Man ARM-W-22
27. STIPANOVIC RD 1983 Function and chemistry of plant trichomes and glands in insect resistance. In PA Hedin, ed, Plant Resistance to Insects. Am Chem Soc Symp 208 pp 69-100
28. KOSMIDOU-DIMITROPOULOU K, JD BERLIN, PR MOREY 1980 Capitate hairs on cotton leaves and bracts. Crop Sci 20: 534-537

29. BRYSON CT, JC McCARTY JR, JN JENKINS, WL PARROTT 1983 Frequency of pigment glands and capitate and covering trichomes in nascent leaves of selected cottons. Crop Sci 23: 369-371
30. MINYARD JP, DD HARDEE, RC GUELDNER, AC THOMPSON, G WIYGUL, PA HEDIN 1969 Constituents of the cotton bud compounds attractive to the boll weevil. J Agric Food Chem 17: 1093-1097
31. SAXENA KN, A BASIT 1982 Inhibition of oviposition by volatiles of certain plants and chemicals in the leafhopper _Amrasca devastans_ (Distant). J Chem Ecol 8: 329-338
32. ELZEN GW, HJ WILLIAMS, SB VINSON 1983 Response by the parasitoid _Campoletis sonorensis_ (Hymenoptera: Ichneumonidae) to synomones in plants: Implications for host habitat location. Environ Entomol 12:1872-1876
33. HEDIN PA, AC THOMPSON, RC GUELDNER, AM RIZK, HS SALMA 1972 Malvaceae: Egyptian cotton leaf essential oil. Phytochemistry 11: 2356-2357
34. GUERRA AA, TN SHAVER 1969 Feeding stimulants from plants for larvae of the tobacco budworm and bollworm. J Econ Entomol 62: 98-100
35. HEDIN PA, LR MILES, AC THOMPSON, JP MINYARD 1968 Constituents of a cotton bud. Formulation of a boll weevil feeding stimulant mixture. J Agric Food Chem 16: 505-513
36. CHAN BG, AC WAISS JR, RG BINDER, CA ELLIGER 1978 Inhibition of Lepidopterous larval growth by cotton constituents. Ent Exp Appl 24: 94-100
37. ELLIGER CA, BG CHAN, AC WAISS JR 1980 Flavonoids as larval growth inhibitors. Structural factors governing toxicity. Naturwissenschaften 67: 358-359
38. SHAVER TN, MJ LUKEFAHR 1969 Effect of flavonoid pigments and gossypol on growth and development of the bollworm, tobacco budworm, and pink bollworm. J Econ Entomol 62: 643-646
39. BINDER RG, BG CHAN 1982 Effects of cyclopropanoid and cyclopropenoid fatty acids on growth of pink bollworm, bollworm, and tobacco budworm. Ent Exp Appl 31: 291-295
40. PANDEY SN, LK SURI 1982 Cyclopropenoid fatty acid content and iodine value of crude oils from Indian cottonseed. J Am Oil Chem Soc 59: 99-101
41. GUINN G, MP EIDENBOCK 1982 Catechin and condensed tannin contents of leaves and bolls of cotton in

relation to irrigation and boll load. Crop Sci 22: 614-616

42. HALLOIN JM 1982 Localization and changes in catechin and tannins during development and ripening of cottonseed. New Phytol 90: 651-657
43. HANNY BW 1980 Gossypol, flavonoid, and condensed tannin content of cream and yellow anthers of five cotton (*Gossypium hirsutum* L.) cultivars. J Agric Food Chem 28: 504-506
44. HEDIN PA, DH COLLUM, WH WHITE, WL PARROT, HC LANE, JN JENKINS 1981 The chemical basis for resistance in cotton to *Heliothis* insects. *In* M Kloza, ed, Regulation of Insect Development and Behaviour, Part II, Wroclaw Tech Univ Press, Wroclaw, Poland pp 1071-1086
45. HOWELL CR, AA BELL, RD STIPANOVIC 1976 Effect of aging on flavonoid content and resistance of cotton leaves to Verticillium wilt. Physiol Plant Pathol 8: 181-188
46. HUNTER RE 1974 Inactivation of pectic enzymes by polyphenols in cotton seedlings of different ages infected with *Rhizoctonia solani*. Physiol Plant Pathol 4: 151-159
47. HUNTER RE, JM HALLOIN, JA VEECH, WW CARTER 1978 Terpenoid accumulation in hypocotyls of cotton seedlings during aging and after infection by *Rhizoctonia solani*. Phytopathology 68: 347-350
48. LANE HC, MF SCHUSTER 1981 Condensed tannins of cotton leaves. Phytochemistry 20: 425-427
49. MEISNER J, M ZUR, E KABONCI, KRS ASCHER 1977 Influence of gossypol content of leaves of different cotton strains on the development of *Spodoptera littoralis* larvae. J Econ Entomol 70: 714-716
50. SCHUSTER MF, HC LANE 1980 Evaluation of high-tannin cotton lines for resistance to bollworms. Proc Beltwide Cotton Prod Res Conf, St Louis, MO pp 83-84
51. SEAMAN F, MJ LUKFAHR, TJ MABRY 1977 The chemical basis of the natural resistance of *Gossypium hirsutum* L. to *Heliothis*. Proc Beltwide Cotton Prod Res Conf, Atlanta, GA pp 102-103
52. SHARMA HC, RA AGARWAL 1981 Behavioural responses to larvae of spotted bollworm *Earias vittella* Fabr. towards cotton genotypes. Indian J Ecol 8: 223-228

53. SHARMA HC, RA AGARWAL 1981 Consumption and utilization of bolls of different cotton genotypes by larvae of *Earias vittella* F. and effect of gossypol and tannins on food utilization. Z Angew Zool 68: 13-38
54. SHARMA HC, RA AGARWAL 1982 Effect of some antibiotic compounds in *Gossypium* on the post-embryonic development of spotted bollworm (*Earias vittella*). Ent Exp Appl 31: 225-228
55. SHARMA HC, RA AGARWAL, M SINGH 1982 Effect of some antibiotic compounds in cotton on post-embryonic development of spotted bollworm (*Earias vittella* F.) and the mechanism of resistance in *Gossypium arboreum*. Proc Indian Acad Sci (Anim Sci) 91: 67-77
56. STIPANOVIC RD, AA BELL, DH O'BRIEN 1980 Raimondal, a new sesquiterpenoid from pigment glands of *Gossypium raimondii*. Phytochemistry 19: 1735-1738
57. STIPANOVIC RD, AA BELL, MJ LUKEFAHR 1977 Natural insecticides from cotton (*Gossypium*). *In* PA Hedin, ed, Host Plant Resistance to Pests. Am Chem Soc Symp Ser 62 pp 197-214
58. ELLIGER CA, BG CHAN, AC WAISS JR 1978 Relative toxicity of minor cotton terpenoids compared to gossypol. J Econ Entomol 71: 161-164
59. VEECH JA 1979 Histochemical localization and nematoxocity of terpenoid aldehydes in cotton. J Nematol 11: 240-246
60. BELL AA, RD STIPANOVIC 1977 Biochemistry of disease and pest resistance in cotton. Mycopathologia 65: 91-106
61. BELL AA, RD STIPANOVIC 1983 Biologically active compounds in cotton: An overview. Proc 7th Cotton Dust Res Conf, San Antonio, TX pp 77-80
62. BATE-SMITH EC 1975 Phytochemistry of proanthocyanidins. Phytochemistry 14: 1107-1113
63. BELL AA 1983 Physiological responses of plant cells to infection. *In* R Moore, ed, Vegetative Compatibility Responses in Plants. Baylor Univ Press, Waco, TX pp 47-70
64. BELL AA 1981 Biochemical mechanisms of disease resistance. Annu Rev Plant Physiol 32: 21-81
65. HALLOIN JM, AA BELL 1979 Production of nonglandular terpenoid aldehydes within diseased seeds and cotyledons of *Gossypium hirsutum* L. J Agric Rood Chem 27: 1407-1409

66. VEECH JA 1978 An apparent relationship between methoxy-substituted terpenoid aldehydes and the resistance of cotton to Meloidogyne incognita. Nematologica 24: 81-87
67. ESSENBERG M, M D'A DOHERTY, BK HAMILTON, VT HENNING, EC COVER, SJ McFAUL, WM JOHNSON 1982 Identification and effects on Xanthomonas campestris pv. malvacearum of two phytoalexins from leaves and cotyledons of resistant cotton. Phytopathology 72: 1349-1356
68. STIPANOVIC RD, GA GREENBLATT, RC BEIER, AA BELL 1981 2-Hydroxy-7-methoxycadalene. The precursor of lacinilene C 7-methyl ether in Gossypium. Phytochemistry 20: 729-730
69. KAUFMAN Z, D NETZER, I BARASH 1981 The apparent involvement of phytoalexins in the resistance response of cotton plants to Fusarium oxysporum F. sp vasinfectum. Phytopath Z 102: 178-182
70. MACE ME 1978 Contributions of tyloses and terpenoid aldehyde phytoalexins to Verticillium wilt resistance in cotton. Physiol Plant Pathol 12: 1-11
71. HARRISON NA, CH BECKMAN 1982 Time/space relationships of colonization and host response in wilt-resistant and wilt-susceptible cotton (Gossypium) cultivars inoculated with Verticillium dahliae and Fusarium f.sp. vasinfectum. Physiol Plant Pathol 21: 193-207
72. MACE ME, AA BELL, RD STIPANOVIC 1978 Histochemistry and identification of flavanols in Verticillium wilt-resistant and -susceptible cottons. Physiol Plant Pathol 13: 143-149
73. WATKINS GM, ed 1981 Compendium of Cotton Diseases. Am Phytopath Soc, St Paul, MN
74. BELL AA 1983 Protection practices in the USA and world. Section B - Diseases. In RJ Kohel, DF Lewis, eds, Cotton. Agron Monograph No 24, Am Soc Agron, Crop Sci Soc Am and Soil Sci Soc Am, Madison, WI
75. CARTER WW 1981 Resistance and resistant reaction of Gossypium arboreum to the Reniform nematode, Rotylenchulus reniformis. J Nematol 13: 368-374
76. VEECH JA 1983 Protection practices in USA and world. Nematodes. In R Kohel, C Lewis, eds, Cotton. Agron Monograph No 24, Am Soc Agron, Crop Sci Soc Am and Soil Sci Soc Am, Madison, WI

77. BRINKERHOFF LA 1970 Variation in Xanthomonas malvacearum and its relation to control. Annu Rev Phytopathol 8: 85-110
78. WILHELM S 1981 Sources and genetics of host resistance in field and fruit crops. In ME Mace, AA Bell, CH Beckman, eds, Fungal Wilt Diseases of Plants, Academic Press, NY pp 299-376
79. BRINKERHOFF LA, LM VERHALEN, R MAMAGHANI, WM JOHNSON 1978 Inheritance of an induced mutation for bacterial blight resistance in cotton. Crop Sci 18: 901-903
80. HEINSTEIN P 1980 Partial purification of an elicitor of Gossypium arboreum phytoalexins from the fungus Verticillium dahliae. Planta Medica 39: 196-197
81. CHARUDATTAN R, JE DEVAY 1981 Purification and partial characterization of an antigen from Fusarium oxysporum f.sp. vasinfectum that cross-reacts with antiserum to cotton (Gossypium hirsutum) root antigens. Physiol Plant Pathol 18: 289-295
82. KEEN NT, M YOSHIKAWA 1983 β-1,3-Endoglucanase from soybean releases elicitor-active carbohydrates from fungus cell walls. Plant Physiol 71: 460-465
83. KEEN NT,M YOSHIKAWA, MC WANG 1983 Phytoalexin elicitor activity of carbohydrates from Phytophthora megasperma f.sp. glycinea and other sources. Plant Physiol 71: 466-71
84. ALBERSHEIM P, AG DARVILL, M McNEIL, B VALENT, MG HAHN, G LYON, JK SHARP, AE DESJARDINS, MW SPELLMAN, LM ROSS, BK ROBERTSEN, P AMAN, L-E FRANZEN 1981 Structure and function of complex carbohydrates active in regulating plant-microbe interactions. Pure Appl Chem 53: 79-88
85. LEE JA 1981 Genetics of D_3 complementary lethality in Gossypium hirsutum and G. barbadense. J Heredity 72: 299-300
86. MACE ME, AA BELL 1981 Flavonol and terpenoid aldehyde synthesis in tumors associated with genetic incompatibility in Gossypium hirsutum x G. gossypioides hybrid. Can J Bot 59: 951-955
87. GERSTEL DU 1954 A new lethal combination in interspecific cotton hybrids. Genetics 39: 628-639
88. LEE JA 1981 A genetical scheme for isolating cotton cultivars. Crop Sci 21: 339-341
89. KOHEL RJ 1973 Genetic nomenclature in cotton. J Heredity 64: 291-295

90. KOHEL RJ 1979 Gene arrangement in the duplicate linkage groups V and IX: Nectariless, glandless, and withering bract in cotton. Crop Sci 19: 831-833
91. LEE JA 1982 Linkage relationships between Le and Gl alleles in cotton. Crop Sci 22: 1211-1213
92. MAHILL JF, DD DAVIS 1978 Influence of male sterile and normal cytoplasms on the expression of bacterial blight in cotton hybrids. Crop Sci 18: 440-443
93. LEE JA 1978 Allele determining rugate fruit surface in cotton. Crop Sci 18: 251-254
94. WILSON FD, JN SMITH 1977 Variable expressivity and gene action of gland-determining alleles in Gossypium hirsutum L. Crop Sci 17: 539-543
95. LEE JA 1977 Inheritance of gossypol level in Gossypium. III Genetic potentials of two strains of Gossypium hirsutum L. differing widely in seed gossypol level. Crop Sci 17: 827-830
96. HANNY BW, WR MEREDITH JR, JC BAILEY, AJ HARVEY 1978 Genetic relationships among chemical constituents in seeds, flower buds, terminals, mature leaves of cotton Crop Sci 18: 1071-1074
97. WHITE WH, JN JENKINS, WL PARROTT, JC McCARTY JR, DH COLLUM, PA HEDIN 1982 Generation mean analyses of various allelochemics in cotton. Crop Sci 22: 1046-1049
98. WHITE WH, JN JENKINS, WL PARROTT, JC McCARTY JR, DH COLLUM, PA HEDIN 1982 Strain and within-season variability of various allelochemics within a diverse group of cottons. Crop Sci 22: 1235-1238
99. BELL AA, RD STIPANOVIC 1977 The chemical composition, biological activity, and genetics of pigment glands in cotton. Proc Beltwide Cotton Prod Res Conf, Atlanta, GA pp 244-258
100. PARNELL FR, HE KING, DF RUSTON 1949 Jassid resistance and hairiness of the cotton plant. Bull Entomol Res 39: 539-585
101. DILDAY RH, TN SHAVER 1981 Seasonal variation in flowerbud gossypol content in cotton. Crop Sci 21: 956-960
102. BELL AA 1982 Plant pest interaction with environmental stress and breeding for pest resistance: Plant diseases. In MN Christiansen, CF Lewis, eds, Breeding Plants for Less Favorable Environments, John Wiley & Sons NY pp 335-363

103. CARTER WW 1982 Influence of soil temperature on *Meloidogyne incognita* resistant and susceptible cotton, *Gossypium hirsutum*. J Nematol 14: 343-346
104. BIRD LS 1980 Breeding for disease and nematode resistance in cotton. *In* Biology and Breeding for Resistance to Arthropods and Pathogens in Agricultural Plants, Texas Agri Exp Sta and Univ of Calif pp 86-100
105. NILES GA 1980 Breeding of cotton. *In* Biology and Breeding for Resistance to Arthropods and Pathogens in Agricultural Plants. Texas Agr Exp Sta and Univ Calif pp 83-85
106. STEWART JM 1981 In vitro fertilization and embryo rescue. Environ Exp Bot 21: 301-315
107. BENEDICT JH, TF LEIGH, AH HYER, PF WYNHOLDS 1981 Nectariless cotton: Effect on growth, survival, and fecundity of *Lygus* bugs. Crop Sci 21: 28-30
108. MEREDITH WR JR 1980 Performance of paired nectaried and nectariless F_3 cotton hybrids. Crop Sci 20: 757-760
109. HYER AH, EC JORGENSON, RH GARBER, S SMITH 1979 Resistance to root-knot nematode-Fusarium wilt disease complex in cotton. Crop Sci 19: 898-901
110. SHEPHERD RL 1982 Genetic resistance and its residual effects for control of the root-knot nematode-Fusarium wilt complex in cotton. Crop Sci 22: 1151-1155
111. BIRD LS 1982 The MAR (Multi-Adversity Resistance) system for genetic improvement of cotton. Plant Disease 66: 172-176

Chapter Eight

BIOREGULATION OF PLANT CONSTITUENTS

H. YOKOYAMA, W. H. HSU, E. HAYMAN, AND S. POLING

United States Department of Agriculture
Agricultural Research Service, WR
Fruit and Vegetable Chemistry Laboratory
263 S. Chester Avenue
Pasadena, California 91106

INTRODUCTION

The synthesis of plant constituents can be regulated. A number of bioregulators can affect the biosynthesis of constituents in plant tissues. Generally, these compounds stimulate the production of the constituents, unlike some of the herbicides such as certain pyridazinone types which inhibit the biosynthetic pathways.[1-3] We will examine the bioregulators and the nature of their effect on plant constituents, particularly the effect on the isoprenoid constituents in plant tissues.

BIOREGULATORS OF PLANT CONSTITUENTS

In 1969, Knpyl[4] had found that the compound chlormequat (2-chloroethyl)-trimethylammonium chloride causes all-*trans* lycopene (ψ,ψ-carotene) to accumulate in detached pumpkin cotyledons. However, no such response was observed when chlormequat was applied to other plant tissues. A major step forward was observed in 1970 with the discovery that another compound, CPTA, 2-(4-chlorophenylthio)-triethylamine hydrochloride, can cause accumulation of lycopene in a wide array of plant tissues and microorganisms.[5] Since the discovery of CPTA, research efforts involved in the development of the bioregulatory agents have generated a substantial amount of information relative to the structure-activity relationships for bioregulation of tetraterpenoids in higher plants and microorganisms and of polyisoprenes in higher plants. Both the *trans* and *cis* biosynthetic pathways can be regulated. The isoprenoid pattern observed is determined essentially by the nature of the bioregulatory agents employed.

GENERAL FORMULA FOR BIOREGULATORS OF ALL-*TRANS* TETRATERPENES

In the regulation of the *trans* tetraterpene systems, the bioregulators have the general formula: $(CH_2H_5)_2NCH_2CH_2R$. In general, the N,N-dimethyl analogs are less effective than the N,N-diethyl compounds. The magnitude of the stimulation, but not the general pattern of tetraterpene response, depends on R. The nature of R can cause modifications in the amount of the individual tetraterpenes or carotenes produced. Thus, when R lacks an aromatic ring, there appears to be less inhibition of the cyclase(s); consequently, more cyclic carotenes accumulate. Examples are given in Tables 1, 2, and 3. All the compounds investigated caused lycopene accumulation in Marsh grapefruit. The untreated fruit had the normal light yellow coloration. After treatment, the coloration of the flavedo ranged from light orange to an intense red. The flavedo of all treated fruits showed lycopene accumulation. Lycopene was not detected in the untreated fruits, and it is not normally seen in mature grapefruit.[6] The biological activity was also correlated with the logarithm of 1-octanol - water partition coefficient (log ρ).

DIETHYLALKYLAMINES

The diethylalkylamines have a fairly consistent response pattern as the length of the alkyl chain was increased (Table 1).[7] The amount of any given carotene remained about the same or increased slightly with diethylpentylamine and diethylhexylamine, while diethylheptylamine caused a larger response, and diethyloctylamine and diethylnonylamine caused very large increases. Lycopene accounted for most of the increase in the total carotene content but the intermediates, ζ-carotene (7,8,7',8' -tetrahydro-ψ,ψ-carotene) and neurosporene (7,8-dihydro-ψ,ψ-carotene) also increased significantly. Of significance are the 3.5 to 4.6-fold increases in the cyclic carotenes, γ-, α- and β-carotene (i.e., β, ψ-, β,ε- and β,β-carotene) caused by diethyloctylamine and diethylnonylamine, respectively. The increases in cyclic carotenes is much larger than that caused by compounds with aromatic rings. Diethylbutylamine has also been observed to cause the development of red coloration in grapefruit but only at higher concentrations and longer treatment periods. The higher members of this series, diethyldecylamine, diethylundecylamine and diethyldodecylamine, caused increasing peel injury as the length of the alkyl chain increased. The coloration of the peel adjacent to the

Table 1. Effect of diethylalkylamines [$(C_2H_5)_2N(CH_2)_nCH_3$] on carotene content of flavedo of Marsh seedless grapefruit (μg/g dry wt)[a]

		Treatment[b] n =				
	Control	4	5	6	7	8
Phytofluene	37.3	38.6	29.0	28.3	27.8	39.5
ζ-Carotene	2.25	2.47	3.13	4.32	14.6	17.8
Neurosporene	1.72	1.28	1.37	1.38	2.94	5.16
Lycopene		1.01	6.99	59.0	143	115
γ-Carotene	0.37	0.30	0.96	1.17	2.95	3.21
α-Carotene	0.54	1.04	1.02	1.26	2.07	2.11
β-Carotene	1.72	1.41	1.20	1.35	4.73	6.75
Total	43.9	46.1	43.7	96.8	197.7	189.5
Log ρ		2.94	3.44	3.94	4.44	4.94

[a] Dried in vacuo at 65°C
[b] Isopropanol solution of diethylalkylamine (0.2M) poured over fruit, postharvest

Table 2. Effect of diethylaminoalkylbenzenes [$(C_2H_5)_2N(CH_2)_nPh$] on carotene content of flavedo of Marsh seedless grapefruit (μg/g dry wt)[a]

	Control	Treatment[b] n = 1	2	3	4	5
Phytofluene	23.1	25.3	29.3	37.9	38.7	86.2
ζ-Carotene	1.13	1.38	3.87	6.77	27.5	53.8
Neurosporene	0.71	0.83	1.85	1.11	7.16	12.0
Lycopene		8.62	60.5	188	104	153
γ-Carotene		0.55	0.77	1.07	0.59	1.67
α-Carotene	0.57	0.77	0.84	0.61	2.19	0.76
β-Carotene	0.95	0.56	0.46	0.95	trace	0.68
Total	26.5	37.8	97.6	235.5	180.1	308.1

[a] Dried in vacuo at 65°C
[b] Isopropanol solution of diethylaminoalkylbenzene (0.2M) poured over fruit, postharvest

damaged area showed color enhancement but to a lessening degree as the alkyl chain increased.

DIETHYLAMINOALKYLBENZENES

The responses of fruits treated with compounds in Table 2 were similar to those of fruits treated with diethylalkylamines.[8] There was a much larger increase in γ-carotene and neurosporene for diethylaminobutylbenzene and diethylaminopentylbenzene while the increases in cyclic carotene was not very large. For the diethylalklylamines lycopene increased to a very high level and then dropped for the last members of the series. The drop in lycopene accumulation was seen with diethylbutylamine but the amount of lycopene increased with diethylpentylamine although remaining less than the maximum for the series. The very large increase in phytofluene (7,8,11,12,7',8'-hexahydro-ψ,ψ-carotene), as well as ζ-carotene, caused δ-diethyl-aminopentylbenzene to give greater increase in the total carotene content. Whether this was caused in part by some side effect of the extensive peel damage or entirely by the compound is not certain. A similar effect has been previously observed with [γ-(diethylamino)-propoxy]-benzene and [σ-(diethylamino)-butoxy]-benzene. In these cases the increases in phytofluene and ζ-carotene were also very large but no great damage was seen. This effect may arise when the diethylamino and phenyl groups of the inducers are separated by a chain of four or five carbon atoms.

SUBSTITUTED DIETHYLAMINOETHYLPHENYLETHERS

Analogs of diethylaminoethylphenylether gave a similar response pattern (Table 3).[9] Neurosporene did not increase as much as previously although ζ-carotene did show a large increase. The cyclic carotenes, with the exception of δ-carotene, did not show a significant increase. There was a general decrease, not previously seen, of all the carotenes with diethylaminoethyl-4-*tert*-butyl-phenylether as compared with diethylaminoethyl-4-*iso*-propylphenylether, instead of the steady increase of ζ-carotene and neurosporene as observed in the other series (Tables 1 and 2). Quite probably, the greater biological activity of those compounds in Table 3 as compared to those in Tables 1 and 2 is due to a higher degree of interaction between the compound and the active site.

Generally, the response increased with increasing concentration to a maximum value and further increases in the concentration of diethylamino-ethyl-4-ethylphenylether from 0.26 to 0.52M reduced the observed response,[8] while treatment with CPTA at 0.18M caused an accumulation of carotenes equal to that in Table 3. Doubling the concentration of diethylaminoethyl-4-methylphenylether while doubling of diethylaminoethyl-4-ethylphenylether had almost no effect. That increasing CPTA concentrations begin to

Table 3. Effect of diethylaminoethylphenylethers [$(C_2H_5)_2NCH_2CH_2OC_6H_4R$] and CPTA on carotene content of flavedo of Marsh seedless grapefruit (μg/g dry wt.)[a]

		Treatment[b] R =					
	Control	H	*p*-Methyl	*p*-Ethyl	*p*-*iso*-Propyl	*p*-*tert*-Butyl	CPTA
Phytofluene	44.9	48.1	50.9	44.9	51.1	33.7	42.6
ζ-Carotene	2.74	5.45	19.0	13.7	20.6	4.57	10.1
Neurosporene	0.33	0.91	0.84	0.93	1.62	0.63	1.21
Lycopene		22.2	249	182	226	54.4	199
γ-Carotene	0.27	1.01	2.48	2.13	2.74	1.04	1.90
α-Carotene	0.35	0.57	0.41	0.38	0.40	0.63	0.23
β-Carotene	1.35	1.26	0.40	0.82	0.70	0.58	0.39
Total	49.9	79.5	323.0	244.9	303.2	95.6	244.4

[a] Dried in vacuo at 65°C
[b] Isopropanol solution of diethylaminoethylphenylether (0.1M), poured over fruit, postharvest

lose effectiveness in inducing larger carotene biosynthesis has also been observed in *Blakeslea trispora*.[10]

DIETHYLAMINOETHYLESTERS

For examination of the β-carotene effect, the 2-diethylaminoethylesters of the shorter chain aliphatic acid series were investigated. As shown in Table 4, these esters caused significant increases in β-carotene, whereas the increase in lycopene was very small.[11] The esters were applied as free amines in isopropanol. The peel remained healthy on all the fruits except for those treated with octanoate and nonanoate which damaged about 30 and 60% of the peel area respectively. The tendency to cause peel damage with increasing lipid solubility has been noted for other inducers.

ESSENTIAL STRUCTURAL FEATURES FOR INDUCTION OF ALL-*TRANS* CAROTENES

As a further elaboration on the general formula $(C_2H_5)_2NCH_2CH_2R$ given for *trans* lycopene inducing bioregulators, none of the following structural alterations totally abolished the activity or changes in the response pattern: [1] Replacement with O-atom of the S-atom connecting the amine portion of the benzene moiety in the CPTA molecule; [2] elimination of the substitution on the benzene ring; [3] elimination of the benzene ring itself; [4] replacement of the ethyl groups of the amine portion by other alkyl groups. However, these alterations change the effectiveness of the compound thus modified. Benzene derivatives are more effective than alkyl derivatives; and the substitutions of certain groups on the benzene ring at the *para* position seem to make the compounds more effective. Table 5 shows that the substitution of a methyl group at the *ortho* position almost completely eliminates the inducing ability of the compound. Substitution at the *meta* position also causes a reduction in the inducing ability. With the methyl group in the *para* position, the compound is a very powerful inducer. The same effect is seen with the strongly electron withdrawing chloro group, as observed with the electron releasing methyl group, for *p*-, *m*-, and *o*-chlorophenoxytriethylamine.[7] Steric hindrance may possibly play an important role in the effectiveness of the position isomers. Also the effectiveness of the compounds

Table 4. Effect of diethylaminoethylesters [$(C_2H_5)_2NCH_2CH_2OOC-R$)] on the carotene content of flavedo of Marsh seedless grapefruit (μg/g dry wt)[a]

		Treatment[b] R =				
	Control	Butyl	4-Pentyl	5-Hexyl	6-Heptyl	7-Octyl
Phytofluene	36.6	35.9	36.6	32.4	31.2	30.7
ζ-Carotene	5.11	10.3	19.4	15.7	13.0	8.48
Neurosporene		0.55		0.61		
Lycopene		9.30	11.4	21.5	8.03	2.79
γ-Carotene	0.33	1.01	5.57	3.54	2.13	2.01
α-Carotene	0.33	1.32	3.68	3.05	6.18	3.22
β-Carotene	1.61	14.2	98.3	58.3	50.5	43.0
Other carotenes	2.40	1.19	2.52	2.16	2.40	2.04
Total carotenes	46.4	73.8	177	137	113	92.2
Total xanthophylls	24.9	18.2	22.0	24.4	22.1	24.3

[a] Dried in vacuo at 65°C
[b] Isopropanol solution of diethylaminoethylester (0.2M) poured over fruit, postharvest

does not depend solely upon the electron withdrawing ability of substituting groups in the benzene moiety.[12] Furthermore, experimental data do not necessarily rule out the possible inductive effect of substituents on the amine portion of the molecule.

It should be emphasized that the variable effectiveness on carotenogenesis among the triethylamine derivatives suggests the possibility that certain compounds did not penetrate well into the fruit flavedo tissue. Thus, the lipophilic-hydrophilic properties of the bioregulators appear to have an influence on the effectiveness of these compounds. These characteristics are reflected in the value of the log ρ, which is the logarithm of the octanol-water partition coefficient of the unionized molecule.[13-16] Log values (Tables 1-4) have proven useful, in combination with considerations of the electronic and steric states of a molecule, as a general rule in the designing of new bioregulators. Thus, there are upper list for log ρ at which peel damage begins to occur. The compounds with log ρ greater than 4.6 probably cause peel damage by disrupting the lipid membranes of the cell. Those compounds with log ρ less than 4.6 show no damage. Some compounds probably interact more strongly at the active site(s) and produce a noticeable effect on carotenogenesis at lower concentrations and for smaller values of log ρ than others. The

Table 5. Effect of ring position of subsitutent in diethylaminoethylphenylether [$(C_2H_5)_2NCH_2CH_2OC_6H_4R$] on the carotene content of the flavedo of Marsh seedless grapefruit (μg/g dry wt)[a]

	Treatment[b] R =			
	Control	_o_-Methyl	_m_-Methyl	_p_-Methyl
Phytofluene	40.2	39.3	36.1	52.7
ζ-Carotene	4.66	3.26	4.00	17.7
Neurosporene	0.26	0.79	1.14	1.77
Lycopene		0.43	24.4	516
γ-Carotene		0.22	0.60	0.83
α-Carotene	0.52	0.34	0.61	0.28
β-Carotene	1.25	0.86	0.59	0.48
Total carotenes	46.9	45.2	67.4	589
Total xanthophylls	21.8	21.3	24.3	25.1

[a] Dried in vacuo at 65°C
[b] Isopropanol solution of diethylaminoethylphenylether (0.1M) poured over fruit, postharvest

optimum value for log ρ appears to be in the range 3.5 to 4.5.

INDUCERS OF _CIS_ CAROTENES

All the compounds reported above cause the accumulation of _trans_ carotenes. They are of the general formula $(C_2H_5)_2NCH_2CH_2R$. The carotene content greatly increases and all-_trans_ lycopene becomes the major constituent unless the compounds are aliphatic esters of diethylaminoethanol. If they are, all-_trans_-β-carotene is the predominant constituent.

Studies of structure-activity relationships of compounds that affect carotenogenesis led to the discovery of a new series of bioregulators causing the accumulation of _cis_ carotenes.[17,18] The natural occurrence of _cis_ and poly-_cis_ carotenes has been observed, but it is not very common.[19,20]

Bioregulators shown in Tables 6, 7, and 8 stimulate the formation of poly-_cis_ carotenes. The N-benzylfurfurylamines are more effective than the corresponding dibenzylamines. The only anomaly is 2-(4-bromophenoxy)ethylfur-

Table 6. Effects of compounds N-R furfurylamine on the carotene content of the flavedo of Marsh seedless grapefruit (μg/g dry wt)[a]

	Control	Treatment[b] R =							
		Benzyl	4-Methyl-benzyl	4-Chloro-benzyl	4-bromo-benzyl	4-Nitro-benzyl	Phen-ethyl	2-Phen-oxy-ethyl	2(4-Bromo)-phenoxy-ethyl
Phytofluene	19.78	22.03	21.17	25.03	31.03	19.44	25.63	31.70	28.32
α-Carotene	0.34	0.24	0.33	0.23	0.20	0.21	0.22	0.30	0.29
β-Carotene	1.54	1.26	1.36	1.31	1.46	0.93	1.48	1.20	0.95
ζ-Carotene	2.58	11.26	6.18	18.71	27.61	2.26	22.79	12.57	14.94
Poly-cis-γ-carotene I		0.95	0.45	2.24	2.72		2.01	1.25	
Proneurosporene		8.41	3.42	15.17	21.42	0.44	18.30	10.56	5.07
γ-Carotene									0.36
cis-Lycopene		7.31	3.81	14.84	23.37	1.13	17.52	8.70	3.78
Neurosporene	1.17					0.72			2.64
Poly-cis-γ-carotene II		4.82	1.68	10.49	14.45		8.45	5.45	
Unknown 453		2.03	1.16	2.05	3.70		3.54	2.49	
Lycopene									32.39
Total carotenes	25.38	70.17	44.79	122.95	172.49	25.83	141.42	94.64	92.73
Total xanthophylls	19.13	22.18	22.40	24.78	30.85	20.62	27.09	24.73	22.14

[a] Dried in vacuo at 65°C
[b] Isopropanol solution of N-R furfurylamine (0.1M) poured over fruit, postharvest

Table 7. Effect of dibenzylamine and substituted dibenzylamines on the carotene content of flavedo of Marsh seedless grapefruit (μg/g dry wt) [a]

	Treatment[b]							
			Substituted dibenzylamines					
	Control	Dibenzylamine	4-Fluro	4-Chloro	4-Bromo	4-Methyl	4-Nitro	4-Cyano
Phytofluene	34.98	36.10	39.21	40.14	37.87	32.30	50.27	43.32
α-Carotene	0.14	0.11	0.10	0.16	0.12	0.11	0.14	0.09
β-Carotene	0.65	0.83	0.95	1.88	1.05	1.00	1.01	0.78
ζ-Carotene	4.37	10.12	12.64	21.16	18.64	14.03	18.71	12.45
Poly-*cis*-γ-carotene I*	0.31 (combined with Proneurosporene*)	0.70	1.35	3.18	3.84	2.50	1.49	0.78
Proneurosporene*		4.91	7.99	16.24	15.03	9.30	13.45	8.28
Prolycopene*	0.55	8.03	14.08	33.10	29.76	21.28	24.95	13.19
cis-Lycopenes*	1.21	5.12	7.58	9.69	9.31	9.45	10.10	6.00
Poly-*cis*-γ-carotene II*	0.87	2.35	5.48	11.94	6.81	5.55	5.91	4.50
Unknown 453	0.48	1.74	1.44	4.18	8.22	4.11	6.40	0.94
Total carotenes	43.56	70.01	90.82	41.67	130.65	99.53	132.43	90.33
Total xanthophylls	25.48	28.76	30.04	39.36	36.58	26.08	39.74	28.47

[a] Dried in vacuo at 65°C
[b] Isopropanol solution of dibenzylamine or substituted dibenzylamine (0.1M) poured over fruit, postharvest
* Tentative assignment of poly-*cis* carotenes in control

furylamine which acted as an inducer of both poly-*cis* carotenes and lycopene. They also inhibited the cyclase(s), a characteristic of lycopene inducers, and prevented the formation of poly-*cis* γ-carotene I and II. N-Methylation of the corresponding secondary amines does not reduce but tend to enhance the activity. This enhancement of activity is especially evident with the N-alkyl,N-methylbenzylamines. While N-n-hexylbenzylamine is only weakly active, N-methyl,N-*n*-hexylamine is a very good inducer. These compounds are not good inhibitors of the cyclase(s) as is the case with the lycopene inducers. In none of the samples was any trace of all-*trans* lycopene detected. Prolycopene was the major constituent but there were relatively large amounts of other poly-*cis* carotenes. In marked contrast, the triethylamine bioregulators primarily stimulated the synthesis of all-*trans* lycopene or β-carotene. Although not determined experimentally, the increase in total xanthophylls observed in the treated fruit could have resulted from the formation of *cis* and poly-*cis* xanthophylls. *cis* Xanthophylls do occur naturally in some plants.[21]

Table 8. Effect of compounds *N*-methyl, *N*-R-benzylamine on the carotene content of the flavedo of Marsh seedless grapefruit (μg/g dry wt)[a]

		Treatment[b] R =				
	Control	*n*-Butyl	*n*-Pentyl	*n*-Hexyl	*n*-Heptyl	*n*-Octyl
Phytofluene	24.86	29.40	24.03	35.20	28.07	34.12
α-Carotene	0.54	0.36	0.23	0.23	0.21	0.31
β-Carotene	0.95	1.01	1.70	1.64	0.77	0.93
ζ-Carotene	2.73	3.85	7.26	21.24	20.86	24.99
Poly-*cis*-γ-carotene 1			0.86	2.94	2.90	3.15
Proneurosporene		1.77	5.01	16.08	16.38	18.80
Prolycopene		2.66	9.45	32.57	36.06	26.44
cis-Lycopenes		3.63	4.75	10.94	13.26	12.33
Poly-*cis*-γ-carotene II		0.39	3.58	11.17	8.03	9.83
Unknown 453		0.20	1.21	2.09	3.58	2.03
Total carotenes	29.04	43.27	58.08	134.10	130.12	132.93
Total xanthophylls	25.83	25.04	27.90	33.85	27.89	27.18

[a]Dried *in vacuo* at 65°C
[b]Isopropanol solution of *N*-methyl, *N*-R-benzylamine (0.1M) poured over fruit, postharvest

BIOREGULATORS OF POLYISOPRENES

A promising approach to improving the yield characteristics of the guayule plant is through the regulation of the polyisoprenoid biosynthetic system to cause an accumulation of increased amounts of rubber. The initial bioregulator which was reported to affect polyisoprenoid biosynthesis was the compound 2-diethylaminoethyl-3,4-dichlorophenylether (DCPTA).[22] DCPTA caused significant increases in rubber content of the guayule plant, (*Parthenium argentatum* Gray) (Table 9). Subsequently, it was found N-*p*-bromobenzylfurfurylamine caused marked increases in the yield of rubber in guayule (Table 10).[23] N-Methylbenzylhexylamine exhibits nearly equal activity in stimulation of rubber synthesis (Table 11).

Effects of bioregulators on the increase of total yield of rubber per plant is limited to a great extent by the availability of parenchyma cells for synthesis and storage of newly induced rubber molecules. Preliminary studies have indicated that bioregulators do affect the activity of cambium, inducing the cambium to produce more bark tissue (storage area) in relation to the wood tissue (non-storage area) (Mehta I, personal communication). There also appears to be an increase in amount of secondary phloem major portion of the bark and no increase in the production of secondary xylem (wood). When the treatment is initiated at an early developmental state with low concentrations of bioregulator, there appears to be a significant increase in the total yield of rubber per plant (Table 12).[24]

Studies were carried out on the rubber composition of guayule plants. Rubber samples isolated from plants treated with DCPTA were compared to those from untreated plants. Nuclear magnetic resonance spectroscopy (^{13}C at 15 MHz) confirm the structural and geometrical purity of rubber isolated from treated guayule plants. The NMR spectra show a complete absence of signals attributable to structural isomers, namely *trans*-1,4 isoprene units, demonstrating that the rubber sample from treated plants is a highly stereospecific polymer composed entirely of *cis*-1,4 isoprene units (Fig. 1). The ^{13}C NMR spectra confirm that improvement of yield in guayule plants is accomplished without altering the microstructure of the rubber.

Table 9. Bioinduction of rubber in stem and branch tissues of guayule (18 months old) by 2-diethylaminoethyl-3,4-dichlorophenylether (DCPTA). The plants were treated with 2000 ppm of bioregulator, 1000 ppm isopropanol and 500 ppm Ortho-X77. All plants were harvested 40 days after treatment. Each result represents the mean of 6 plants.

	Rubber content (mg/g dry wt.)	
Strain	Control	Created
212	58	146
228	61	100
230	42	77
234	52	92
239	57	122
241	48	122
242	51	108

The molecular weight distribution of rubber samples from treated plants was compared to those from untreated plants. The distribution of molecular weights of the polyisoprene chain in rubber from treated guayule plant is identical to that in rubber from untreated plant, and both are unimodal (Fig. 2).

MODE OF ACTION OF BIOREGULATORS

The fact that carotenogenic microorganisms such as *Blakeslea trispora* and *Phycomyces blakesleeanus* respond to treatment provided an opportunity to study the mechanism of action of the bioregulators in tetraterpenoid formation. Studies conducted using the mold *B. trispora* indicated that the bioregulators act at the enzyme level by inhibiting the cyclase(s); the transformation of the acyclic lycopene to the moncyclic λ-carotene and the bicyclic β-carotene was inhibited (Fig. 3).[10] These studies also suggested that the bioregulators induce (derepress) a gene(s) regulating the synthesis of a specific enzyme or enzymes in the primary biosynthetic pathway of carotenoids and thus increase net synthesis of carotenoids. The nullifying action of cycloheximide on the effect of the bioregulatory agent in carotenoid synthesis indicated that the latter compound acts as an inducer (derepressor) of enzyme synthesis rather than as an activator pre-existing enzymes(s) (Fig. 4). Once the enzymes participating in

Table 10. Bioinduction of rubber in stem and branch tissues of guayule (18 months old) by N-p-bromobenzylfurfurylamine. The plants were treated with 2000 ppm of bioregulator, 1000 ppm of isopropanol and 500 ppm of Ortho-X77. All plants were harvested 40 days after treatment. Each result represents the mean of 6 plants.

Strain	Rubber content (mg/g dry wt.)	
	Control	Treated
228	58	146
239	5	121
234	54	162
89	72	207

carotenoid synthesis have been formed, cycloheximide does not affect their activity. Cycloheximide is known to act at the ribosomal protein level to inhibit the formation of messenger DNA which, in turn, inhibits protein synthesis. The relative effectiveness of individual bioregulator as a cyclase inhibitor and as an enzyme inducer could lead to variations in the carotenoid response pattern amongst the different bioregulators.

The mode of action of the poly-*cis* inducers is probably gene derepression, similar to that of the lycopene inducers. However, these compounds probably derepress a recessive gene controlling the biosynthesis of the poly-*cis* carotenoids, whereas lycopene inducers derepress the dominant genes giving rise to the normal all-*trans* carotenoids. The existence of this recessive gene can only be postulated as present in grapefruit and other citrus fruits. Impor-

Table 11. Bioinduction of rubber in stem and branch tissues of guayule (10 months old) by N-methylbenzylhexylamine. The plants were treated with 2000 ppm of bioregulator, 100 ppm of isopropanol and 500 ppm of Ortho-X77. All plants were harvested 40 days after treatment. Each result represents the mean of 5 plants.

Strain	Rubber content (mg/g dry wt.)	
	Control	Treated
593	39	131

Table 12. Effects of several bioregulators on rubber content of guayule

	Treatment	Rubber content[a]
Control		15.5 ± 3.9
125 ppm	2-Diethylaminoethyl-3,4-dichlorophenylether	23.5 ± 6.2
250 ppm	2-Diethylaminoethyl-3,4-dichlorophenylether	23.6 ± 6.5
500 ppm	2-Diethylaminoethyl-2,4-dichlorophenylether	24.3 ± 4.2
1000 ppm	N-methylbenzylhexylamine	26.4 ± 7.4
500 ppm	2-Diethylaminoethyl-3,4-dimethylphenylether	29.6 ± 8.4

[a]Grams of rubber per plant ± the standard deviation

tant genetic studies have shown such a gene exists in the tomato fruit. Tangerine tomato fruits which are homozygous for the recessive allele t, accumulate the orange pigment prolycopene at the expense of lycopene. Red fruit carry the dominant allele t+. The same situation could exist in other fruit and explain the action of the poly-*cis* inducers. This interpretation lends support to the postulated parallel pathways for all-*trans* and poly-*cis* carotenoids. If this hypothesis is true, then one could expect to find

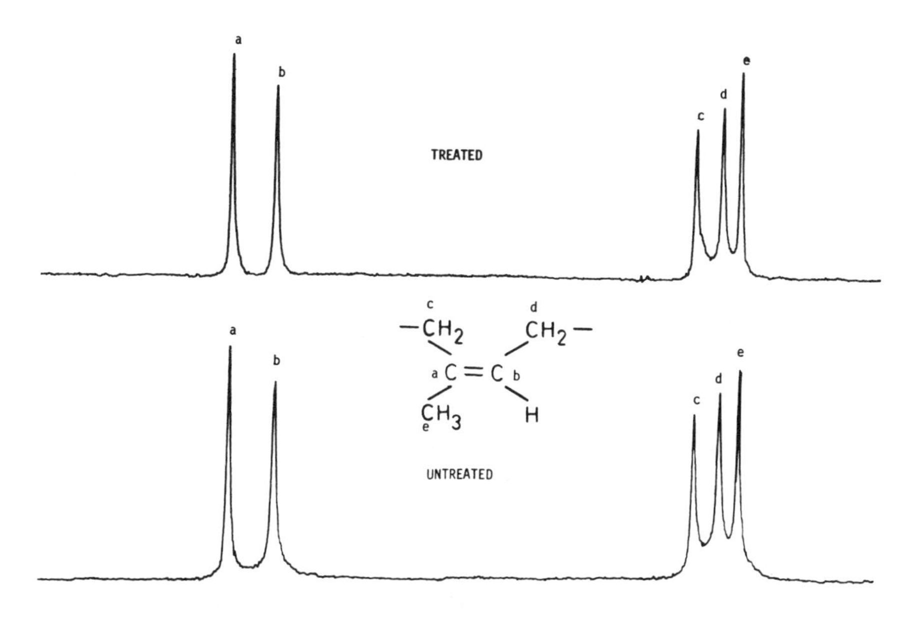

Figure 1. Comparison of ^{13}C-NMR spectra of rubber from DCPTA treated and untreated guayule plants.

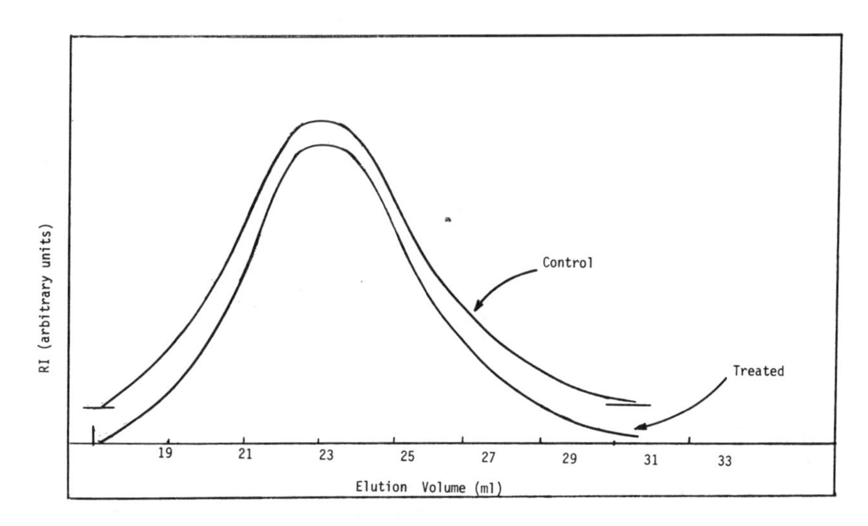

Fig. 2. Gel permeation chromatography (GPC) of guayule rubber. The GPC analysis was conducted using a Waters model 6000 chromatograph with differential refractive index detector, solvent system of tetrahydrofuran (at 28°C) stabilized with 200 ppm of 2,6-di-tert-butyl-4-ethylphenol and a set of 5 μ Styragel columns each with a nominal porosity of 10^6, 10^5, 10^3, and 500 Å.

poly-cis ζ-carotene, -phytofluene, and -phytoene present in the fruit. The poly-cis inducers appear to differ from the lycopene inducers in another respect. In addition to gene derepression, the lycopene inducers inhibit the cyclase(s), causing lycopene to accumulate at the expense of the cyclic carotenes. The accumulation of significant amounts of poly-cis γ-carotene I and II indicates that the cyclase(s) are not inhibited by the poly-cis inducers and whose only function appears to be to derepress the recessive gene.

The mode of action of the bioregulators which induce the synthesis of rubber in the guayule plant is similar to that of the lycopene inducers, namely gene derepression. Studies have shown that DCPTA induces the synthesis of enzyme proteins in guayule plant tissues.[25]

DCPTA causes the de novo synthesis of new rubber molecules in plant tissues. Studies show that newly induced rubber molecules formed after the guayule plants are treated with DCPTA are stored in parenchymatous cells

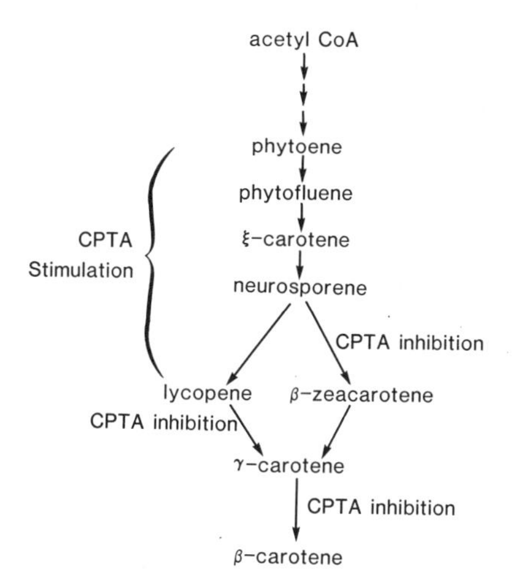

Fig. 3. The possible sites of CPTA action on carotenoid biosynthesis in mated strains of Blakeslea trispora.

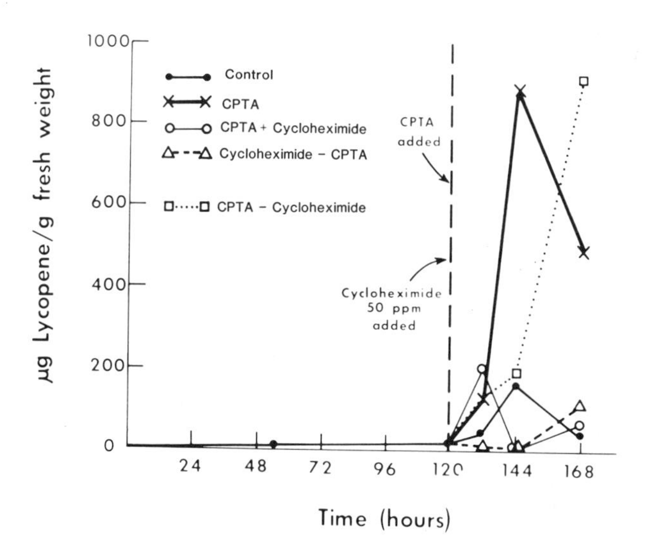

Fig. 4. Effect of cycloheximide and CPTA on lycopene biosynthesis in mated strains of Blakeslea trispora.

which did not have notable quantity of rubber before the treatment (Bauer T., R. Glaeser, and H. Yokoyama unpublished observation). No increase of rubber was noted in cells which contain rubber prior to the treatment. Additionally, gel permeation chromatography studies indicate that DCPTA induce the formation of new rubber molecules rather than a chain extension of rubber molecules at the surface of existing rubber particles.[23] Results of GPC strongly suggest no changes in the molecular weight distribution pattern or rubber samples from treated guayule plants.

SUMMARY

Bioregulators can affect the biosynthesis of plant constituents, particularly the isoprenoid compounds. Generally, the effect is one of stimulation of the synthesis of the isoprenoids ranging from the lower terpenoids to the polyisoprenoids. Both the *cis* and *trans* systems are affected. The mode of action appears to be due to regulation of gene expression. The bioregulators act directly or indirectly as derepressors.

REFERENCES

1. KUMMEL HW, LH GRIMME 1975 The inhibition of carotenoid biosynthesis in green algae by Sandoz 6706: accumulation of phytoene and phytofluene in *Chlorella fusca*. Z Naturforsch 30c: 333-335
2. URBACH D, M SCHANKA, W URBACH 1976 Effect of substituted pyridazinone herbicides and of difunone (EMD-IT 5914) on carotenoid biosynthesis in green algae. Z Naturforsch 31c: 652-655
3. EDER FA 1979 Pyridazinones, their influence on the biosynthesis of carotenoids and metabolism of lipids in plants. Z Naturforsch 34c: 1052-1054
4. KNYPL JS 1969 Accumulation of lycopene in detached cotyledons of pumpkin treated with (2-chloroethyl) -trimethylammoniumchloride. Naturwissenschaften 56: 572-573
5. COGGINS C, GL HENNING, H YOKOYAMA 1970 Lycopene accumulation induced by 2-(4-chlorophenylthio)-triethylamine hydrochloride. Science 168: 1589-1590
6. YOKOYAMA H, M WHITE 1965 Carotenoids in the flavedo of Marsh seedless grapefruit. J Agr Food Chem 65:693-696
7. POLING SM, WJ HSU, H YOKOYAMA 1975 Structural-activity

relationships of chemical inducers of carotenoid biosynthesis. Phytochemistry 144: 1933-1938
8. POLING SM, WJ HSU, H YOKOYAMA 1973 New chemical inducers of carotenoid biosynthesis. Phytochemistry 12: 2665-2667
9. YOKOYAMA H, C DeBENEDICT, C COGGINS, GL HENNING 1972 Induced color changes in grapefruit and orange. Phytochemistry 11: 1721-1723
10. HSU WJ, H YOKOYAMA, C COGGINS 1972 Carotenoid biosynthesis in _Blakeslea trispora_. Phytochemistry 11: 2985-2990
11. POLING SM, WJ HSU, FJ KOEHRM, H YOKOYAMA 1977 Chemical induction of β-carotene biosynthesis. Phytochemistry 16: 551-555
12. HUS WJ, SM POLING,, C DeBENEDICT, C RUDASH, H YOKOYAMA 1975 Chemical inducers of carotenogenesis. J Agr Food Chem 23: 831-834
13. HANSCH C, RM MIUR, T FUJITA, PP MALONEY, F GEIGER, M STREICH 1963 The correlation of biological activity of plant growth regulators and chloromycetin derivatives with Hammett constants and partition coefficients. J Am Chem Soc 85: 2817-2824
14. HANSCH C, T FUJITA 1964 Analysis. A method for the correlation of biological activity and chemical structure. J Am Chem Soc 86: 1616-1626
15. FUJITA T, J IWASA, C HANSCH 1964 A new substituent constant, derived from partition coefficients. J Am Chem Soc 86: 5175-5180
16. HANSCH C, JE QUINLAN, GL LAWRENCE 1967 The linear free-energy relationship between partition coefficients and aqueous ability of organic liquids. J Org Chem 33: 347-350
17. POLING SM, WJ HSU 1980 Chemical induction of poly-_cis_ carotenoid biosynthesis. Phytochemistry 19: 1677-1680
18. POLING SM, WJ HSU, H YOKOYAMA 1982 Synthetic bioregulators of poly-_cis_ carotenoid biosynthesis. Phytochemistry 21: 601-604
19. ZECHMEISTER L 1962 _Cis-trans_ isomeric carotenoids, vitamin A and arylpolyenes. Academic Press, New York
20. GOODWIN TW 1971 Biosynthesis in Carotenoids (O Isler,ed) Birkhauser-Verlag, Basel, pp 578-629
21. WEEDON BCL 1971 Stereochemistry in Carotenoids (O Isler,ed) Birkhauser-Verlag, Basel, pp 268-319
22. YOKOYAMA H, E HAYMAN, W HSU, SM POLING, A BAUMAN 1977 Bioinduction of rubber in the guayule plant. Science 197: 1076-1078

23. YOKOYAMA H, E HAYMAN, WJ HSU, SM POLING 1983 Bioregulation of rubber synthesis in the guayule plant *In* L Nickell, ed Plant Growth Regulatory Chemicals CRC Vol 1, Press, Boca Raton FL pp 59-70
24. HAYMAN E, H YOKOYAMA 1983 The effect of bioregulators on the accumulation of rubber in guayule. J Agr Food Chem 31: 1120-1121
25. BENEDICT C, PH REIBACH, S MADHAVAN, RV STIPANOVIC, H YOKOYAMA 1983 Effect of 2-(3,4-dichlorophenoxy)-triethylamine on the synthesis of *cis* polyenes in guayule plant (*Parthenium argentatum* Gray). Plant Physiol 72: 879-899

Chapter Nine

PRODUCTION OF RESINS BY ARID-ADAPTED ASTEREAE

JOSEPH J. HOFFMANN, BARBARA E. KINGSOLVER,
STEVEN P. MCLAUGHLIN, AND BARBARA N. TIMMERMANN

Bioresources Research Facility
Office of Arid Lands Studies
University of Arizona
Tucson, Arizona 85719

INTRODUCTION

The arid regions of the world and their native flora remain a largely unexplored frontier in the search for new plant chemicals. This search has been revived in recent years because, in addition to the scientific gains to be made through the study of the production and ecological significance of plant secondary metabolites, many of these compounds have practical value as well.

As we approach the end of a long era of abundant, inexpensive petroleum products, our extensive reliance on non-renewable resources is becoming less desirable and, in fact, conspicuously short-sighted. In response to the well-publicized need for a transition towards a sustainable resource base, many researchers have looked increasingly to natural products.[1] Calvin and others have proposed the use of "petroleum plantations" for the cultivation of plants that produce highly reduced, hydrocarbon-like compounds, which could serve as substitutes for crude oil.[2] An approach that appears more feasible, for the immediate future at least, is the use of plant-derived compounds as chemical intermediates for industry, or as replacements for specialty chemicals and other commodities that are currently petroleum-derived, or otherwise non-renewable.[3] In either case, the success of any endeavors to develop a renewable, biomass-based industry will depend ultimately on a more thorough knowledge of plants and their potentially useful chemical constituents. Although the literature shows a scarcity of screening programs of this type, the few surveys that have been conducted for the purpose of identifying potential chemical crops have revealed the existence of a considerable battery of possibilities.[4] Of special interest are the surveys of the economic botany of arid-adapted plants conducted by Duisberg and Hay[5] and, more recently, by McLaughlin and Hoffmann, who concentrated specifically on petroleum-like phytochemicals.[6]

It is particularly appropriate to conduct investigations of this type on arid-adapted species. In practical terms, the arid lands of the world are vast and underutilized, and the development of crops that could be grown in these regions would be an accomplishment of considerable economic importance.

From a scientific point of view, the most persuasive reasons for seeking new compounds from arid-adapted plant species are ecological and phytochemical ones. The stresses imposed by a xeric environment are diverse and extreme, and the evolutionary responses of desert plants to these stresses have produced an impressive array of phytochemicals.[7]

The most universal adversities to life in the desert are heat, or in some cases extreme fluctuations between heat and cold, and a shortage of water. Of the many adaptations to water-stress and/or temperature-stress that have been identified, many are phytochemical. Resins, waxes, and other impermeable coatings, for example, are believed by many to have the function of conserving water by lowering the rate of transpiration.[8-10] In another adaptive strategy, complex sugars such as pentosans may retard desiccation by imbibing water; sharp increases in the production of pentosans and hexosans in times of water-stress has been reported in guayule and several other species.[5,11]

Predation pressure has also played an extremely

important evolutionary role in the desert. Owing to the scarcity of water, and to the relatively slow and energetically costly regeneration of biomass, herbivory is particularly devastating to desert plants. Thus, phytochemical adaptation in these species has also included the development of a tremendous variety of insecticidal, antimicrobial, and other chemicals with anti-feedant or cytotoxic properties. Compounds of this type that have been identified include chromenes, prenylated quinones, and sesquiterpenes esterified with phenolic acids, all of which repel phytophagous insects and in some cases inhibit larval development.[7] Synergisms and other complex interactions have also been reported. In the creosotebush, for example, phenol oxidase in the leaves appears to interact with the resin on the leaf surfaces to produce proteolytic effects in insect herbivores.[10] The evidence indicates that among the practical benefits of the study of secondary chemicals produced by arid-adapted plants may be the identification of new sources of natural pesticides and antibiotics.

Thus, in spite of their low biomass productivity, many xerophytic plant species may have potential as important crops for the future because of the complex, highly reduced organic chemicals they produce. The allocation of a large proportion of the plant's energy to the production of secondary chemicals rather than to biomass appears to be a relatively common phenomenon among xerophytic species; because of this, these plants have value not only for their economic potential, but as subjects for scientific inquiry as well. The ecological significance and the anatomical and biochemical mechanisms for this unusual pattern of energy allocation are not well understood, and clearly warrant further investigation.

This paper is intended to present an overview of recent investigations of one category of plant chemicals, diterpene resins, which are produced by a number of species native to the arid regions of the southwestern U.S. In particular, we will concentrate on our work with the genus *Grindelia*. While little is known about the biology of resin production or its significance in the herbaceous species, we will present our data on *Grindelia* in the context of the previous literature on related or convergent resin-producing species, in the hope that some general conclusions may be drawn.

RESINOUS PLANTS OF NORTH AMERICA

The discovery of a category of arid-adapted, resinous plants concentrated in the tribe Astereae was an unexpected outcome of a recent survey of McLaughlin and Hoffmann.[6] The survey analyzed more than two hundred desert species for their economic potential as producers of energy and petroleum-replacement products. The screening included more than four hundred plant collections representing considerable taxonomic diversity: 195 species and varieties in 107 genera, 35 families. The study emphasized plants that were known to produce latex and later, as the trend was recognized, those that had resinous exudates.

An important outcome of the survey was the discovery that the highest cyclohexane extracts (representing the proportion of hydrocarbon-like materials in the plant) were produced by resinous species of the Compositae in the tribe Astereae. Some members of the genera *Grindelia*, *Chrysothamnus*, *Xanthocephalum*, *Gutierrezia*, *Baccharis*, and *Haplopappus* produced extractables exceeding 10% of the plant's dry weight (Table 1). Latex-bearing plants, by contrast, contained between 4 and 8% extractables, and plants producing neither latex nor resinous exudates consistently had less than 4% extractables.

The most highly productive of the resinous species appear to be members of the genus *Grindelia*, and in particular *G. camporum*, in which the crude resin content can be as high as 20% of the dried biomass. Chemical analysis of the resins coating the leaves, stems and involucres of *Grindelia* and most other resinous Astereae showed the extracts to be composed chiefly of diterpene acids.

Chemistry of the Tribe Astereae

The phytochemistry of the tribe Astereae is not well known; terpenoid, alkaloid, flavonoid or acetylenic chemistry has been partially described for about 10% of the species.[12,13] Some chemical results are available for the genera *Grindelia*, *Chrysothamnus*, "*Xanthocephalum* complex," *Haplopappus*, *Solidago*, *Psiadia*, and *Baccharis*.

Reports of diterpenoids in members of the tribe Astereae have been growing rather rapidly in the chemical

Table 1. Laboratory analyses of Southwestern Compositae, tribe Astereae. Values are percentage of whole-plant dry weight extracted with cyclohexane (CH) and ethanol. Adapted from Reference 6.

Taxa	Extractables	
	CH	EtOH
Baccharis glutinosa Pers.	1.7	9.5
B. sarothroides Gray*	4.8	18.0
" "	4.9	14.0
Chrysothamnus nauseosus ssp. bigelovii(Gray)Hall	15.1	20.8
" " " " "	16.0	12.8
C. nauseosus ssp. consimilis (Greene) Hall	7.9	16.4
" " " "	10.4	14.1
C nauseosus ssp. gnaphalodes (Greene) Hall	4.9	17.1
C. nauseosus ssp. junceus (Greene) Hall	6.2	15.8
C. nauseosus ssp. latisquameus (Gray) Hall	2.5	9.7
C. paniculatus (Gray) Hall	9.7	10.7
" " "	18.3	14.3
C. viscidiflorus ssp. stenophyllus (Gray) Hall	22.6	17.2
Conyza coulteri Gray	1.7	5.2
Erigeron neomexicanus Gray	2.2	8.1
Grindelia aphanactis Rydb.	7.2	9.2
G. camporum Greene	13.0	11.8
" "	5.5	6.8
G. robusta Nutt.	8.4	12.2
G. squarrosa (Pursh)Dunal.	13.8	9.9
G. squarrosa (Pursh) Dunal.	8.8	6.9
Gutierrezia lucida Greene	5.6	25.2
" " "	3.9	9.6
G. microcephala (DC.) Gray	9.9	20.4
Haplopappus acradenius (Greene) Blake	7.5	19.4
H. heterophyllus (Gray) Blake	9.5	12.5
H. laricifolius Gray	2.4	16.0
H. linearifolius DC.	11.9	20.6
Heterotheca grandiflora Nutt.	3.6	11.0
H. psammophila Wagenk.	5.4	7.6
" "	2.2	13.8
Lessingia germanorum Cham.	9.5	12.5
Machaeranthera tanacetifolia (HBK) Nees.	2.6	14.0
Solidago altissima L.	5.6	13.6
S. missouriensis Nutt.	6.2	15.1
S. wrightii Gray	5.0	22.0
Xanthocephalum gymnospermoides (Gray) B.& H.	12.1	14.8
" " " "	4.6	13.3

*Many species were collected several times, but only highest and lowest percentages are shown.

literature during the last few years; among these reports, the presence of the grindelane-type diterpenes has been noted in Grindelia, Chrysothamnus, Haplopappus, and Solidago.[14-21]

Of the large genus *Grindelia*, the chemistry of only fourteen species has been investigated. Eight of these species were found to contain the flavonoids apigenin, luteolin, kaempferol and quercetin, along with their methyl ethers and glycosides.[22-24] Simple phenolic acids have been isolated from three *Grindelia* species, and the essential oils borneol, terpineol, α-pinene and β-pinene were characterized from *Grindelia robusta* and *G. squarrosa*.[23] Numerous C-10 acetylenic compounds were isolated mainly from the roots of several *Grindelia* species, and sterols, sesquiterpenoids, and unusual tropones were found in their above-ground parts.[19,25,26] The labdane grindelic acid and many other related diterpenes have been isolated from seven *Grindelia* species, including *G. camporum*, *G. robusta*, and *G. squarrosa*.[16,17,23,27-29]

The Genus *Grindelia*

Because of its high resin production and other characteristics, including a marked variation in both quantity and composition of resin produced by different populations, *Grindelia* serves as an excellent model for the study of resin production in arid lands. Most of our recent work has concentrated on *G. camporum*, an arid-adapted, herbaceous perennial.

The majority of North American *Grindelia* species occur in the Southwest; the other two centers of diversity are the Pacific Northwest and Central Mexico. Of the many southwestern *Grindelia* species, only three are widespread: *G. camporum*, *G. aphanactis*, and *G. squarrosa*, all of which grow in arid or semi-arid environments. *G. camporum* is native to the Central Valley of California, where it is active during the hot, rainless summer months. It is not only xerophytic but appears to be halotolerant as well, as it is often found in saline flats, alluvial soils of alkaline streams, and near salt lakes and springs.

G. camporum is one of several *Grindelia* species in which both diploid and tetraploid populations are known.[30] The basic chromosome number of the genus is six, without exception. All *Grindelia* species appear to be obligate outcrossers, and plants at the same ploidy level cross readily. The principal barriers to hybridization appear to be geographic and ecological, rather than cytological or physiological.

RESIN PRODUCTION IN GRINDELIA

Resin Chemistry and Variation

Our phytochemical investigations of *G. camporum* have shown that the resin is composed of acidic and neutral fractions.[16] Twelve labdane-type diterpene acids have been isolated from the acid fraction, accounting for more than 90% of that portion of the resin, and are shown in Figure 1. The remainder of the acidic fraction appears to be comprised of trace amounts of modified labdane diterpene acids such as chrysolic acid, which has been reported in *Chrysothamnus paniculatus*, another resinous member of the Astereae whose chemical composition closely resembles that of *G. camporum*.[15] A major difference is that 3,4,5-tricaffeoylquinic acid and 3,4-, 3,5-, and 4,5-dicaffeoylquinic acids constituted the major part of the non-carbohydrate portion of a methanol extract obtained from *C. paniculatus*, after the resins were recovered with ethyl acetate;[31] these compounds are present in *G.* camporum in only trace amounts.

A substantial part (25%) of the non-acidic fraction of *G. camporum* resin was made up of the naturally-occurring methyl esters of grindelic acid and its 6,8(17)-diene; 7α,8α-epoxy; 7-hydroxy-8(17)dehydro; 6-hydroxy; 6-oxo; 17-acetoxy and 17-propionyloxy derivatives. The remainder of the neutrals were found to be a mixture of the flavonoids apigenin 4'-methyl ether, quercetin and its 3,3'-dimethyl ether, and kaempferol 3,7-dimethyl ether, a mixture of C_{25} to C_{35} paraffin waxes and an unidentified mixture of wax esters and plant sterols.

A number of different solvent systems were used during the course of our research on *Grindelia*. Ethyl acetate was used for the phytochemical studies reported here; this solvent extracts more of the phenolic materials than others we used. Dichloromethane appears to be the most efficient solvent for the recovery of the resin acids, although a drawback is that it also extracts more of the pigments and chlorophyll than does cyclohexane, which was used for the initial screening.

We have observed a large amount of variability in the composition and quantity of resin produced by *G. camporum*. The proportion of the dry biomass that can be extracted with dichloromethane varies between populations and even between

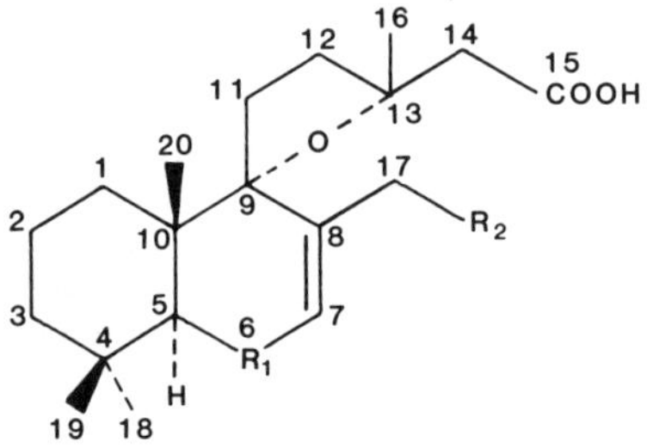

	R_1	R_2	
1	CH_2	H	grindelic acid
2	C=O	H	6-oxo-grindelic acid
3	CHOH	H	6-hydroxy grindelic acid
4	CH_2	OMe	17-methoxy grindelic acid
5	CH_2	OCOMe	17 acetoxy grindelic acid
6	CH_2	$OCOCH_2Me$	17 propionyloxy grindelic acid
7	CH_2	$OCOCH(Me)_2$	17 isobutyryloxy grindelic acid
8	CH_2	$OCOCH_2CH(Me)_2$	17-isovaleryloxy grindelic acid

9 — 7a,8a,-epoxy grindelic acid

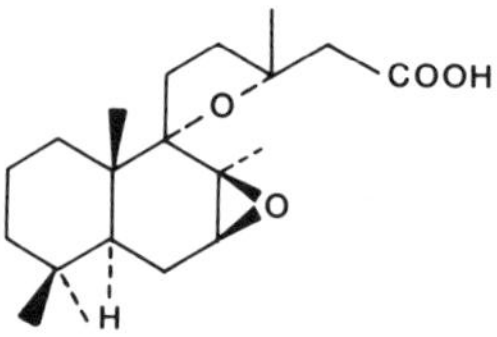

10 — 7-hydroxy-8(17)-dehydro grindelic acid

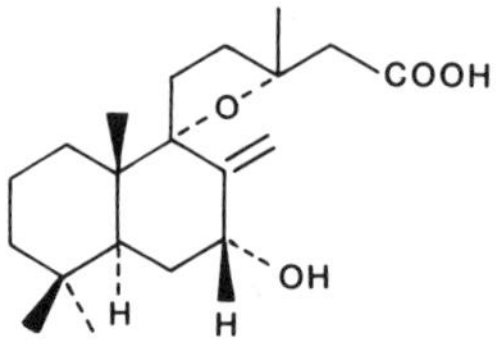

11 — 6,8(17)-diene grindelic acid

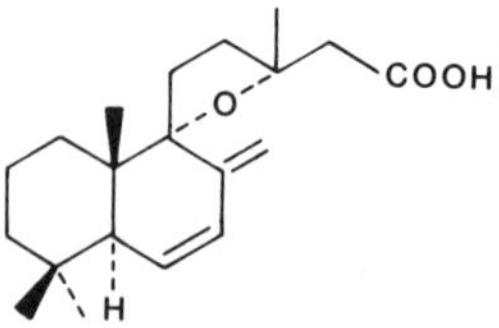

12 — strictanonoic acid

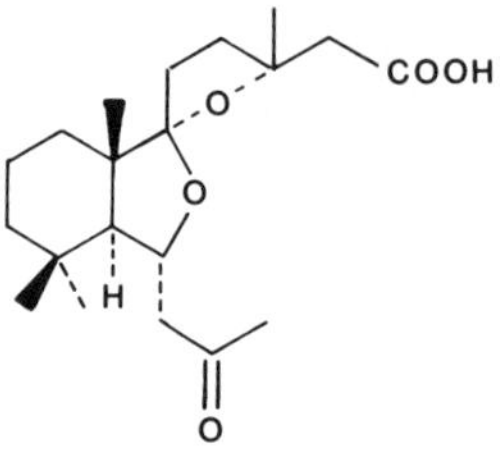

Fig. 1. Twelve grindelane diterpenoids isolated from G. camporum (adapted from reference 16).

individuals from the same population; the quantity generally falls within a range of 5 to 18%. The neutral component, which is separated from the resin acid fraction by partitioning with sodium carbonate, varies from 20 to 45% of the resin, while the resin acid fraction varies, correspondingly, from 80 to 55%. Moreover, the relative amounts of the various diterpene acids also vary within the species; grindelic acid, for example, has been found to comprise anywhere from 40 to 80% of the resin acid fraction.

This variation in resin yield and composition has yet to be explained. Our preliminary genetic studies have indicated that the broad-sense heritability of resin production is fairly high. However, in addition to genotype, factors such as population structure, phenology, climate and soil type may contribute to the variation. These latter factors may also explain the low correlation we have observed between resin yields in wild plants and those under cultivation.

Anatomy of Resin Production

Our preliminary anatomical studies have revealed the presence of two types of resin-producing structures in *Grindelia camporum*: multicellular *resin glands*, which occur in shallow pits on the surfaces of the stems, leaves, and phylaries (bracts surrounding the flower heads), and multicellular *resin ducts* which occur in the leaf mesophyll and stem cortex (Figures 2 and 3). Similar structures have been reported in other *Grindelia* species, although their functional significance has not previously been investigated.[32,33]

The resin glands in *G. camporum* are most abundant on the involucres of the flower heads (see frontispiece); they are less densely distributed on the leaves, and occur fairly infrequently on the stems.

A partitioning study was conducted in order to determine the relationship, if any, between resin production and distribution of resin glands. Mature *G. camporum* plants, grown in our experimental plots, were harvested after the onset of flowering. The plant biomass was separated into stems, leaves, and capitula. Each fraction was washed with dichloromethane, then separately ground and analyzed for crude resin content, proportion of resin acids and

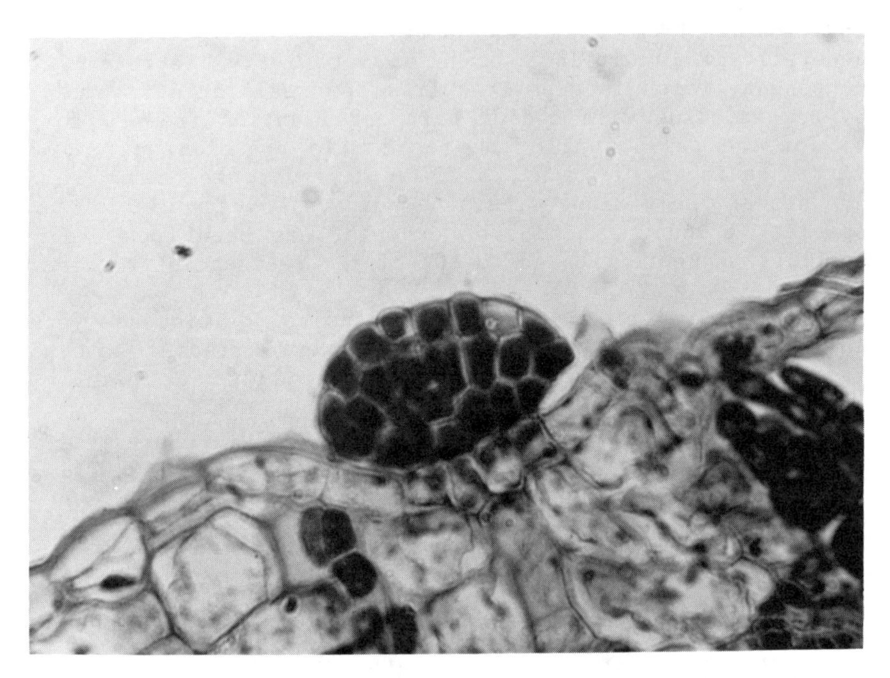

Fig. 2. Resin gland on surface of G. camporum leaf, shown in cross section.

grindelic acid content. Results are shown in Table 2. Although the flower heads accounted for only 25% of the total biomass, they contained 57% of the total crude resin, 71% of the resin acids and 91% of the grindelic acid.

Surface washes, accomplished by dipping plant material briefly into dichloromethane, removed most of the crude resin acid and nearly all of the grindelic acid. It is therefore assumed that the resin glands are responsible for most of the resin production.

Although this work is preliminary, it is possible to draw from it two conclusions that may shed some light on the function of resin-producing structures, and the function of resin. First, we now have evidence that most of the resin produced by G. camporum occurs on the plant surface, as part of the cuticular layer. The ecological implications of this finding will be discussed in the next section. Secondly, there is a positive correlation between the density of resin glands on a plant part and the quantity of resin it yields. Grindelia flowers were found to contain 20 to 30% resin while leaves and stems had less than 10%. This could indicate that resin glands are directly and quantitatively responsible for resin production. It could also imply that grindelic acid is a precursor to the oxygenated compounds in the plant, or it could indicate that grindelic acid is metabolized at differing rates in different plant parts. Further study is

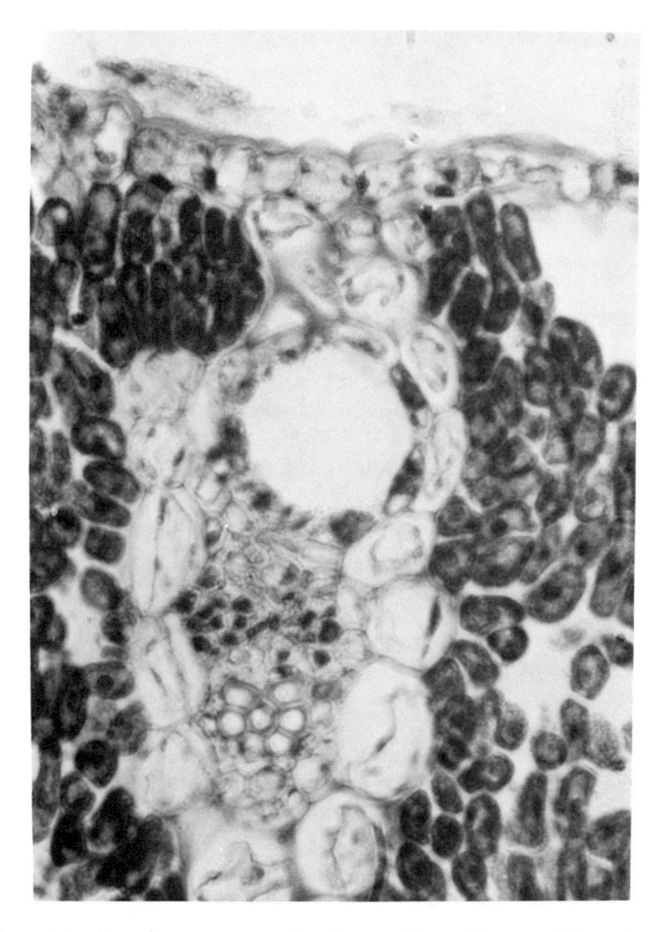

Fig. 3. Resin duct, consisting of a lumen lined with small, densely staining cells. Resin ducts are found in all above-ground organs of G. camporum.

Table 2. Yields of crude resin, resin acids and grindelic acid from partitioned plants. Fractions were surface-extracted with dichloromethane (DCM) followed by grinding and extraction with ethyl acetate (EtOAc).

Fraction	Crude resin	Resin acids in crude resin	Grindelic acid in resin acids
		%	
Flowers			
DCM wash	24.6	79.3	41.2
EtOAc extract	4.7	47.8	17.9
Leaves			
DCM wash	10.1	54.2	10.4
EtOAc extract	4.6	6.1	1
Stems			
DCM wash	2.8	49.4	9.9
EtOAc extract	0.7	28.8	1
Whole plant*			
DCM wash	10.1	68.7	32.0
EtOAc extract	2.8	26.2	13.6
Total	12.9	59.5	30.2

*Weighted average based on prior estimates of 25% flowers, 25% leaves, and 50% stems in above-ground dry weight.

needed to determine with certainty the mechanisms of synthesis, metabolism, and function of resin acids.

RESINS AS A PHYTOCHEMICAL ADAPTATION

A strong body of observation supports the hypothesis that resins in herbaceous plants represent a phytochemical adaptation to an arid environment. The accumulation of resins, particularly of the diterpene type, is well known among members of the Pinaceae but is uncommon in flowering plants; the relatively few species known to produce resins are predominantly xerophytic.

Consistent with this trend, a survey of North American species having conspicuously resinous leaves showed that this feature is associated almost exclusively with arid-adapted plants.[6] Although the resins produced by these plants are apparently similar in function, they vary somewhat in composition. Diterpene resins appear to be fairly common in arid-adapted, New World members of the Astereae tribe of the Compositae, including the genera _Grindelia_, _Chrysothamnus_, _Gutierrezia_, and _Xanthocephalum_, as previously discussed. The resin is accumulated in ray cells, fibers and vessel elements within the wood,[34] in resin canals, or ducts in the stems and leaves, or in multicellular, glandular trichomes.[35] Resinous exudates also occur in the Zygophyllaceae, a plant family that is almost universally important in arid lands; the major components responsible for the conspicuous coating of resin on the leaves of the North American desert shrub *Larrea tridentata*, and its Argentine analogue *L. cuneifolia*, are phenolic resins.[36]

Several xerophytic shrubs of western Australia are also noted for their resin production. The secretion of diterpene resin acids has been reported in *Newcastelia viscida*, *Beyeria viscosa*, and *Eremophila fraseri*.[37-39] In latter species, distinctive geographical differences in resin composition were noted. Although this variation has not been explained, it may be an indication of phytochemical response to varying degrees of environmental stress.

Three prevailing theories exist for the explanation of the ecological significance of resins in plants; all of these theories are presented in the evolutionary context of the arid environment. The most obvious explanation, perhaps, is that when secreted onto the leaf surface, these compounds make the cuticle less permeable to water and, as a consequence, protect the plant from excessive water loss. The presence of resins and other impermeable compounds is physiologically important, since cutin, the principal polymeric constituent of plant cuticles, is permeable to water.[40] Although the direct relationships between cuticle thickness, composition, and xerophytic adaptation are not well understood, many workers have noted that plants in arid zones possess thick cuticles. Experimental evidence of the role of cuticle thickness in water conservation was provided by Hall and Jones, who found that cuticular transportation in clover leaves was significantly increased when part of the cuticle was removed.[41] As further evidence of this phenomenon, intraspecific variation in both the thickness and the composition of the cuticle can often be correlated with water economy. Many species produce a waxy, glaucous coating on the leaf surface, called a "bloom," when they are subjected to water stress. Sorghum leaves in this condition have a lower rate of cuticular transpiration, increased water-use efficiency, and higher water potential than do "bloomless" leaves under similar conditions of stress.[42,43]

Another, closely related, explanation of the adaptive significance of resins occurring on leaf surfaces is that they increase the reflectance of the leaves, thereby reducing the amount of solar radiation penetrating the surface and, as a consequence, lowering the internal temperature. Support for this hypothesis comes from the work of B. Dell, who measured the reflectance spectrum from the leaves of *Beyeria viscosa*, a resin-producing shrub native to the arid regions of western Australia.[9] Radiant heat on *Beyeria* leaves causes a gradual softening of the resin, which becomes mobile and eventually forms a continuous, glossy coating on the abaxial surface of the leaf. While the presence of this coating does not appear to alter the wavelengths of light available to the plant for photosynthesis, it does reduce the overall amount of light, and heat, entering the leaf.

A third hypothesis regarding the ecological significance of resins emphasizes the importance of these compounds as plant defense substances. Terpenoid compounds have been implicated as feeding deterrents in a number of instances.[44-46] This evidence indicates that diterpenes are probable sources of the pest resistance observed in many plant species in which they occur. Two grindelic acid diterpenoids (18-hydroxygrindelic acid and 18-succinyloxygrindelic acid) isolated from *Chrysothamnus nauseosus* were found to exhibit significant antifeeding behavior against third-instar larvae of the Colorado potato beetle.[20] In addition, a series of diterpene acids including 6-hydroxygrindelic acid, isolated from *Grindelia humilis*, showed deterrent activity towards the aphid *Schizaphis graminum*.[29]

Separate discussions of these hypotheses explaining the value of resins in xeric adaptation do not imply that these functions are mutually exclusive; on the contrary, all three types of adaptation are likely to occur, often within the same plant. The phenolic resin produced by *Larrea*, for example, is reported to be an antidesiccant, to reduce herbivory, and to have ultraviolet screening properties.[10]

It is highly probable that the diterpene resins found in *Grindelia* and other Astereae serve a similar diversity of functions. The observation that most of the resinous material in these plants is cuticular, along with the fact that most of the Astereae species that have resinous cuticles do occur in xeric and/or saline habitats, provides at least strong circumstantial evidence of the role of resins in drought resistance. in addition, reports that many diterpene and other resins are toxic to insects are consistent with the observation that *Grindelia* and other resinous Astereae grown in our experimental plots over a period of years have shown an outstanding resistance to herbivory.

The concentration of resins in the leaves and, particularly, in the capitula of *Grindelia* and other resinous species we have examined presumably reflects the importance of protecting these plant parts from excessive desiccation and insect damage. This may enable the plant to produce at least some seeds, thus maintaining some reproductive potential even during periods of extreme and prolonged drought. The metabolic costs of producing the large quantities of resin observed in these species are quite high; however, this unusual allocation of energy appears to confer on these species the selective advantage of relative freedom from competition. *G. camporum*, for example, flowers in late summer in the Central Valley of California, and *Chrysothamnus paniculatus* flowers late in summer in the Mohave Desert; few other plants in these regions are physiologically active during this time.

CONCLUSIONS

Although the adaptive significance of resins in xerophytic plant species is of scientific interest, the majority of attention that will ultimately be given to these compounds will probably be directed at their practical applications. The diterpene resins produced by *Grindelia*, for example, have chemical and physical properties nearly identical with those of the resins used in the naval stores industry; several applications of the resin from *G. squarrosa* have, in fact, been patented.[47-49]

Resins are useful for a variety of industrial applications including adhesives, tackifiers, and paper sizings; their derivatives are widely used in the production of synthetic polymers. The world naval stores market, currently at a volume of approximately 700 thousand metric tons per year, is growing rapidly. Resin supplies, meanwhile, are diminishing steadily because of problems associated with traditional means of obtaining pine resins.

For this reason, there is currently considerable interest in alternative resin sources.[50-52] Consequently, a degree of significance has been attached to the discovery of several members of the Astereae in the arid Southwest that produce large amounts of resinous exudates. Among these species, Grindelia camporum appears to be a promising candidate for development as a commercial resin crop for arid lands; positive features of the plant are its high resin content, the quality of resin produced, and the plant's tolerance of xeric conditions.[53] Our preliminary agricultural data indicate that it can be grown on less than 75 cm water, and can produce market-quality resin acid at a cost of approximately 22¢ per pound.[54] The quality of this resin acid, in its purified form, is virtually identical to a commercial brand of hydrogenated wood rosin that currently sells for more than $1.00 per pound (A.D. Little, private communication). High-value commodities such as this may provide the economic incentives necessary for the commercial development of new crops providing a renewable, biomass-based chemical industry.

The elucidation of the ecological roles of resins in arid-adapted plants, however, is far from complete. Basic research of this type will be critically important in the future, for a number of reasons. Studies of resin biosynthesis, transport, and metabolism may provide answers to many important questions in plant physiology and chemistry, and could also lead to significant taxonomic advances. In addition to the scientific gains to be made, however, a better understanding of phytochemical adaptation in xerophytic species will have far-reaching practical benefits. A knowledge of the relationships between environmental factors and phytochemical responses such as resin production will eventually lead to an ability to manipulate the production of the desired products in a controlled setting. It will also provide valuable information for plant breeding and varietal improvement.

Furthermore, an understanding of the ecological roles of resins and other plant chemicals may in many cases shed light on their potential for practical applications. The broad range of softening points observed in the resin of Beyeria, for example, which results in a gradually decreasing viscosity and the coating of the leaf surface in response to increasing temperature, is precisely the same physical property that is desirable in commercial resins. The slow-melt response, as opposed to a sharply defined melting point, is the characteristic of resin that is of primary importance in its multitude of industrial applications.

Plant chemicals that deter herbivory offer another example of an adaptation that has obvious practical value. Pesticides produced by plants are growing in demand, not only because of their derivation from a renewable source, but also because they are often highly specific, and consequently cause a minimum of ecological disturbance.

The development of new crops for arid and marginal lands will carry with it the necessity for new pesticides that will act against the insects and pathogens endemic to these regions. It will also create the necessity for a better understanding of the relationships between temperature, water-stress, and the production of biomass and secondary chemicals. The pursuit of this development in a context that integrates both scientific and practical concerns will require the combined efforts of researchers in many disciplines, and will offer both challenges and the potential for great rewards.

ACKNOWLEDGEMENTS

This work was funded in part by a research agreement with the Diamond Shamrock Corporation.

REFERENCES

1. Solar Energy Research Institute 1981 A New Prosperity: Building a Sustainable Energy Future. Brick House Publishing, Andover, Mass
2. CALVIN M 1979 Petroleum plantations for fuel and materials. BioScience 29: 533-538
3. PALSSON BO, S FATHI-AFSHAR, DF RUDD, EN LIGHTFOOT 1981 Biomass as source of chemical feedstocks: an economic evaluation. Science 213: 513-517
4. BUCHANAN RA, IM CULL, FH OTEY, CR RUSSELL 1978 Hydrocarbon- and rubber-producing crops. Econ Bot 32: 131-153
5. DUISBERG, PC, JL HAY 1971 Economic botany of arid

regions. In WG McGinnies, BJ Goldman, and P Paylore, eds, Food, Fiber and the Arid Lands, University of Arizona Press, Tucson, pp 247-270.

6. McLAUGHLIN SP, JJ HOFFMANN 1982 Survey of biocrude-producing plants from the southwest. Econ Bot 36: 323-339

7. RODRIGUEZ E 1983 Cytotoxic and insecticidal chemicals of desert plants. In PA Hedin, ed, Plant Resistance to Insects Am Chem Soc Symp Ser, Vol 208, Am Chem Soc, Washington, DC pp 291-302

8. SHANTZ HL, RL PIEMEISEL 1924 Indicator significance of the natural vegetation of the southwestern desert region. J Agr Res 28: 721-801

9. DELL B 1977 Distribution and function of resins and glandular hairs in Western Australian plants. J Royal Soc West Aust 59: 119-123

10. RHOADES DF 1977 Integrated antiherbivore, antidesiccant and ultraviolet screening properties of creosotebush resin. Biochem Syst Ecol 5: 281-290

11. TRAUB HP, McSLATTERY, WL McRARY 1940 The effect of moisture stress on nursery-grown guayule with reference to changes in reserve carbohydrates. Am J Bot 33: 699-705

12. HEGNAUER R 1964 Chemotaxonomie der Pflanzen. Band 3 Dicotyledoneae: Acanthaceae-cyrillaceae. Birkhauser-Verlag, Stuttgart

13. HERZ W 1977 Astereae, chemical review. In V. Heywood, JB Harborne and BL Turner, eds, Biology and Chemistry of the Compositae. Academic Press, New York, London pp 567-576

14. HOFFMANN JJ, SP McLAUGHLIN, SD JOLAD, KH SCHRAM, MS TEMPESTA, RB BATES 1982 Constitutents of *Chrysothamnus paniculatus* (Compositae) 1: Chrysothame, a new diterpene, and 6-oxogrindelic acid. J Org Chem 47: 1725-1727

15. TIMMERMANN BN, JJ HOFFMANN, SD JOLAD, KH SCHRAM, RE KLENCK, RB BATES 1982 Constituents of *Chrysothamnus paniculatus* (Compositae) 2: Chrysolic acid, a new labdane-derived diterpene with an aromatic B-ring. J Org Chem 47: 4114-4116

16. TIMMERMANN BN, DJ LUZBETAK, JJ HOFFMANN, SD JOLAD, KH SCHRAM, RB BATES, RE KLENCK 1983 Grindelane diterpenoids from *Grindelia camporum* and *Chrysothamnus paniculatus*. Phytochemistry 22: 523-525

17. GUERREIRO E, J KAVKA, J SAAD, M ORIENTAL, O GIORDANO

1981 Acidos diterpénicos en *Grindelia pulchella* y *G. chiloensis*. Rev Latinoamer Quím 12: 77-81

18. BOHLMANN F, U FRITZ, H ROBINSON, RM KING 1979 Isosesquicaren aus *Haplopappus tenuisectus*. Phytochemistry 18: 1749-1750
19. BOHLMANN F, M AHMED, N BORTHAKUR, M WALLMEYER, J JAKUPOVIC, RM KING, H ROBINSON 1982 Diterpenes related to grindelic acid and further constituents from *Grindelia* species. Phytochemistry 21: 167-172
20. ROSE A 1980 Grindelane diterpenoids from *Chrysothamnus nauseosus*. Phytochemistry 19: 2689-2693
21. GUTIERREZ A, J OBERTI, H JULIANI 1981 Constituyentes del *Solidago chilensis*. Anales Asoc Quím Argent 69: 27
22. WAGNER H, M IYENGAR, O SELIGMANN, L HORHAMMER, W HERZ 1972 Chrysoeriol-7-glucuronid in *Grindelia squarrosa*. Phytochemistry 11: 2350
23. PINKAS M, N DIDRY, M TORCK, L BÉZANGER, JC CAZIN 1978 Recherches sur les polyphenols de quelques especes de *Grindelia*. Ann Pharm Franc 36: 97-104
24. RUIZ SO, E GUERREIRO, OS GIORDANO 1981 Flavonoides en tres especies del género *Grindelia*. Anales Asoc Quím Argent 69: 293-295
25. BOHLMANN F, W THEFELD, C ZDERO 1970 Über die Inhaltsstoffe von *Grindelia*-Arten. Chem Ber 103: 2245-2251
26. BOHLMANN F, T BURKHARDT, C ZDERO 1973 Naturally Occurring Acetylenes. Academic Press, London
27. MANGONI L, M BELARDINI 1962 Constituents of *Grindelia robusta*. Gazz Chim Ital 92: 983-994
28. BRUUN T, L JACKMAN, E STENHAGEN 1962 Grindelic and oxygrindelic acids. Acta Chem Scand 16: 1675-1681
29. ROSE A, K JONES, W HADDON, D DREYER 1981 Grindelane diterpenoid acids from *Grindelia humilis*: feeding deterrency of diterpene acids toward aphids. Phytochemistry 20: 2249-2253
30. DUNFORD MP 1964 A cytogenetic analysis of certain polyploids in *Grindelia* (Compositae). Am J Bot 51:41-56
31. TIMMERMANN BN, JJ HOFFMANN, SD JOLAD, KH SCHRAM, RE KLENCK, RB BATES 1983 Constituents of *Chrysothamnus paniculatus* (Compositae) 3: 3,4,5-tricaffeoylquinic acid (a new shikimate prearomatic) and 3,4-, 3,5-, and 4,5-dicaffeoylquinic acids. J Nat Prod 46: 365-368

32. FORLIARD N 1969 Contribution a l'etude de la composition chimique de l'herbe de Grindelia. J Pharm Belg 24: 397-414
33. HOHMANN VB 1967 Botanisch-varenkundliche Untersuchungen innerhalb der Gattung Grindelia. Plant medica 15: 255-263
34. CARLQUIST S 1960 Wood anatomy of Astereae (Compositae). Tropical Woods 113: 54-84
35. METCALFE CR, L CHALK 1950 Anatomy of the dicotyledons. Oxford Univ Press
36. MABRY TJ, DR DIFEO JR, M SAKAKIBARA, CF BOHNSTEDT JR, D SEIGLER 1977 The natural products chemistry of Larrea. In TJ Mabry, JH Hunziker, DR Difeo Jr, eds, Creosotebush: Biology and Chemistry of Larrea in New World Deserts Dowden Hutchinson and Ross, Stroudsburg, PA pp 115-134
37. DELL B, AJ McCOMB 1975 Glandular hairs, resin production, and habitat of Newcastelia viscida E. Pritzel (Dicrastylidaceae). Aust J Bot 23: 373-390
38. DELL B, AJ McCOMB 1974 Resin production and glandular hairs in Beyeria viscosa (Labill.) Miq (Euphorbiaceae). Aust J Bot 25: 195-210
39. DELL B 1975 Geographical differences in leaf resin components of Eremophila fraseri F. Muell. (Myoporaceae) Aust J Bot 23: 889-897
40. SCHONHERR J 1976 Water permeability of cuticular membranes. In OL Lange, L Kappen, ED Schulze, eds, Water and Plant Life Springer-Verlag, Berlin pp 148-159
41. HALL DM, RL JONES 1961 Physiological significance of surface wax on leaves. Nature (London) 191: 95-96
42. CHATTERTON NJ, WW HANNA, JB POWELL, DR LEE 1975 Photosynthesis and transpiration of bloom and bloomless sorghum Can J Plant Sci 55: 641-643
43. BLUM A 1975 Effect of the Bm gene on epicuticular wax and the water relations of Sorghum bicolor L. (Moench). Israel J Bot 24: 50-51
44. MABRY TJ, JE GILL 1979 Sesquiterpene lactones and other terpenoids In GA Rosenthal, DH Janzen, eds, Herbivores: Their interaction with Secondary Plant Metabolites Academic Press New York, London pp 501-537
45. ELLIGER CA, DF ZINKEL, GB CHAN, AC WAISS JR 1976 Diterpene acids as larval growth inhibitors. Experientia 32: 1364-1366
46. BURNETT WC, JR, SB JONES JR, TJ MABRY, WG PADOLINA

1974 Sesquiterpene lactones: Insect feeding deterrence in *Vernonia*. Biochem Syst Ecol 2: 25-29

47. McNAY RE 1964 Emulsion polymerization. US Patent 3,157,608
48. McNAY RE, WR PETERSON 1964 Treatment of synthetic rubber. US Patent 3,157,609
49. McNAY RE, WR PETERSON 1965 Method of sizing cellulose fibers with resinous material from the plant *Grindelia* and products thereof. US Patent 3,186,901
50. ZINKEL DF 1975 Naval stores: silvichemicals from pine. *In* TE Timell, ed, Applied Polymer Symp No 28, John Wiley & Sons, New York pp 309-327
51 ZINKEL DF 1981 Turpentine, rosin, and fatty acids from conifers. *In* IS Goldstein, ed, Organic Chemicals from Biomass. Chap 9 CRC Press Boca Raton, FL
52. HANNUS K 1976 Lipophilic extractives in technical foliage of pine (*Pinus sylvestris*). *In* TE Timell, ed, Applied Polymer Symp No 28, John Wiley & Sons, New York pp 485-501
53. HOFFMANN JJ 1983 Arid lands plants as feedstocks for fuels and chemicals. Crit Rev Plant Sci 1: 95-116
54. McLAUGHLIN SP 1982 Plant terpenoids as economic products from arid lands. Paper presented to 1982 Annual Meet Soc Econ Bot, June 14-17, Tuscaloosa, AL

Chapter Ten

CHANGES IN THE LEVELS OF PLANT SECONDARY METABOLITES UNDER WATER AND NUTRIENT STRESS

JONATHAN GERSHENZON

Department of Botany
University of Texas
Austin, Texas 78712

INTRODUCTION

Increasing interest in the stress physiology of higher plants has, in the last few years, greatly expanded our knowledge concerning the effects of stresses such as drought, freezing, nutrient deficiencies, disease and insect attack on plant metabolism. This work, however, has concentrated on the influences of stress on the basic processes of growth and development (primary metabolism). Very little attention has been paid to the changes in secondary metabolism that may be occurring. Information on the effects of stress conditions on secondary metabolites has come mainly from research efforts to maximize the yield of active constituents from herb, spice and drug plants and from attempts to reduce the levels of toxins in certain food and forage crops. Advances in our understanding of the functional roles of secondary metabolites in plants, however, and new information about how pest resistance is altered by stress should stimulate additional interest in the responses of secondary metabolites to stress.

The idea that plant secondary compounds have important functions is now widely accepted. In the two decades since the appearance of the classic papers of Fraenkel[1] and Ehrlich and Raven,[2] a large body of evidence has been obtained which suggests that secondary compounds often serve as defenses against predators and pathogens. Investigations of the response of these constitutents to stress, therefore, may help in explaining patterns of herbivore utilization of plants and the incidence of fungal invasion, as well as in evaluating the importance of defense under different conditions.

A variety of stresses has recently been shown to influence the resistance of plants to herbivores.[3,4] For example, nitrogen deficiency increases the resistance of a wide range of crop plants to insects,[4] while drought makes some plants more susceptible to insect attack.[5,6] Changes in resistance of this sort have usually been ascribed to changes in plant water content, alterations in the level of soluble nitrogen compounds, or shifts in the sugar-starch balance. Clearly, though, alterations in the levels of secondary compounds caused by stress could also have a strong influence on plant resistance to herbivory.

In this chapter I will discuss the effects of two stresses that affect plants, water deficits (water stress) and nutrient deficiency (nutrient stress). Omission of other stress factors is not meant to minimize the importance of their effects on secondary metabolism. Low light and low temperature, for instance, have been shown to have significant influences on levels of secondary compounds [7-12] and fungal infection has long been known to induce the synthesis of particular secondary compounds called phytoalexins in certain plants. Recent evidence demonstrates that herbivory can also trigger the production of secondary metabolites.[13-15]

This chapter will review the effects of water deficits and nutrient deficiencies on the production of different classes of secondary metabolites, suggest physiological mechanisms by which these changes might occur, and discuss the adaptive significance of the patterns observed.

CYANOGENIC GLYCOSIDES

Water Stress

Reports that certain forage grasses containing cyanogenic glycosides become toxic to livestock during periods of drought[16-19] suggest that water deficits can lead to increased concentrations of cyanogenic glycosides in some plants. Controlled experiments with cassava (Manihot esculenta),[20] grain sorghum (Sorghum bicolor)[21] and Sudangrass (S. sudanense)[17,22] have confirmed this suggestion, showing increases in cyanogenic glycoside accumulation of as much as 75% under water stress. Since cyanogenic glycoside concentrations are highest in young leaves,[17,22] it is possible that water stress acts simply by inhibiting leaf expansion and therefore preventing the dilution of cyanogen content caused by growth. Even mild water deficits can have a very strong effect on leaf expansion.[23,24]

In wild plants, times or places of low moisture availability may be correlated with higher cyanogenic glycoside content, but this is not always the case, perhaps because of the overriding influence of other environmental factors. Drought increases the level of cyanogenic glycosides in white clover (Trifolium repens)[25] and arrowgrass (Triglochin maritima)[19] and Amelanchier alnifolia was found to have a higher concentration of cyanogenic glycosides in drier sites than in wetter ones.[26] However, leaf cyanogen content was significantly lower during the dry season than during the wet season in individuals of Acacia farnesiana in Costa Rica[27] and the cyanogenic potential of bracken (Pteridium aquilinum) fronds from open sites was 40-50% less than that found in shady sites.[28] In the chaparral shrub Heteromeles arbutifolia, native to the Mediterranean type climate of California, cyanogenic glycoside concentration in the leaves is maintained at high levels through the spring and most of the summer, but decreases during the hot, dry months of late summer and fall, reaching a low in December before the start of the rainy season.[29] This late season decline may be a response to increasing water stress or to some other environmental variable such as decreasing nitrogen availability (see below) or lowered predation pressure. On an annual basis, however, cyanogenic glycoside levels in H. arbutifolia during a dry year were at times nearly double those measured during a wet year.[29]

Nutrient Stress

Nitrogen fertilization increased the concentration of cyanogenic glycosides in leaves of sorghum,[21] Sudangrass [17,22,30-34] and cassava,[35] as much as eight-fold in some cases. Although these investigations were not designed to examine the effects of nutrient stress, since fertilization usually increased overall growth (though growth data were not reported in every study) the unfertilized controls must have been nitrogen-limited to some degree, and therefore were by definition[36] under nitrogen stress. A reduction in cyanogenic glycoside concentration under low nitrogen supply would not be surprising since cyanogenic glycoside molecules contain nitrogen.

In *H. arbutifolia*, leaf cyanogenic glycoside accumulation is directly correlated with nitrogen availability, with the highest percentages of cyanogens being found during the warm, moist months of spring and summer when nitrogen is considered to be most available.[29] In studies with other nutrients, the concentration of cyanogenic glycosides in *Sorghum* spp. increased under low phosphorus levels,[17,31,34] but was not affected significantly by low potassium supplies.[17,30,34]

GLUCOSINOLATES AND OTHER SULFUR COMPOUNDS

Water Stress

Although there has been very little formal study of the influence of water stress on glucosinolate production, growers of cruciferous vegetables have long known that plants grown with a restricted water supply tend to be smaller and more flavorful than those grown under well-watered conditions. Controlled experiments on cabbage (*Brassica oleracea capitata*) and watercress (*Rorippa nasturtium-aquaticum*) showed that water deficits increase the concentration of the characteristic flavor components of these plants, isothiocyanates (the hydrolysis products of glucosinolates) and other sulfur-containing secondary compounds such as alkyl sulfides.[37] These changes were too great to be explained by a dilution effect in plants grown with adequate water.[37] In late season plantings of cabbage, glucosinolate concentration, as measured by levels of thiocyanate, was more than twice as high in non-irrigated plants as in irrigated ones.[38] There were no significant differences between these treatments, however, in early season plantings.

Two additional studies may be relevant here, though water deficits were not directly manipulated in either one. Investigations of the effects of different crop spacings on the flavor of cabbages and Brussels sprouts demonstrated that closer spacing usually resulted in higher percentages of isothiocyanates and other sulfur compounds.[39,40] These increases, which were sometimes over ten-fold, could be a result of increased competition for water or nutrients. Field experiments with the Rocky Mountain crucifer *Cardamine cordifolia*, which normally grows in moist, shaded localities, showed that when plants were exposed to direct sun by severe pruning of over-hanging branches, there was a doubling in the average percentage of isothiocyanate-yielding glucosinolates in the leaves.[41] This change might be due to the effects of increased water stress, although higher light intensity or higher temperatures could also be controlling factors.

Nutrient Stress

Since glucosinolates contain both nitrogen and sulfur, one might expect that deficiencies of these nutrients would reduce glucosinolate production. Low sulfur conditions do cause a decrease in glucosinolate accumulation, a fact that has been appreciated for many years. Experiments conducted in the 1940s revealed that sulfur-deficient black mustard, *Brassica nigra*, was almost completely without its characteristic flavor or pungency.[42,43] The volatile sulfur content of *B. nigra* grown in nutrient solution without sulfur, measured after hydrolysis of the glucosinolates, was at least an order of magnitude below controls grown in complete nutrient solution.[42,43]

A series of more recent studies carried out by Freeman and Mossadeghi[44-46] has confirmed that low sulfur supply leads to a reduction in the concentration of glucosinolates and other sulfur-containing secondary compounds (such as alkyl sulfides and mercaptans) in a number of vegetable crops, including cabbage, radish (*Raphanus sativus*), white mustard (*Sinapis alba*), garlic

(Allium sativum), onion (A. cepa) and wild onion (A. vineale). For example, the amount of the glucosinolate sinalbin, p-hydroxybenzylglucosinolate, in Sinapis alba seed (measured by the gas chromatographic peak area of its principal hydrolysis product, p-hydroxybenzylisothiocyanate) was directly related to the amount of sulfur in the applied nutrient medium, increasing over 18-fold as the sulfur concentration was raised from 0 to 3 milliequivalents/liter.[46] In many of these studies, almost no glucosinolates or other sulfur-containing secondary compounds were detected when sulfur was not provided. Results of additional studies on Brassica nigra leaves[47] and B. napus seed meal[48,49] follow the same pattern, but in kale (a cultivar of B. oleracea) glucosinolate concentrations were unaffected by low sulfur supply.[50]

In contrast to sulfur deficiency, a deficiency of nitrogen has been shown to either increase glucosinolate concentrations in plants or have no effect on them. Under low nitrogen conditions, the glucosinolate content in various organs of Brassica species (B. napus,[48] B. nigra,[42,47] B. oleracea[50]) increased up to 120% in some experiments, but did not significantly change in others.[48,50]

It is unclear why nitrogen deficiencies do not lead to decreased glucosinolate content, as sulfur deficiencies do. Perhaps, since glucosinolates contain a much smaller proportion of the total nitrogen in the plant than they do of the total sulfur, deficiencies of nitrogen need not inhibit the rate of glucosinolate synthesis.

The altered levels of glucosinolates found in nutrient deficient plants may be of real significance in plant defense. A recent investigation showed that the larvae of the lepidopteran Spodoptera eridania (the southern armyworm), a generalist feeder, grew more rapidly on Brassica nigra raised under low sulfur conditions (decreased levels of glucosinolates) than on controls, but did not complete development on B. nigra raised under low nitrogen conditions (increased levels of glucosinolates).[47]

ALKALOIDS

Water Stress

The influence of environmental factors on alkaloid production has been frequently studied because many plants containing alkaloids are (or have been) important as sources of pharmaceuticals or as forage for livestock. For years it has been observed that alkaloid-bearing plants are often more potent in dry periods than in wet periods. For instance, wild hemlock (*Conium maculatum*) is most toxic during hot and dry seasons[16] while cinchona (*Cinchona ledgeriana*) produces no quinine during the rainy season.[51] In lupines (*Lupinus*, spp.)[51] and opium poppies (*Papaver somniferum*),[51,52] there is a general tendency for alkaloid concentrations to be higher during dry years than wet ones. In field-grown *P. somniferum*, a very wet period during the ripening of the capsule coincided with a sharp decline in morphine and codeine content.[53] *Papaver somniferum* in the drier conditions of a nearby greenhouse, however, did not show this decrease.

Controlled experiments have confirmed these basic trends. Water stress has been shown to increase alkaloid percentages in a variety of plants including *Nicotiana rustica*[54] and *N. tabacum*[51,55-58] (nicotine and other tobacco alkaloids), *Datura innoxia*[54] and *Hyoscyamus muticus*[59] (tropane alkaloids), *Solanum aviculare*[60] and *Lycopersicon esculentum*[61] (steroidal glycoalkaloids), *Lobelia sessifolia*[54] (piperidine derivatives), *Senecio longilobus*[62] and *Lolium multiflorum* x *Festuca arundinacea*[63] (pyrrolizidine alkaloids), and *Phalaris aquatica*[64] and *P. tuberosa*[65] (indolealkylamines). Alkaloid increases reported in these studies ranged from 5% to 500%.

The mechanisms by which water deficits increase alkaloid levels have been the subject of some speculation. One possible explanation is that water deficits raise the levels of amino acids and amides which can serve as biosynthetic precursors to alkaloids. Water stressed plants are well known to accumulate high levels of free amino acids, especially proline, and the amides glutamine or asparagine.[66,67] This accumulation has been attributed to a decline in protein synthesis[24,66] accompanied in some cases by accelerated protein breakdown.[68,69] In the study on the Lolium-Festuca hybrid mentioned above,[63] the amount of free amino acids appeared to have increased under water stress, since the protein content decreased while the total nitrogen level remained essentially unchanged. But, the decline in protein content apparently took place well before alkaloid concentrations actually increased, suggesting that amino acid accumulation was not the sole factor responsible for alkaloid build-up.[63]

Many processes involved in plant growth, including cell enlargement, cell wall synthesis and protein synthesis, are extremely sensitive to water deficits.[23,24,66] If water stress inhibited growth, but had comparatively little effect on alkaloid production, this could provide another way to explain the greater alkaloid concentrations observed under stress.

To decide which, if either, of these two explanations might be correct, it is necessary to compare the total yields of alkaloids in stressed plants with those of non-stressed controls. If water deficits act by raising the supply of alkaloid precursors, an increase in total alkaloids would be expected, whereas if greater alkaloid content is the result of growth being inhibited more than alkaloid synthesis, one would expect the total amount of alkaloids to remain constant or decline.

Total alkaloid yields are usually not reported, but can be computed from measurements of alkaloid concentration and plant weight. Only two of the references cited in this paper provide sufficient information for this calculation. These two show opposing trends (Table 1), suggesting that different mechanisms for the accumulation of alkaloids under water stress may operate in different species. In the Festuca-Lolium hybrid (Table 1B), the decrease in total alkaloid production is consistent with the second explanation (growth being affected more than alkaloid synthesis), however, as Kennedy and Bush point out,[63] plant dry weight does not decline in synchrony with increased alkaloid concentration in this study, but rather precedes it. Therefore, at least in this species, neither proposed mechanism alone may satisfactorily explain the accumulation of alkaloids. Detailed investigations of alkaloid metabolism under water stress are clearly needed.

Table 1. Computation of the effect of water stress on total alkaloid production. Total alkaloids were calculated as the product of dry matter harvested and the percentage of alkaloid.

Nicotiana tabacum[58]

Treatment	Yield of cured leaves lbs/acre	Nicotine %	Total nicotine lbs/acre
High irrigation	1649	1.76	29.0
Moderate irrigation	1622	2.10	34.1
No irrigation	1334	4.02	53.6

Festuca arundinacea x *Lolium multiflorum*[63]

Treatment	Forage dry wgt at 12 wk harvest g	N-Acetyl-loline and N-formyl-loline at 12 wk harvest %	Total alkaloid in forage at 12 wk harvest mg
Control	7.8	0.32	25.0
Moderate stress	1.4	0.52	7.3
Severe stress	0.3	1.11	3.3

Nutrient Stress

The relationship between nitrogen supply and alkaloid production has been examined frequently in the course of research efforts to increase the yield of pharmaceuticals from some medicinal plants and reduce the toxicity of certain forage grasses. These studies were generally intended to test the effects of applying additional nitrogen on alkaloid levels rather than to evaluate the impact of nitrogen deficiences. However, whenever growth measurements are reported in these investigations, these show that nitrogen supplements significantly increased plant growth and therefore, the unfertilized controls, like those in the cyanogenic glycoside work, must have been under nitrogen stress.[36] Studies concerning the effects of nitrogen fertilization on alkaloid content have given somewhat inconsistent results (Table 2), but in a majority of cases additional nitrogen was found to raise the alkaloid concentration, with increases ranging from 25% to 300%. Deficiencies of nitrogen then may be generally assumed to reduce alkaloid content.

Table 2. Effects of nitrogen fertilization on alkaloid concentration.

Family and Species	Alkaloid	Change in Concentration*	Reference
Papveraceae			
Papaver somniferum	Benzylisoquinoline	+	70,71
Fabaceae			
Lupinus, sp.	Quinolizidine	+	51,72
Lupinus angustifolius		+	71
Zygophyllaceae			
Peganum harmala	β-Carboline	+	73,74
Euphorbiaceae			
Ricinus communis	Simple pyridine	+	51,71
Apocynaceae			
Catharanthus roseus	Indole-monoterpene	0	71
Vinca perenne		0	71
Solanaceae			
Atropa belladonna	Tropane	+	75
		+,0	76
		0	77,78
Datura innoxia	Tropane	+	76
D. stramonium		+,0	75,76,79
Hyoscyamus muticus	Tropane	+,-	75
		+,0	76
		+	59,80
		0	78
Nicotiana tabacum	Tobacco alkaloids	+	71,72,76,81,83
		+,0	82
N. rustica		+	71
Solanum tuberosum	Steroidal glycoalkaloids	+	84
(aerial parts)		-	79
		+,-	71
(tubers)		+	76
		-	84
Campanulaceae			
Lobelia inflata	Piperidine derivatives	-	75
Lobelia, sp.		-	76
Poaceae			
Festuca arundinacea	Diazaphenanthrene	+	85,86
F. arundinacea x *Lolium multiflorum*	Pyrrolizidine	-	63
Hordeum vulgare	Indolealkylamines	+	71
Phalaris aquatica	Indolealkylamines	+,0	64
P. tuberosa =		+,-	65
(*P. aquatica*)		+,0	87
P. arundinacea		+,0	88

* (+) = increase in concentration, (-) = decrease, (0) = no significant change. If different phases of a single study gave different results, these are reported as "+,0" or "+, -".

The promotion of alkaloid production by supplementary nitrogen is scarcely surprising since nitrogen is by definition a part of every alkaloid molecule. Added nitrogen increases levels of soluble nitrogen compounds, particularly glutamine and asparagine,[71] from which nitrogen can be readily transferred to amino acids and other alkaloid precursors. Nitrogen can also stimulate alkaloid production by increasing the activity of the enzymes involved in biosynthesis.[51]

As is clear from Table 2, there is sometimes considerable intra-specific variability in the effect of added nitrogen on alkaloid concentrations. Part of this can be attributed to the fact that different experiments used different genetic lines which may vary in their response to added nitrogen.[64,88] Apparent intra-specific variability could also have been caused by failures to adequately control for diurnal, developmental and seasonal variation[76,89] and differences in experimental conditions, such as light and temperature regimes,[65,87] initial soil fertility, the time course of nitrogen application or the amount of fertilizer applied. In some situations, high nitrogen supplements could actually have suppressed alkaloid production by inhibiting root growth,[71] since roots are considered to be the principal sites of alkaloid biogenesis in a number of species.[72,90]

The effect of nitrogen supply on plant alkaloid content may depend in part on the biosynthetic origin of the alkaloids being produced. Most alkaloids are synthesized from amino acids, but the entire carbon skeleton of the steroidal glycoalkaloids of Solanum (except for the sugars) and a significant portion of the carbon atoms of the complex indole alkaloids of Catharanthus and Vinca are of terpenoid origin. The biosynthesis of these alkaloids, therefore, might not be affected by nitrogen availability in the same fashion as the production of alkaloids derived from amino acids. As shown in Table 2, neither Solanum tuberosum, Catharanthus roseus nor Vinca perenne consistently show increases in alkaloid concentration in response to added nitrogen.

The relative amount of nitrogen in the alkaloids themselves could also have an important influence on whether or not alkaloid content rises with increased nitrogen supply. In general, the species that show consistent increases in alkaloid concentration in response to added nitrogen are those that produce alkaloids relatively rich in nitrogen (top half of Table 3). In species where nitrogen fertilization decreases or has no consistent effect on alkaloid content, on the other hand, the alkaloids synthesized usually have larger ratios of carbon to nitrogen (bottom half of Table 3). It seems reasonable that, in the presence of adequate nitrogen, alkaloid biosynthesis should be limited principally by the supply of fixed carbon. Under these conditions, the production of alkaloids relatively rich in nitrogen (requiring proportionally less carbon) could be stimulated to a greater degree than that of alkaloids with a relatively larger carbon requirement. Studies with plants that produce alkaloids of more than one type with significantly different carbon to nitrogen ratios, such as species of *Duboisia*,[51] could help to test this idea.

The effect of supplemental nitrogen on plant alkaloid levels could also vary with the type of environment to which each plant is adapted. Species of infertile habitats often have low nutrient requirements and inherently low growth rates and do not show dramatic responses in growth to sudden nutrient abundance. In such species,

Table 3. Relationships between the carbon:nitrogen ratios of some alkaloids and how their concentrations in plants change with added nitrogen.

Species	Representative alkaloid	Carbon:nitrogen ratio	General effect of added nitrogen*
Papaver somniferum	Morphine	16	+
Lupinus, spp.	Sparteine	7.5	+
Peganum harmala	Harmine	6.5	+
Ricinus communis	Ricinine	4	+
Nicotiana tabacum	Nicotine	5	+
Festuca arundinacea	Perloline	10	+
Hordeum vulgare	Gramine	5.5	+
Catharanthus roseus	Vinblastine	11.5	0
Atropa belladonna	Hyoscyamine	17	variable
Datura, spp.	Scopolamine	17	variable
Hyoscyamus muticus	Hyoscyamine		variable
Solanum tuberosum	Solanine	45	variable
Lobelia, spp.	Lobeline	22	-
Festuca arundinacea x *Lolium multiflorum*	N-Acetyl-loline	5	-
Phalaris aquatica	N,N-dimethyl-tryptamine	6	variable

*See footnote to Table 2 for explanation of symbols.

alkaloid biogenesis could be relatively unaffected by changes in the amount of nitrogen present, although there is no evidence available on this point. Almost all of the alkaloid-containing species studied are plants of nutrient-rich, disturbed habitats.

Increases in nitrogen supply lead to increased concentrations of alkaloids in most of the species studied. An explanation of the adaptive significance of this trend must consider the importance of nitrogen to both plants and their herbivores. Nitrogen is frequently a major growth-limiting nutrient for both these groups.[92] Many plants, particularly short-lived species and those of disturbed or unpredictable environments, may be adapted to take advantage of any local abundance of nitrogen by absorbing as much as possible and storing the excess for later growth and reproduction.[93,94] Since nitrogen is also an extremely valuable resource for herbivores, it may be necessary for plants to store surplus nitrogen in a protected form by incorporating it into defensive compounds, such as alkaloids or cyanogenic glycosides or non-protein amino acids, in order to avoid increased predation.[93] Radiolabeling evidence suggests that nitrogen stored in this manner is not metabolically inaccessible to the plant, but can be readily mobilized when needed.[95,96] Thus, the increased accumulation of alkaloids observed following nitrogen fertilization may simply represent a secured storage form of this valuable nutrient. Nitrogen deficiencies, then, could be expected to deplete such storage pools, leading to lower alkaloid concentrations.

Other nutrients can also have an influence on alkaloid levels in plants. In contrast to nitrogen, fertilization with potassium usually decreases alkaloid concentration,[63,76,83,85,88,97,98] while potassium deficiences lead to increased alkaloid content.[51,72,84] The magnitude of these changes average 15-30%, considerably less than those usually resulting from varying nitrogen supply. The effects of added phosphorus are extremely variable and show no clear trend.[51,73,74,76,85,88,90,99] Deficiencies of potassium and phosphorus (and all other elements except molybdenum and nitrogen) cause increases in the levels of soluble nitrogen compounds,[67] which may provide added substrate for alkaloid biosynthesis.

PHENOLIC COMPOUNDS

Water Stress

Plant phenolics appear to exhibit a variety of responses to water deficits, but unfortunately, very few controlled studies have been carried out. The levels of several simple cinnamic and benzoic acid derivatives in wheat (*Triticum aestivum*) declined noticeably under drought conditions, although this was not precisely quantified.[100] In *Helianthus annuus* however, concentrations of chlorogenic and isochlorogenic acid (caffeic acid esters of quinic acid) increased dramatically, up to six-fold, under NaCl-induced water stress.[101] Experiments on radish (*Raphanus sativus*) and cucumber (*Cucumis sativus*) seedlings showed that water stress increased cotyledon flavonoid content in cucumbers after eight days of growth, but had no effect on that of radishes.[102] Under enhanced levels of UV-B irradiation, however, the reverse was true; the flavonoid content of radishes rose under water stress, but that of cucumbers showed no change.[102]

Several correlations between field moisture conditions and levels of phenolic compounds have been reported, but these also show no definite trend. For instance, there was no obvious connection between the catechin and condensed tannin contents of cotton (*Gossypium hirsutum*) and the number of days following the most recent irrigation.[103] The condensed tannin content in the leaves of *Heteromeles arbutifolia*, however, was higher during a wet year than during a dry one.[29] In bracken (*Pteridium aquilinum*) the condensed tannin content of fronds was lowest in midsummer, possibly due to the greater water stress prevailing at that time,[104] but another study on bracken showed that fronds from sunny areas, which could have suffered greater water stress, had higher tannin concentrations than those from shady areas.[28] The same situation was also observed in plum (*Prunus domestica*),[105] suggesting that tannin synthesis might depend on the amount of photosynthate available, but this possibility has not been directly tested. Thus, no clear relationship between water stress and phenolic content has been established and this area requires considerable further study.

Water deficits could alter the levels of phenolic compounds in several ways. A decrease in phenolic production might result from a decrease in the activity of key enzymes in phenolic biosynthesis, such as phenylalanine ammonia-lyase,[106] or from a reduction in the supply of substrate[28,29] due to the inhibition of photosynthesis.[68,107-109] Water stress could lead to increased levels of phenolic compounds, on the other hand, by reducing the rate of growth leading to a build up of lignin precursors, which are then metabolically diverted to various phenolics.[101]

Nutrient Stress

The pattern of phenolic response to nutrient deficiences is less ambiguous than that seen for water stress. Deficiencies of nitrogen, phosphorus, potassium and sulfur usually result in greater concentrations of phenolic compounds (Table 4). Some of the increases measured in these studies are quite striking. For example, caffeoylquinic acid levels in sunflower (*Helianthus annuus*)[101,112] or tobacco (*Nicotiana tabacum*)[113] can increase as much as five or ten-fold under nitrogen deficiency, while anthocyanin content of tomato (*Lycopersicon esculentum*) under phosphorus-deficient conditions can be up to 13 times the normal amount.[117] However, most observed increases are much smaller.

Callus and cell suspension cultures have been used to study the effects of nutrients on phenolic synthesis in a number of cases. These provide carefully controlled systems for studying the influence of factors such as nutrients, growth hormones or metabolic inhibitors on the production of secondary compounds, although there is some question as to how closely their behavior models that of intact plants. Experiments with cell suspension cultures of Paul's Scarlet Rose (*Rosa* sp.),[134,136,137] *Acer pseudoplatanus*,[135,138] tobacco[133] and periwinkle (*Catharanthus roseus*) cells[130,131] have shown that accumulation of phenolic substances can be very sensitive to nutrient supply. Total phenolic levels often increase substantially when the medium becomes depleted of nitrogen[135] or phosphorus.[130,131] Additional amounts of these nutrients usually stimulate growth and suppress the formation of phenolics.[130,133-136,138]

The dramatic increases in plant phenolic levels reported under nutrient stress suggest that, in at least some cases, elevated phenolic concentrations are a result of de novo synthesis rather than simply a relative slowing of primary metabolism. Nutrient shortages could promote

Table 4. Effects of low nutrient levels on plant phenolic content.

Type of compound	Species	Effect of low* N	P	K	S	Reference
Caffeoylquinic acids	*Fagopyrum esculentum*	0	0			110
	Helianthus annuus	+	+	+	+	101,111
	Nicotiana tabacum	+		+	+	113-115
Coumarins	*Helianthus annuus*	0		+	0	
	Nicotiana tabacum	+				113
Anthocyanins	*Brassica oleracea*			+		116
	Fragaria, sp.		+	+		116
	Gossypium hirsutum		+			116
	Hordeum vulgare	+	+	+		116
	Impatiens, sp.	+				116
	Lactuca sativa	+				116
	Lycopersicon esculentum	0	+	+		116,117
	Malus, sp.	+	+	+		116
	Pyrus Communis		+			116
	Raphanus sativus	+	+	-		116
	Saxifraga crassifolia	+				118
	Solanum melongena	-				116
	Sorghum bicolor	+	+			119
	Spirodela oligorrhiza		+			116
	Vitis vinifera			+		116
	Zea mays	+,-	+	+,-		116,120
Flavonoid glycosides	*Nicotiana tabacum*	+		+		114
	Fagopyrum esculentum	0	0			110
	Hydrangea macrophylla	+				121
	Lycopersicon esculentum	-	-	-		121
	Nicotiana tabacum	+,-				115,121
	Malus robusta	+	0	0	+	122
Isoflavones	*Trifolium subterraneum*	+	+		+	123-125
Tannins	*Blechnum brasiliense*	+	0	-	-	126
	Lespedeza cuneata	0	0	+		127
Total phenolics	*Helianthus annuus*	(general deficiency +)				128
	Lycopersicon esculentum	+				129
	Nicotiana tabacum	-	+			115,121
	CELL, CALLUS AND ORGAN CULTURES					
Cinnamoyl putrescines	*Nicotiana tabacum*	+				130,131
Anthocyanins	*Brassica oleracea*	+				132
Polyphenols	*Catharanthus roseus*	+	+			133
	Paul:s Scarlet Rose	+				134
Total phenolics	*Acer pseudoplatanus*	+				135
	Nicotiana tabacum		+			130,131
	Paul:s Scarlet Rose	+				134,136,137

* (+) = increase in concentration, (-) = decrease, (0) = no change

the synthesis of phenolic compounds by increasing the activity of enzymes such as phenylalanine ammonia- lyase (PAL)[121,139,140] or by increasing the supply of precursors.[141] Many environmental factors have been reported to influence the levels of PAL,[142] the enzyme catalyzing the deamination of phenylalanine to *trans*-cinnamic acid. Nutrient deficiences increase PAL activity in apples, *Malus*, sp. (nitrogen and potassium deficiencies),[143] in *Acer pseudoplatanus* cell suspension cultures (nitrogen deficiency)[135] and in tobacco cell cultures (phosphorus deficiency),[130] and therefore could stimulate phenolic synthesis in this manner.

Nutrient stress could increase the supply of phenolic precursors, such as phenylalanine, by several mechanisms. Reductions in plant growth under stress could lead to a build-up of substrate for lignin synthesis, as already mentioned, which could be shunted into the production of secondary phenolic compounds. When nitrogen is in short supply, phenylalanine could be directly deaminated to reclaim its nitrogen for other purposes.[101] The *trans*-cinnamic acid remaining is then available for conversion to more complex phenolics. Finally, as the rate of protein synthesis slows under conditions of nitrogen deficiency, the unused carbohydrate could be diverted to phenolic synthesis. In general, conditions unfavorable for protein synthesis in plants result in the increased production of phenolic compounds,[141] a point graphically illustrated by recent investigations with cell suspension cultures.[134,136-138]

Regardless of the physiological mechanisms involved, the accumulation of large amounts of phenolics under nutrient stress may have some role in the adaptation of plants to stress situations. Several possible roles can be proposed.

1) *Growth Inhibition*. A reduction in growth rate could be an important mechanism for survival under conditions of low nutrient availability.[91,144] Several types of phenolic compounds have the potential to serve as internal growth inhibitors, acting to slow down various aspects of plant metabolism.[95,96,101,120,144] Flavonoids, for example, may serve as cofactors of indoleacetic acid oxidase or inhibitors of mitochondrial ATP production, although it has not yet been conclusively demonstrated

that they perform these functions in intact plants.[121,146,147]

2) Allelopathy. Many phenolics have been suggested to function as allelopathic agents.[148,149] For plants under nutrient stress, it may be selectively advantageous to produce higher levels of phenolics in order to more effectively inhibit the growth of neighboring individuals, as this could reduce competition for the limited supply of nutrients. *Helianthus annus* litter from plants grown under nutrient deficiencies was shown to be more inhibitory to the germination of *Amaranthus retroflexus* seeds than that from control plants,[128] although no evidence was presented which indicates that this interaction is ecologically significant.

3) Defense against predators and pathogens. Numerous reports have implicated phenolic compounds in plant defense.[150] The synthesis of increased concentrations of phenolics during periods of nutrient deficiency may be an adaptation to reduce damage at a time when the cost of replacing the foliage lost to herbivores and fungi is considerably higher than it would be under conditions of adequate nutrient supply.[151] According to this hypothesis, plants that normally occupy infertile habitats could have continuously high levels of phenolic compounds. Tropical rain forest trees growing on nutrient-poor, sandy soils have been shown to have significantly higher concentrations of phenolic substances than those growing on soils with greater nutrient reserves and are therefore thought to be better defended against herbivory.[152,153]

4) Storage. Nutrient deficiences, by reducing growth to a greater extent than photosynthesis,[154,155] may result in a temporary accumulation of unused carbohydrate.[92] The stepped-up production of phenolics under such conditions can be viewed as a method for storing surplus fixed carbon in a form that is better defended against predation than the ordinary storage carbohydrates would be (although phenolics are more oxidized than carbohydrates and so not as rich in stored energy). It may be more critical to store reserves in a well-protected form when unfavorable growing conditions prevent them from being quickly replaced. This proposal is analogous to the suggestion made earlier that high concentrations of alkaloids function as a secure storage form for surplus nitrogen. Like alkaloids, phenolic monomers can apparent-

ly be readily degraded by plants, so carbon stored in them could be mobilized when needed.[156] However, much less is known about the ability of plants to catabolize polymeric phenolics, such as condensed tannins.[156]

The increased amounts of phenolic compounds produced under conditions of nutrient scarcity may therefore act in a number of different ways enabling plants to survive these unfavorable circumstances. Various classes of phenolic substances could, of course, perform different functions in different plants and particular compounds may serve in more than one capacity at any given time, such as storage and defense.

TERPENOIDS

Terpenoids display such a diversity of behavior under stress that it is necessary to focus on individual groups of compounds to find consistent patterns of response.

Monoterpenoids in Mints

There has been considerable long-term interest in the effects of environmental factors on essential oil production in mints, because these aromatic, monoterpenoid-rich oils are important in flavoring food, in perfume and in the cosmetics industry. Water deficits in mints have been shown to increase the concentration of essential oils in several controlled studies. In marjoram, _Majorana hortensis_, for instance, the essential oil content increased under low relative humidity,[157] while data calculated from a study on peppermint, _Mentha piperita_,[158] demonstrate that irrigation decreased the concentration of essential oil per plant. In _Satureja douglasii_, water stress increased leaf monoterpenoid concentration slightly in a control chamber study,[159] although measurement of natural populations showed that high monotepenoid concentrations were usually associated with conditions of less water stress, probably because of the dominating influence of other factors, such as light intensity and temperature.[160] In field experiments, correlations between seasonal moisture availability and monoterpenoid content have not shown a consistent trend.[157,161]

Water stress could increase monoterpenoid concentration by inhibiting leaf growth to a greater extent than monoterpenoid production. In mints, monoterpenoids are found on the leaf surface in glandular hairs that are initiated very early in the development of the leaf.[162-164] Moderate water deficits reduce the initiation of glandular hairs and the concomitant synthesis of monoterpenoids in young leaves, but appear to have more substantial effects on subsequent leaf expansion, resulting in a higher density of glandular hairs and a greater monoterpenoid content in leaves produced under conditions of water stress.[159,161] This is another example of a physiological mechanism for the accumulation of increased amounts of secondary compounds under stress in which increased concentrations are the result of alterations in the basic course of growth and development, rather than the direct influence of stress on secondary metabolism.

Water deficits can also have an impact on the composition of essential oils produced by mints. No mention has been made of changes in the composition of plant secondary metabolites up to this point because, in the case of phenolics, alkaloids and glucosinolates, compositional changes occurring under stress have been reported only rarely and are usually much less important than overall quantitative changes. However, the monoterpenoid composition of some mints can be drastically altered by water stress. For example, during dry periods, the essential oil of *Thymus serrulatus* is composed principally (60-70%) of the phenolic monoterpenoids thymol and carvacrol. At wet times of the year, however, the oil is very rich in linalool, an acyclic monoterpene, and the phenolic compounds make up only 15-20% of the total.[165]

It has been suggested on several occasions that monoterpenoids function to reduce transpiration by inducing stomatal closure or by providing a vapor shield which keeps leaf temperatures low.[166-170] Although this idea has not been widely accepted, it could provide a rationale for the increases in monoterpenoid concentration observed under water stress, since a reduction in transpiration could certainly be of adaptive significance when water deficits are severe. Rovesti[170] claimed that the compositional changes observed in the monoterpenoids of *Micromeria* under water stress (an increase in the percentage of camphor and a decrease in the percentage of menthone) favor a more rapid reduction in transpiration under these conditions, since camphor has a lower vapor pressure than menthone. Monoterpenes in most mints, however, seem unlikely to be involved in regulating transpiration since they are localized in glandular hairs and are not volatilized to any great extent unless these hairs are broken off.[160,171] This does not preclude the involvement of other secondary terpenoid compounds in reducing water loss, however. The terpenoid-containing resin found on the leaf surfaces of certain desert plants in western Australia, for example, passes out of the glandular hairs where it is made and forms a layer on the leaf surface with breaks for the stomatal apertures.[172] This resin could function in reducing transpiration by decreasing leaf temperature or by cutting down water loss through the cuticle.[172] It is worth mentioning that the plant hormone generally implicated in stomatal closure in water-stressed plants, abscisic acid, is a sesquiterpene and many monoterpenoid-rich essential oils contain small concentrations of sesquiterpenes, which could induce stomatal closure[173] if they ever contact the stomata.

The effects of nutrient deficiencies on monoterpenoid levels in mints have not been studied, but attempts have been made to increase the yield of essential oil by fertilizing. The results, though, have been rather ambiguous; additions of nitrogen, phosphorus or potassium increased the concentration of essential oil in some studies and had no significant effect in others.[76,157,158]

Oleoresin in conifers

It has been appreciated for a long time that various types of stress, including water and nutrient stress, predispose trees to insect and fungal attack.[174-176] Stressed trees could become more vulnerable to attack because of altered carbohydrate or nitrogen levels, which increase their nutritional value, or because of decreased amounts of defensive compounds. The terpenoid-containing oleoresin of conifers, which consists chiefly of monoterpenes and diterpene acids, is known to serve as a defense against insect and fungal invasion.[177]

There are several indications that water and nutrient stress can reduce the production of conifer oleoresin, although this question needs much further study. Nutrient deficiencies were shown to decrease the levels of oleoresin volatiles in one study with pines.[178] Trees from nitrogen-deficient soils had lower concentrations of essential oil in their foliage and, consequently, the feeding activity and survival of several species of defoliating insects were considerably enhanced. Fertilization, on the other hand, was reported to increase resin content,[179] with urea (a nitrogen source) being more effective than gypsum (calcium sulfate).[180]

Work with many conifers, including species of _Pinus_, _Abies_ and _Pseudotsuga_, has established that water deficits reduce the rate of oleoresin flow from wounds and thus decrease the resistance of trees to bark beetles.[176,181-184] However, it is not clear if water stress reduces the actual amount of oleoresin in the tree or just the rate at which it is forced out of a wound, since oleoresin exudation pressure is very sensitive to sudden changes in turgor pressure.[182,185,186] Decreased flow in water-stressed trees could be simply due to reductions in turgor. Nevertheless, since stressed trees are apparently not able to exude resin for as long a period of time as healthy ones,[183] the actual amount of oleoresin present in these trees could be lower.

As in mints, water stress can cause changes in the composition of monoterpenoids in conifers. Monoterpenoid composition is under strong genetic control in conifers, with each tree often having its own unique mixture of components,[187] but developmental, seasonal and environmental factors can sometimes cause extensive modification.[188-191] Water deficits in loblolly pine (_Pinus taeda_) increase the relative percentage of α-pinene in xylem oleoresin from 79% to 89%, while decreasing the concentration of β-pinene, camphene, myrcene and limonene.[192] In another study, the total proportion of monoterpene hydrocarbons in _P. taeda_ xylem oleoresin increased slightly under water stress, while the relative amount of diterpene acids decreased.[193] Such changes affect the viscosity[194] and volatility of the resin and therefore could alter its defensive properties. In young foliage of Douglas fir (_Pseudotsuga menziesii_) the relative amount pinene also increased under water stress, while that of α-pinene and bornyl acetate decreased.[195] Western spruce budworm larvae that were reared on these water-stressed trees had a higher survivorship and weighed more than those reared on control trees.[195]

Other Terpenoids

Information concerning the influence of water and nutrient stresses on the production of several other types of terpenoids is also available. The sesquiterpene lactone and diterpene content of the leaves of the desert sunflower Helianthus ciliaris doubled from 0.40% to 0.79% under moderate water stress (J. Gershenzon and E. Lee, unpublished results). Measurements of leaf growth and total sesquiterpene and diterpene yield showed that this increase was due to a substantial decrease in leaf weight, while terpenoid production remained almost constant, analogous to the situation for monoterpenoids in mints. Like mint monoterpenoids, the sesquiterpene lactones in Helianthus are also found in glandular hairs on the leaf surface (J. Gershenzon and G. Kreitner, unpublished observations).

The tropical leguminous tree Hymenaea courbaril produces a leaf resin consisting of a mixture of sesquiterpene hydrocarbons. Water deficits had no effect on total resin concentration, computed on a dry weight basis, but definite changes in sesquiterpene composition were noted when leaves of the same age in stressed and non-stressed trees were compared.[196] Water stress, which retarded leaf development, appeared to have also slowed down the rates of normal developmental changes[197] in sesquiterpene composition in the leaves.[196] Water deficits have been said to suspend leaf aging because of their effects on various physiological parameters.[23] Drought-stressed leaves of Panicum maximum, for example, show high nitrogen levels and high photosynthetic rates on rewatering, comparable with those of much younger leaves from unstressed plants.[198,199] Perhaps the changes seen in the sesquiterpene composition of H. courbaril under water stress are also a consequence of a delay in the aging process.

Very little is known about the effects of stress on the production of steroids, triterpenes and related compounds. Fertilization with nitrogen, phosphorus or potassium increased the cardiac activity of species of Digitalis in most cases,[76] indicating that nutrient deficiencies could depress the synthesis of cardenolides in these plants. The effects of fertilization on the saponin content of Saponaria species, however, were variable.[76] In cotton (Gossypium hirsutum) seeds, the level of gossypol, a phenolic sesquiterpene dimer, was positively correlated with increased rainfall,[200] so water stress may inhibit the production of this toxic pigment.

Turning to polyisoprenoids, the environmental factors influencing rubber production have, quite predictably, been the subject of considerable investigation. In the United States, attention has focused on guayule (Parthenium argentatum), a shrub native to the deserts of northern Mexico and west Texas, that has been a commercially important source of rubber in the past and is currently generating renewed interest. Rubber in guayule accumulates during those times of the year that are least favorable for vegetative growth, suggesting that stress may generally promote its production.[201] Water deficits[202,206] and low temperatures[7,203] do lead to substantial increases in rubber content, up to eight-fold in some cases. The effects of nutrient deficiencies, though, are less clear. Low levels of nitrogen or phosphorus usually decrease rubber content,[207,208] but when nutrient supplies to actively growing plants are suddenly restricted, there is a rapid rise in rubber concentration.[203] Changes in potassium or sulfur levels have no significant effect on rubber content.[207]

The effects of stress have also been studied in some other rubber-bearing plants. In the Russian dandelion (Taraxacum kok saghyz), nitrogen stress might increase rubber production, since heavy applications of nitrogen reduce rubber content.[203] Addition of phosphorus, on the other hand, results in greater rubber content, though only in wet years.[203] The principal source of natural rubber at present is the Amazonian tree Hevea brasiliensis. The flow of latex obtained by tapping the trunk of this species is very responsive to changes in turgor pressure,[209] like the exudation of conifer oleoresin. Water deficits reduce latex flow, but increase the concentration of rubber in the latex, since there is less dilution.[209,210] The effect of water deficits on rubber synthesis itself, however, does not appear to have been studied. Low levels of nitrogen or phosphorus, like water deficits, increase latex yield.[211,212]

In general, terpenoids seem to exhibit a variety of responses to water and nutrient stresses, with no trend readily apparent. However, there are some differences in the behavior of most herbs and shrubs, on the one hand, and trees, on the other. In many herbs and shrubs (mints, *Helianthus*, *Parthenium*) terpenoids tend to accumulate under stress conditions, particularly water deficits, while in trees (conifers, *Hymenaea*), stress usually causes a reduction in terpenoid levels or has no significant effect.

SUMMARY OF CHANGES

From this survey of most of the major types of secondary metabolites, it is clear that water and nutrient stress frequently have significant effects on the amounts of these compounds present in plants. A summary of the major patterns of change discussed in this paper is presented in Table 5. Except for nitrogen and sulfur deficiencies which, as would be expected, decrease the production of most kinds of secondary compounds containing these elements, stress usually results in increased concentrations of secondary metabolites. This finding is somewhat surprising and worth emphasizing because it seems to have been completely overlooked until very recently.[25] Extensive further study, though, is certainly required to more fully substantiate it.

There are large gaps in our knowledge of what happens to plant secondary metabolites under stress. Evidence available for many classes of compounds is sketchy and there appears to be no information at all concerning such constituents as non-protein amino acids, polyacetylenes, quinones and many kinds of di- and triterpenoids. In addition, study has been limited to a small group of predominantly temperate species, most of which are commercially important and may have undergone heavy selection pressure by humans for certain desirable traits, so that their behavior under stress may not be representative of that of wild plants in general. It could also be useful to know something about the effects of stress on secondary metabolism in lower plants. Interestingly, many fungi and bacteria show responses very similar to those of higher plants, producing antibiotics and other secondary compounds at a much greater rate when the culture medium is deficient in nitrogen or phosphorus.[213-216]

More investigation is also needed into the physiological mechanisms responsible for altered

Table 5. Summary of the major effects of water and nutrient stress on the concentrations of secondary metabolites in higher plants

Class of secondary metabolite	Type of stress*				
	Water deficits	Nutrient deficiencies -N	-P	-K	-S
Cyanogenic glycosides	+	-			
Glucosinolates and other sulfur compounds	+	+			-
Alkaloids	+	-	?	+	
Phenolics	?	+	+	+	+
Terpenoids Herbs and shrubs	+	?	+	?	
Trees	-	-			

*(+) = increase in concentration, (-) = decrease, (?) = no clear trend

concentrations of secondary metabolites. Only meager evidence is available for most of the postulated mechanisms discussed in this paper. Further advances in our undestanding of stress physiology and the regulation of secondary metabolism are required before we will be able to explain, in any sort of detail, how concentrations of secondary compounds change under stress conditions.

ACCUMULATION UNDER STRESS: POSSIBLE EXPLANATIONS

The widespread increases in the concentrations of plant secondary metabolites under stress may be just unavoidable consequences of changes in the primary processes of growth and development with no particular functional importance to the plant. Water deficits and nitrogen deficiencies cause a very rapid reduction in growth rate[24,68,91,154,155] and deficiencies of other nutrients, such as phosphorus and potassium, have similar if less pronounced effects.[91] If these stresses had, for whatever reason, only minor effects on the actual biosynthesis of secondary compounds, increased concentrations of these compounds would result. In the case of mild nitrogen shortages, photosynthesis may continue at nearly normal rates even though growth is limited.[91,92] Excess carbohydrate could then build up and be channeled into the production of secondary metabolites. These changes are not the result of any direct influence of stress on secondary metabolism and could conceivably have little significance in the survival of plants under stress.

Water deficits and nutrient deficiencies of at least moderate severity, however, could be expected to have a direct impact on the synthesis of secondary compounds because of their inhibition of photosynthesis.[68,91,107-109] A reduction in the supply of fixed carbon would presumably force a plant to cut back on its biosynthetic out-put in some way. In theory, a plant could respond to a reduced supply of photosynthate: 1) by sharply curtailing its production of secondary metabolites, while continuing to allocate resources to primary metabolic activities at the previous rate, 2) by cutting back on resources for growth and development while continuing to synthesize secondary metabolites at pre-stress levels, or 3) by reducing both primary and secondary metabolic expenditures to varying degrees. Plants can exhibit a good deal of plasticity in their allocation of resources to different functions under changing environmental conditions.[217-219]

As noted in this chapter, many species respond to water and nutrient stress by slowing their growth while continuing to produce secondary metabolites at a rate that results in higher concentrations of these compounds. The continued synthesis of secondary compounds under adverse conditions, though, is extremely difficult to explain from an evolutionary perspective, unless these compounds have a critical function at these times, since it would seem that plants under stress should be selected to channel their limited resources directly into growth, maintenance or reproduction. The accumulation of secondary metabolites under stress then must be an adaptive response to conditions under which the functions of these compounds become more important. One can envision a number of possible functions for secondary compounds. As discussed earlier in the section on phenolics, several of these functions, including protected storage of reserves, growth inhibition, allelopathy, and defense against predators and pathogens, could become more critical under stress.

The most generally accepted role for plant secondary metabolites is that they serve as defensive compounds. The presence of high concentrations of these constituents under stress, therefore, might be a result of natural selection for increased resistance to herbivory or disease at a time when damage could cause a particularly great loss of fitness.[151,220] According to recent theoretical treatments, the production of secondary compounds depends on a number of factors, including 1) the cost of synthesis and storage, 2) the risk of attack and 3) the value of the plant parts protected.[25,93,220] Under stress, the relative importance of these factors could change. The cost of producing secondary metabolites should be comparatively higher under stress due to the reduced availability of resources. The value of plant parts should also increase, since their relative replacement costs would be greater and recovery from any damage suffered is likely to be slow. The risk of attack could be greater under stress too, although there is not much information available on this point. It has been suggested that drought increases the nutritional value of plants for insects in some instances[5,6] and that nutrient deficiencies and water stress increase the susceptibility of plants to fungal infection in others.[221-223] In summary, the cost of producing defensive secondary compounds in plants should increase under stress, but so should the value of the organs to be protected and, possibly, the chance of being attacked. Janzen[224] likened the production of secondary compounds to buying an insurance policy. If there is a greater risk of damage, or the value of the property to be protected is greater, it is worth paying an increased premium to get more protection.

One way plants could mitigate the higher costs associated with the synthesis of defensive secondary metabolites under stress is to increase the synthesis of cheaper protective compounds,[25] ones that make less of a demand on scarce resources. Under nitrogen stress, for example, plants that make both nitrogenous and non-nitrogenous defenses could increase their relative production of the latter.[92] When overall carbon supply is limiting, one might expect an increased investment in defenses that are more effective at lower concentrations, though possibly not as active against as broad of a range of herbivores (toxins _sensu_ Feeny[225] or qualitative defenses _sensu_ Rhoades[226]). However, there is as yet no real evidence for this type of response.

It is difficult to test the validity of predictions about the cost of producing secondary metabolites or the value of different organs to the plant, since there is no easy way to quantify these concepts, let alone know how they might be altered by stress. Nevertheless, these ideas do help to explain how the possession of increased concentrations of defensive secondary compounds could be of selective value to plants under stress.

The accumulation of secondary metabolites may be part of a suite of adaptations to unfavorable conditions. Plants that have evolved the ability to grow under one of a number of different types of stress frequently share some basic characteristics that are said to increase their fitness under these conditions: a slow rate of growth, xeromorphism, long-lived leaves, a small relative investment in reproduction and increased resistance to predation.[144] It seems reasonable to suggest that plants which normally grow under relatively benign conditions could respond to stress by temporarily adopting some of these traits. The production of high levels of secondary metabolites is one way of increasing resistance to predation.

A reduction in growth rate, although usually considered to be an unfavorable consequence of stress rather than a way of adjusting to it, may make plants less susceptible to the effects of stress. It is interesting to note that there is a strong association between the increased production of secondary compounds and periods of reduced growth, even in situations where no stress is readily apparent. This has been shown for a number of species already mentioned: *Brassica oleracea capitata*[38] (glucosinolates), *Datura metel*[227] (tropane alkaloids), *Nicotiana rustica* callus cultures[228] (nicotine), various Solanaceae[229] (alkaloids), *Catharanthus roseus* cell cultures[99] (alkaloids and phenolics), *Pisum arvense*[230] (anthocyanins), *Hordeum vulgare*[145] (flavonoids), *Heteromeles arbutifolia*[231] (phenolics), *Pinus monticola*[232] (monoterpenes) and *Hevea brasiliensis*[203] (rubber). A negative correlation between growth and the production of secondary metabolites is also seen in many cultures of microorganisms where antibiotics are synthesized following growth during a sharply defined period known as the idiophase. Addition of a carbon source, such as glucose, will frequently stimulate further growth and repress antibiotic synthesis.[214,215,233,234] The enhanced production of secondary compounds in these studies during periods of reduced growth may be a consequence of the fact that both a slow rate of growth and increased production of secondary compounds are adaptive responses to stress. Alternatively, increased production of secondary compounds could necessarily result in a slower growth rate because of resource limitations.

In conclusion, it is worth mentioning that the study of stress and its influences on plant secondary metabolism has a broad range of practical applications. Advances in our understanding of the environmental factors regulating the synthesis of secondary compounds should be useful in 1) maximizing the harvest of usable products from herb, spice and drug plants, 2) decreasing livestock losses caused by ingestion of toxic forage by anticipating dangerous conditions in advance and 3) increasing agricultural efficiency in general. Growing crops under moderate water or nutrient stress may reduce pest problems sufficiently (due to increased concentration of defensive secondary metabolites), so that any reduction in yield will be more than compensated for by savings on fertilizer, irrigation and pesticides.

ACKNOWLEDGMENTS

I thank Hector E. Flores, Laurence H. Hurley and Connie Nozzolillo for suggesting and providing reference material and Jerry J. Brand, Douglas A. Gage, Tom J. Mabry, Diane L. Marshall and Carl D. Schlichting for critical review of the manuscript.

REFERENCES

1. FRAENKEL G 1959 The raison d'être of secondary plant substances. Science 129: 1466-1470
2. EHRLICH PR, PH RAVEN 1965 Butterflies and plants: a study in co-evolution. Evolution 18: 586-608
3. KOGAN M, J PAXTON 1983 Natural inducers of plant resistance to insects. In PA Hedin, ed, Plant Resistance to Insects, American Chemical Society, Washington, DC pp 153-171
4. TINGEY WM, SR SINGH 1980 Environmental factors influencing the magnitude and expression of resistance. In FG Maxwell, PR Jennings, eds,

Breeding Plants Resistant to Insects, John Wiley & Sons, New York pp 89-113

5. WHITE TCR 1974 A hypothesis to explain outbreaks of looper caterpillars, with special reference to populations of Selidosema suavis in a plantation of Pinus radiata in New Zealand. Oecologia 16: 279-302
6. WHITE TCR 1976 Weather, food and plagues of locusts. Oecologia 22: 119-134
7. BONNER J 1943 Effects of temperature on rubber accumulation by the guayule plant. Bot Gaz 105: 233-243
8. BURBOTT AJ, WD LOOMIS 1967 Effects of light and temperature on the monoterpenes of peppermint. Plant Physiol 42: 20-28
9. HAMADA N 1983 The effect of temperature on lichen substances in Ramalina subbreviuscula (lichens). Bot Mag (Tokyo) 96: 121-126
10. HANSON AD, KM DITZ, GW SINGLETARY, TJ LELAND 1983 Gramine accumulation in leaves of barley grown under high-temperature stress. Plant Physiol 71: 896-904
11. TSO TC, MJ KASPERBAUER, TP SOROKIN 1970 Effect of photoperiod and end-of-day light quality on alkaloids and phenolic compounds of tobacco. Plant Physiol 45: 330-333
12. VOIRIN B, P LEBRETON 1972 Influence de la temperature sur le metabolisme des flavonoides chez Asplenium trichomanes. Phytochemistry 11: 3435-3439
13. GREEN TR, CA RYAN 1972 Wound-induced proteinase inhibitor in plant leaves: a possible defense mechanism against insects. Science 175: 776-777
14. LOPER GM 1968 Effect of aphid infestation on the coumestrol content of alfalfa varieties differing in aphid resistance. Crop Sci 8: 104-106
15. SCHULTZ JC, IT BALDWIN 1982 Oak leaf quality declines in response to defoliation by gypsy moth larvae. Science 217: 149-151
16. BLOHM H 1962 Poisonous Plants of Venezuela. Harvard University Press, Cambridge, Massachusetts
17. BOYD FT, OS AAMODT, G BOHSTEDT, E TRUOG 1938 Sudan grass management for control of cyanide poisoning. J Amer Soc Agron 30: 569-582
18. KINGSBURY JM 1964 Poisonous Plants of the United

States and Canada. Prentice-Hall, Englewood Cliffs, New Jersey

19. MAJAK W, RE McDIARMID, JW HALL, AL VAN RYSWYK 1980 Seasonal variation in the cyanide potential of arrowgrass (Triglochin maritima). Can J Plant Sci 60: 1235-1241
20. BRUIJN GH DE 1973 In B Nestel, R MacIntyre, eds, Chronic Cassava Toxicity, International Development Research Centre, Ottawa pp 43-48
21. NELSON CE 1953 Hydrocyanic acid content of certain sorghums under irrigation as affected by nitrogen fertilizer and soil moisture stress. Agron J 45: 615-617
22. ECK HV 1976 Hydrocyanic acid potentials in leaf blade tissue of eleven grain sorghum hybrids. Agron J 68: 349-351
23. BEGG JE, NC TURNER 1976 Crop water deficits. Adv Agron 28: 161-217
24. HSIAO TC, E ACEVEDO, E FERERES, DW HENDERSON 1976 Stress metabolism, water stress, growth, and osmotic adjustment. Phil Trans R Soc Lond B 273: 479-500
25. RHOADES DF 1979 Evolution of plant chemical defense against herbivores. In GA Rosenthal, DH Janzen, eds, Herbivores, Their Interaction with Secondary Plant Metabolites, Academic Press, New York pp 3-54
26. MAJAK W, DA QUINTON, K BROERSMA 1980 Cyanogenic glycoside levels in Saskatoon serviceberry. J Range Management 33: 197-199
27. JANZEN DH, ST DOERNER, EE CONN 1980 Seasonal constancy of intra-population variation of HCN content of Costa Rican Acacia farnesiana foliage. Phytochemistry 19: 2022-2023
28. COOPER-DRIVER G, S FINCH, T SWAIN, E BERNAYS 1977 Seasonal variation in secondary plant compounds in relation to the palatability of Pteridium aquilinum. Biochem Syst Ecol 5: 177-183
29. DEMENT WA, HA MOONEY 1974 Seasonal variation in the production of tannins and cyanogenic glucosides in the chaparral shrub, Heteromeles arbutifolia. Oecologia 15: 65-76
30. CLARK RB, HJ GORZ, FA HASKINS 1979 Effects of mineral elements on hydrocyanic acid potential in sorghum seedlings. Crop Sci 19: 757-761
31. HARMS CL, BB TUCKER 1973 Influence of nitrogen fertilization and other factors on yield, prussic

acid, nitrate, and total nitrogen concentrations of Sudangrass cultivars. Agron J 65: 21-26

32. JUNG GA, B LILLY, SC SHIH, RL REID 1964 Studies with Sudangrass. I. Effect of growth stage and level of nitrogen fertilizer upon yield of dry matter; estimated digestibility of energy, dry matter and protein; amino acid composition; and prussic acid potential. Agron J 56: 533-537
33. McBEE GG, FR MILLER 1980 Hydrocyanic acid potential in several sorghum breeding lines as affected by nitrogen fertilization and variable harvests. Crop Sci 20: 232-234
34. PATEL CJ, MJ WRIGHT 1958 The effect of certain nutrients upon the hydrocyanic acid content of Sudangrass grown in nutrient solution. Agron J 50: 645-647
35. LANCASTER PA, JE BROOKS 1983 Cassava leaves as human food. Econ Bot 37: 331-348
36. GREENWOOD EAN 1976 Nitrogen stress in plants. Adv Agron 28: 1-35
37. FREEMAN GG, N MOSSADEGHI 1971 Water regime as a factor in determining flavor strength in vegetables. Biochem J 124: 61F-62F
38. BIBLE BB, H-Y JU, C CHONG 1980 Influence of cultivar, season, irrigation, and date of planting on thiocyanate ion content in cabbages. J Amer Soc Hort Sci 105: 88-91
39. MACLEOD AJ, ML NUSSBAUM 1977 The effects of different horticultural practices on the chemical flavor composition of some cabbage cultivars. Phytochemistry 16: 861-865
40. MACLEOD AJ, HE PIKK 1978 A comparison of the chemical flavor composition of some Brussels sprouts cultivars grown at different crop spacings. Phytochemistry 17: 1029-1032
41. LOUDA SM, JE RODMAN 1983 Ecological patterns in the glucosinolate content of a native mustard, *Cardamine cordifolia*, in the Rocky Mountains. J Chem Ecol 9: 397-422
42. EATON SV 1942 Volatile sulfur content of black mustard plants. Bot Gaz 104: 82-89
43. EATON SV 1942 Influence of sulfur deficiency on metabolism of black mustard. Bot Gaz 104: 306-315
44. FREEMAN GG, N MOSSADEGHI 1970 Effect of sulfate nutrition on flavor components of onion (*Allium cepa*). J Sci Fd Agric 21: 610-615

45. FREEMAN GG, N MOSSADEGHI 1971 Influence of sulfate nutrition on the flavor components of garlic (Allium sativum) and wild onion (A. vineale). J Sci Fd Agric 22: 330-334
46. FREEMAN GG, N MOSSADEGHI 1972 Influence of sulfate nutrition on flavor components of three cruciferous plants: radish (Raphanus sativus), cabbage (Brassica oleracea capitata) and white mustard (Sinapis alba). J Sci Fd Agric 23: 387-402
47. WOLFSON JL 1982 Developmental responses of Pieris rapae and Spodoptera eridania to environmentally induced variation in Brasssica nigra. Environ Entomol 11: 207-213
48. JOSEFSSON E 1970 Glucosinolate content and amino acid composition of rapeseed (Brassica napus) meal as affected by sulfur and nitrogen nutrition. J Sci Fd Agric 21: 98-103
49. JOSEFSSON E, L-Å APPELQVIST 1968 Glucosinolates in seed of rape and turnip rape as affected by variety and environment. J Sci Fd Agric 19: 564-570
50. McDONALD RC, TR MANLEY, TN BARRY, DA FORSS, AG SINCLAIR 1981 Nutritional evaluation of kale (Brassica oleracea) diets. 3. Changes in plant composition induced by soil fertility practices, with special reference to SMCO and glucosinolate concentrations. J Agric Sci 97: 13-23
51. WALLER GR, EK NOWACKI 1978 Alkaloid Biology and Metabolism in Plants. Plenum Press, New York
52. BUNTING ES 1963 Changes in the capsule of Papaver somniferum between flowering and maturity. Ann Appl Biol 51: 459-471
53. HOFMAN PJ, RC MENARY 1979 Variations in morphine, codeine and thebaine in the capsules of Papaver somniferum L. during maturation. Aust J Agric Res 31: 313-326
54. SOKOLOV VS 1959 The influence of certain environmental factors on the formation and accumulation of alkaloids in plants. Symp Soc Expt Biol 13: 230-257
55. AVEYTAN EM, OK KAZANCHYAN 1973 Effect of watering conditions on the chemical composition of tobacco leaves and its smoke. Tabak 51-52 (cited in Chem Abst 81:10998t)
56. CAMPBELL RB, GT SEABORN 1972 Yield of flue-cured

tobacco and levels of soil oxygen in lysimeters with different water table depths. Agron J 64: 730-733

57. MANDY G, SA KISS 1975 The relation between the water supply and the nicotine content of tobacco leaves, and its biochemical interpretation. Bot Kozl 62: 113-116 (cited in Chem Abst 85: 59850k)
58. VAN BAVEL CHM 1953 Chemical composition of tobacco leaves as affected by soil moisture conditions. Agron J 45: 611-614
59. AHMED ZF, IR FAHMY 1949 The effect of environment on the growth and alkaloidal content of Hyoscyamus muticus L. II. J Am Pharm Assoc Sci Ed 38: 484-487
60. ISMAILOV NM, SM ASLANOV 1965 Effect of soil dryness and increased temperature on the accumulation of glycoalkaloids and soluble sugars in Australian nightshade (Solanum aviculare). Izv Akad Nauk Azerb SSR, Ser Biol Nauk 10-15 (cited in Chem Abst 65: 4589c)
61. AZIZBEKOVA ZS, S ASLANOV, ZA BABAEVA 1973 Effect of sodium chloride on the levels of glycoalkaloids and free amino acids in tomatoes. Dokl Akad Nauk Azerb SSR 29: 58-61 (cited in Chem Abst 80: 104457x)
62. BRISKE DD, BJ CAMP 1982 Water stress increases alkaloid concentrations in threadleaf groundsel (Senecio longilobus). Weed Sci 30: 106-108
63. KENNEDY CW, LP BUSH 1983 Effect of environmental and management factors on the accumulation of N-acetyl and N-formyl loline alkaloids in tall fescue. Crop Sci 23: 547-552
64. BALL DM, CS HOVELAND 1978 Alkaloidal levels in Phalaris aquatica L. as affected by environment. Agron J 70, 977-981
65. WILLIAMS JD 1972 Effects of time of day, moisture stress, and frosting on the alkaloid content of Phalaris tuberosa. Aust J Agric Res 23: 611-621
66. HSIAO TC 1973 Plant responses to water stress. Annu Rev Plant Physiol 24: 519-570
67. STEWART GR, F LAHRER 1980 Accumulation of amino acids and related compounds in relation to environmental stress. In BJ Mifflin, ed, The Biochemistry of Plants, Vol. 5, Amino acids and derivatives, Academic Press, New York pp 609-635
68. BRADFORD KJ, TC HSIAO 1982 Physiological responses to moderate water stress. In OL Lange, PS Nobel, CB Osmond, H Ziegler, eds, Physiological Plant

Ecology II, Water Relations and Carbon Assimilation, Springer-Verlag, Berlin pp 264-324

69. NAYLOR AW 1972 Water deficits and nitrogen metabolism. In Kozlowski, ed, Water Deficits and Plant Growth, Vol. III, Academic Press, New York pp 241-254
70. ANNETT HE 1920 Factors influencing alkaloidal content and yield of latex in the opium poppy (Papaver somniferum). Biochem J 14: 618-636
71. NOWACKI E, M JURZYSTA, P FORSKI, D NOWACKA, GR WALLER 1976 Effect of nitrogen nutrition on alkaloid metabolism in plants. Biochem Physiol Pflanzen 169: 231-240
72. MOTHES K 1960 Alkaloids in the plant. In RHF Manske, ed, The Alkaloids, Chemistry and Physiology, Vol VI, Academic Press, New York pp 1-29
73. NETTLESHIP L, M SLAYTOR 1974 Adaptation of Peganum harmala callus to alkaloid production. J Exp Bot 25: 1114-1123
74. SASSE F, U HECKENBERG, J BERLIN 1982 Accumulation of beta-carboline alkaloids and serotonin by cell cultures of Peganum harmala L. I. Correlation between plants and cell cultures and influence of medium constituents. Plant Physiol 69: 400-404
75. JAMES GM 1947 Effects of manuring on growth and alkaloid content of medicinal plants. Econ Bot 1: 230-237
76. FLÜCK H 1954 The influence of the soil on the content of active principles in medicinal plants. J Pharm Pharmacol 6: 153-163
77. SCHERMEISTER LJ, FA CRANE, RF VOIGT 1960 Nitrogenous constituents of Atropa belladonna L. grown on different sources of externally supplied nitrogen. J Am Pharm Assoc Sci Ed 49: 698-705
78. BREWER WR, LD HINER 1950 Cultivation studies of the solanaceous drugs II. The effect of nutritional and soil reaction fertilizers on the production yields and total alkaloidal content of Atropa belladonna and Hyoscyamus niger. J Am Pharm Assoc Sci Ed 39: 586-591
79. NOWACKI E, M JURZYSTA, P GORSKI 1975 Effect of availability of nitrogen on alkaloid synthesis in Solanaceae. Bull Acad Pol Sci Ser Sci Biol 23: 219-225
80. SCHERMEISTER LJ, RF VOIGT, FT MAHER 1950 The

influence of varying nitrogen levels on hydroponic growth and alkaloid production in Hyoscyamus muticus L. J Am Pharm Assoc Sci Ed 39: 669-672
81. MOTHES K 1928 Pflanzenphysiologische Untersuchungen über die Alkaloide. I. Das Nikotin im Stoffwechsel der Tabakpflanze. Planta 5: 563-615
82. CAMPBELL CR, JF CHAPLIN, WH JOHNSON 1982 Cultural factors affecting yield, alkaloids, and sugars of close-grown tobacco. Agron J 74: 279-283
83. LAMARRE M 1983 Influence de la fertilization N, P et K sur la composition chimique du tabac a cigarette. Can J Plant Sci 63: 523-529
84. AHMED SS, K MÜLLER 1979 Seasonal changes and the effect of nitrogen-and potash-fertilization on the solanine and alpha-chaconine content in various parts of the potato plant. Z Pflanzenernaehr Bodenkd 141: 275-279
85. GENTRY CE, RA CHAPMAN, L HENSON, RC BUCKNER 1969 Factors affecting the alkaloid content of tall fescue (Festuca arundinacea Schreb.). Agron J 61: 313-316
86. BUSH LP, RC BUCKNER 1973 Tall fescue toxicity. In AG Matches, ed, Antiquality Components of Forage, Crop Sci Soc Am Spec Publ Series 4: 99-110
87. MOORE RM, JD WILLIAMS, J CHIA 1967 Factors affecting concentrations of dimethylated idolealkylamines in Phalaris tuberosa L. Aust J Biol Sci 20: 1131-1140
88. MARTEN GC, AB SIMONS, JR FRELICH 1974 Alkaloids of reed canarygrass as influenced by nutrient supply. Agron J 66: 363-368
89. MIKA ES 1962 Selected aspects on the effect of environment and heredity on the chemical composition of seed plants. Lloydia 25: 291-295
90. JAMES WO 1953 Alkaloid formation in plants. J Pharm Pharmacol 5: 809-822
91. CHAPIN FS III 1980 The mineral nutrition of wild plants. Annu Rev Ecol Syst 11: 233-260
92. MATTSON WJ JR 1980 Herbivory in relation to plant nitrogen content. Ann Rev Ecol Syst 11: 119-161
93. MOONEY HA, SL GULMON, ND JOHNSON 1983 Physiological constraints on plant chemical defenses. In PA Hedin, ed, Plant Resistance to Insects, American Chemical Society, Washington, DC pp 21-36
94. McNEIL S, TRE SOUTHWOOD 1978 The role of nitrogen

in the development of insect plant relationships. In JB Harborne, ed, Biochemical Aspects of Plant and Animal Coevolution, Academic Press, London pp 77-98

95. SEIGLER DS 1977 Primary roles for secondary compounds. Biochem Syst & Ecol 5: 195-199
96. SEIGLER D, PW PRICE 1976 Secondary compounds in plants: primary functions. Amer Nat 110: 101-105
97. JAMES WO 1950 Alkaloids in plants. In RHF Manske, HL Holmes, eds, The Alkaloids, Chemistry and Physiology, Vol. 1, Academic Press, New York pp 15-90
98. SMOLENSKI SJ, FA CRANE, RF VOIGT 1967 Effects of the ratio of calcium to potassium in the nutrient medium on the growth and alkaloid production of Atropa belladonna. J Pharm Sci 56: 599-602
99. CAREW DP, RJ KRUEGER 1977 Catharanthus roseus tissue culture: the effects of medium modifications on growth and alkaloid production. Lloydia 40: 326-336
100. TSAI S-D, GW TODD 1972 Phenolic compounds of wheat leaves under drought stress. Phyton 30: 67-75
101. DEL MORAL R 1972 On the variability of chlorogenic acid concentration. Oecologia 9: 289-300
102. TEVINI M, W IWANZIK, AH TERAMURA 1983 Effects of UV-B radiation on plants during mild water stress II. Effects on growth, protein and flavonoid content. Z Pflanzenphysiol 110: 459-467
103. GUINN G, MP EIDENBOCK 1982 Catechin and condensed tannin contents of leaves and bolls of cotton in relation to irrigation and boll load. Crop Sci 22: 614-616
104. TEMPEL AS 1981 Field studies of the relationship between herbivore damage and tannin concentration in bracken (Pteridium aquilinum Kuhn). Oecologia 51: 97-106
105. HILLIS WE, T SWAIN 1959 The phenolic constituents of Prunus domestica. II. The analysis of tissues of the Victoria plum tree. J Sci Fd Agric 10: 135-144
106. BARDZIK JM, HV MARSH JR, JR HAVIS 1971 Effects of water stress on the activities of three enzymes in maize seedlings. Plant Physiol 47: 828-831
107. BERRY JA, WJS DOWNTON 1982 Environmental regulation of photosynthesis. In Govindjee, ed, Photosynthesis, Vol. II, Development, Carbon

Metabolism, and Plant Productivity, Academic Press, New York pp 263-343

108. BOYER JS 1976 Water deficits and photosynthesis. In TT Kozlowski, ed, Water Deficits and Plant Growth, Vol. IV, Academic Press, New York pp 153-190
109. KLUGE M 1976 Carbon and nitrogen metabolism under water stress. In OL Lange, L Kappen, E-D Schulze, eds, Water and Plant Life, Problems and Modern Approaches, Springer-Verlag, Berlin pp 243-252
110. KRAUSE J, H REZNIK 1972 Der Einfluss der Phosphat- und Nitratversorgung auf den Phenylpropanstoffwechsel in Buchweizenblättern (Fagopyrum escultentum Moench). Z Pflanzenphysiol 68: 134-143
111. KOEPPE DE, LM SOUTHWICK, JE BITTELL 1976 The relationship of tissue chlorogenic acid concentrations and leaching of phenolics from sunflowers grown under varying phosphate nutrient conditions. Can J Bot 54: 593-599
112. LEHMAN RH, EL RICE 1972 Effect of deficiencies of nitrogen, potassium and sulfur on chlorogenic acids and scopolin in sunflower. Amer Midl Nat 87: 71-80
113. ARMSTRONG GM, LM ROHRBAUGH, EL RICE, SH WENDER 1970 The effect of nitrogen deficiency on the concentration of caffeoylquinic acids and scopolin in tobacco. Phytochemistry 9: 945-948
114. CHOUTEAU J, J LOCHE 1965 Incidence de la nutrition azotée de la plante de tabac sur l'accumulation des composés phénoliques dans les feuilles. CR Acad Sci Paris 260: 4586-4588
115. WENDER SH 1970 Effects of some environmental stress factors on certain phenolic compounds in tobacco. Recent Adv Phytochem 3: 1-29
116. NOZZOLILLO C 1978 The effects of mineral nutrient deficiencies on anthocyanin pigmentation in vegetative tissues. Phytochem Bull 11
117. ULRYCHOVA M, V SOSNOVA 1970 Effect of phosphorus deficiency on anthocyanin content in tomato plants. Biol Plant 12: 231-235
118. EBERHARDT F, W HAUPT 1959 Über Beziehungen zwischen Anthocyanbildung und Stickstoffumsatz. Planta 53: 334- 338
119. DOAK KD, PR MILLER 1968 Influence of mineral nutrition on pigmentation in sorghum. Agron J 60: 430-432

120. LAWANSON AO, BB AKINDELE, PB FASALOJO, BL AKPE 1972 Time-course of anthocyanin formation during deficiencies of nitrogen, phosphorus and potassium in seedlings of Zea mays Linn. var. E.S.1. Z Pflanzenphysiol 66: 251-253
121. MC CLURE JW 1975 Physiology and functions of flavonoids. In JB Harborne, TJ Mabry, H. Mabry, eds, The Flavonoids, Chapman and Hall, London pp 970-1055
122. HUTCHINSON A, CD TAPER, GHN TOWERS 1959 Studies of phloridzin in Malus. Can J Biochem Physiol 37: 901-910
123. ROSSITER RC 1969 Physiological and ecological studies on the oestrogenic isoflavones in subterranean clover (T. subterraneum L.). VII. Effects of nitrogen supply. Aust J Agric Res 20: 1043-1051
124. ROSSITER RC, NJ BARROW 1972 Physiological and ecological studies on the oestrogenic isoflavones in subterranean clover (T. subterraneum L.). IX. Effects of sulfur supply. Aust J Agric Res 23: 411-418
125. ROSSITER RC, AB BECK 1966 Physiological and ecological studies on the oestrogenic isoflavones in subterranean clover (T. subterraneum L.). II. Effects of phosphate supply. Aust J Agric Res 17: 447-456
126. LAURENT S, M-F LEFEBVRE 1980 Etude de l'effet de différentes carences en éléments minéraux sur la teneur en tanoïdes de gamétophytes de Filicinées. Bull Soc Bot Fr 127: 119-127
127. WILSON CM 1955 The effect of soil treatments on the tannin content of Lespedeza sericea. Agron J 47: 83-86
128. HALL AB, U BLUM, RC FITES 1982 Stress modification of allelopathy of Helianthus annuus L. debris on seed germination. Am J Bot 69: 776-783
129. SARHAN ART, Z KIRÁLY 1981 Tomatine and phenol production associated with control of fusarial wilt of tomato by the NO_3-nitrogen, lime, and fungicide integrated systems. Acta Phytopathol Acad Sci Hung 16: 133-135
130. KNOBLOCH K-H, J BERLIN 1981 Phosphate mediated regulation of cinnamoyl putrescine biosynthesis in cell suspension cultures of Nicotiana tabacum. Planta Med 42: 167-172

131. KNOBLOCH K-H, G BEUTNAGEL, J BERLIN 1981 Influence of accumulated phosphate on culture growth and formation of cinnamoyl putrescines in medium-induced cell suspension cultures of Nicotiana tabacum. Planta 153: 582-585
132. SZWEYKOWSKA A 1959 The effect of nitrogen feeding on anthocyanin synthesis in isolated red cabbage embryo. Acta Soc Bot Pol 28: 539-549
133. KNOBLOCH K-H, J BERLIN 1980 Influence of medium composition on the formation of secondary compounds in cell suspension cultures of Catharanthus roseus (L.) G. Don. Z Naturforsch 35c: 551-556
134. AMORIM HV, DK DOUGALL, WR SHARP 1977 The effect of carbohydrate and nitrogen concentration on phenol synthesis in Paul's Scarlet Rose cells grown in tissue culture. Physiol Plant 39: 91-95
135. WESTCOTT RJ, GG HENSHAW 1976 Phenolic synthesis and phenylalanine ammonia-lyase activity in suspension cultures of Acer pseudoplatanus L. Planta 131: 67-73
136. DAVIES ME 1972 Polyphenol synthesis in cell suspension cultures of Paul's Scarlet Rose. Planta 104: 50-65
137. NASH DT, ME DAVIES 1972 Some aspects of growth and metabolism of Paul's Scarlet Rose cell suspensions. J Exp Bot 23: 75-91
138. PHILLIPS R, GG HENSHAW 1977 The regulation of synthesis of phenolics in stationary phase cell culture of Acer pseudoplatanus L. J Exp Bot 28: 785-794
139. HAHLBROCK K, H GRISEBACH 1979 Enzymic controls in the biosynthesis of lignin and flavonoids. Annu Rev Plant Physiol 30: 105-130
140. HANSON KR, EA HAVIR 1981 Phenylalanine ammonia-lyase. In EE Conn, ed, The Biochemistry of Plants, Vol. 7, Secondary Plant Products, Academic Press, New York pp 577-625
141. MARGNA U 1977 Control at the level of substrate supply--an alternative in the regulation of phenylpropanoid accumulation in plant cells. Phytochemistry 16: 419-426
142. CAMM EL, GHN TOWERS 1973 Phenylalanine ammonia lyase. Phytochemistry 12: 961-973
143. TAN SC 1980 Phenylalanine ammonia-lyase and the

phenylalanine ammonia-lyase inactivating system: effects of light, temperature and mineral deficiencies. Aust J Plant Physiol 7: 159-167

144. GRIME JP 1977 Evidence for the existence of three primary strategies in plants and its relevance to ecological and evolutionary theory. Amer Nat 111: 1169-1194

145. PODSTOLSKI A, J SZNAJDER, G WICHOWSKA 1981 Accumulation of phenolics and growth rate of barley seedlings (Hordeum vulgare L.). Biol Plant 23: 120-127

146. STENLID G 1970 Flavonoids as inhibitors of the formation of ATP in plant mitochondria. Phytochemistry 9: 2251-2256

147. STENLID G 1976 Effects of substituents in the A-ring on the physiological activity of flavones. Phytochemistry 15: 911-912

148. RICE EL 1974 Allelopathy. Academic Press, New York

149. WILSON RE, EL RICE 1968 Allelopathy as expressed by Helianthus annuus and its role in old-field succession. Bull Torr Bot Club 95: 432-448

150. LEVIN DA 1971 Plant phenolics: an ecological perspective. Amer Nat 105: 157-181

151. JANZEN DH 1974 Tropical blackwater rivers, animals, and mast fruiting by the Dipterocarpaceae. Biotropica 6: 69-103

152. GARTLAN JA, DB McKEY, PG WATERMAN, CN MBI, TT STRUHSAKER 1980 A comparative study of the phytochemistry of two African rain forests. Biochem Syst Ecol 8: 401-422

153. MCKEY D, PG WATERMAN, CN MBI, JS GARTLAN, TT STRUHSAKER 1978 Phenolic content of vegetation in two African rain forests: ecological implications. Science 202: 61-64

154. CLARKSON DT, JB HANSON 1980 The mineral nutrition of higher plants. Annu Rev Plant Physiol 31: 239-298

155. INGESTAD T, A-B LUND 1979 Nitrogen stress in birch seedlings I. Growth technique and growth. Physiol Plant 45: 137-148

156. BARZ W, W HOESEL 1979 Metabolism and degradation of phenolic compounds in plants. In T Swain, JB Harborne, CF Van Sumere, eds, Recent Advances in Phytochemistry Vol. 12, Plenum Press, New York pp 339-369

157. FLÜCK H 1963 Intrinsic and extrinsic factors

affecting the production of secondary plant products. In T Swain, ed, Chemical Plant Taxonomy, Academic Press, London pp 167-186

158. CLARK RJ, RC MENARY 1980 The effect of irrigation and nitrogen on the yield and composition of peppermint oil (Mentha piperita L.). Aust J Agric Res 31: 489-498
159. GERSHENZON J, DE LINCOLN, JH LANGENHEIM 1978 The effect of moisture stress on monoterpenoid yield and composition in Satureja douglasii. Biochem Syst Ecol 6: 33-43
160. LINCOLN DE, JH LANGENHEIM 1978 Effect of light and temperature on monoterpenoid yield and composition in Satureja douglasii. Biochem Syst Ecol 6: 21-32
161. FLÜCK H 1955 The influence of climate on the active principles in medicinal plants. J Pharm Pharmacol 7: 361-383
162. AMELUNXEN F 1964 Electronenmikroskopische Untersuchungen an den Drüsenhaaren von Mentha piperita L. Planta Med 12: 121-139
163. AMELUNXEN F 1965 Electronenmikroskopische Untersuchungen an den Drüsenschuppen von Mentha piperita L. Planta Med 13, 457-473
164. DEMENT WA, BJ TYSON, HA MOONEY 1975 Mechanism of monoterpene volatilization in Salvia mellifera. Phytochemistry 14: 2555-2557
165. HEGNAUER R 1966 Chemotaxonomie der Pflanzen, Vol. IV, Birkhauser Verlag, Basel
166. AUDUS LJ, AH CHEETHAM 1940 Investigations on the significance of ethereal oils in regulating leaf temperatures and transpiration rates. Ann Bot 4: 465-483
167. FRIES N, K FLODIN, J BJURMAN, J PARSBY 1974 Influence of volatile aldehydes and terpenoids on the transpiration of wheat. Naturwissenschaften 61: 452-453
168. HAFEZ MGA 1958 Effects of some essential oil vapors on the stomata of Eupatorium and Mentha. Plant Physiol 33: 177-181
169. HAFEZ MGA 1958 Effects of rosemary and thyme oil vapors on the stomata of cherry laurel. Plant Physiol 33: 181-185
170. ROVESTI P 1952 Les essences extraites des variétés physiologiques de Micromeria abyssinica et de M. biflora. Industr Parfumerie 17: 165-168
171. TYSON BJ, WA DEMENT, HA MOONEY 1974 Volatilization

of terpenes from Salvia mellifera. Nature 252: 119-120
172. DELL B, AJ McCOMB 1978 Plant resins--their formation, secretion and possible functions. Adv Bot Res 6: 277- 316
173. WELLBURN AR, AB OGUNKANMI, R FENTON, TA MANSFIELD 1974 All-trans-farnesol: a naturally occurring antitranspirant? Planta 120: 255-263
174. MATTSON WJ, ND ADDY 1975 Phytophagous insects as regulators of forest primary production. Science 190: 515-522
175. RUDNEW DF 1964 Physiologischer Zustand der Wirtspflanze und Massenvermehrung von Forstschadlingen. Z Angew Entomol 53: 48-68
176. STARK RW 1965 Recent trends in forest entomology. Ann Rev Entomol 10: 303-324
177. BERRYMAN AA 1972 Resistance of conifers to invasion by bark beetle-fungus associations. BioScience 22: 598-602
178. GRIMAL'SKII VI 1961 Zool Zh 40: 1656 (cited in MATTSON WJ, ND ADDY 1975 Phytophagous insects as regulators of forest primary production. Science 190: 515-522)
179. OLDIGES H 1958 Waldbodendängung und Schädlingsfauna des Kronenraumes. Allgem Forstz 10: 1-2
180. MAARSE H, RE KEPNER 1970 Changes in composition of volatile terpenes in Douglas fir needles during maturation. J Agric Fd Chem 18: 1095-1101
181. FERRELL GT 1978 Moisture stress threshold of susceptibility to fir engraver beetles in pole-size white firs. Forest Sci 24: 85-92
182. RUDINSKY JA 1966 Host selection and invasion by the Douglas-fir beetle, Dendroctonus pseudotsugae Hopkins, in coastal Douglas-fir forests. Can Entomol 98: 98-111
183. SMITH RH 1966 Resin quality as a factor in the resistance of pines to bark beetles. In HD Gerhold, EJ Schreiner, RE McDermott, JA Winieski, eds, Breeding Pest-Resistant Trees, Pergamon Press, Oxford pp 189-196
184. VITE JP 1961 The influence of water supply on oleoresin exudation pressure and resistance to bark beetle attack in Pinus ponderosa. Contrib Boyce Thompson Inst 21: 37-66
185. LORIO PL JR, JD HODGES 1968 Oleoresin exudation

pressure and relative water content of inner bark as indicators of moisture stress in loblolly pines. Forest Sci 14: 392-398
186. LORIO PL JR, JD HODGES 1968 Microsite effects on oleoresin exudation pressure of large loblolly pines. Ecology 49: 1207-1210
187. SMITH RH 1964 Variation in the monoterpenes of Pinus ponderosa Laws. Science 143: 1337-1338
188. ADAMS RP 1970 Seasonal variation of terpenoid constituents in natural populations of Juniperus pinchotii Sudw. Phytochemistry 9: 397-402
189. ADAMS RP, A HAGERMAN 1976 A comparison of the volatile oils of mature versus young leaves of Juniperus scopulorum: chemosystematic significance. Biochem Syst Ecol 4: 75-79
190. HANOVER JW 1966 Genetics of terpenes. I. Gene control of monoterpene levels in Pinus monticola Dougl. Heredity 21: 73-84
191. HRUTFIORD BF, SM HOPLEY, RI GARA 1974 Monoterpenes in Sitka spruce: within tree and seasonal variation. Phytochemistry 13: 2167-2170
192. GILMORE AR 1977 Effects of soil moisture stress on monoterpenes in loblolly pine. J. Chem Ecol 3: 667-676.
193. HODGES JD, PL LORIO JR 1975 Moisture stress and composition of xylem oleoresin in loblolly pine. Forest Sci 21: 283-290
194. RUNCKEL WJ, IE KNAPP 1946 Viscosity of pine gum. Ind Eng Chem 38: 555-556
195. CATES RG, RA REDAK, CB HENDERSON 1983 Patterns in defensive natural products chemistry: Douglas-fir and western spruce budworm interactions. In PA Hedin, ed, Plant Resistance to Insects, American Chemical Society, Washington, DC pp 3-19
196. LANGENHEIM JH, WH STUBBLEBINE, CE FOSTER 1979 Effect of moisture stress on composition and yield in leaf resin of Hymenaea courbaril. Biochem Syst Ecol 7: 21-28
197. LANGENHEIM JH, WH STUBBLEBINE, DE LINCOLN, CE FOSTER 1978 Implications of variation in resin composition among organ, tissues and populations in the tropical legume Hymenaea. Biochem Syst Ecol 6: 299-313
198. LUDLOW MM, TT NG 1974 Water stress suspends leaf aging. Plant Sci Lett 3: 235-240
199. WILSON JR, TT NG 1975 Influence of water stress on

parameters associated with herbage quality of *Panicum maximum* var. *trichoglume* Aust J Agric Res 26: 127-136
200. PONS WA JR, CL HOFFPAUIR, TH HOPPER 1953 Gossypol in cottonseed. Influence of variety of cottonseed and environment. J Agric Food Chem 1: 1115-1118
201. MITCHELL JW, AG WHITING, HM BENEDICT 1945 Rubber content, stem anatomy, and seed production as related to rate of vegetative growth in guayule. Bot Gaz 106, 341-349
202. BENEDICT HM, WL MCRARY, MC SLATTERY 1947 Response of guayule to alternating periods of low and high moisture stresses. Bot Gaz 108: 535-549
203. BONNER J, AW GALSTON 1947 The physiology and biochemistry of rubber formation in plants. Bot Rev 13: 543-596
204. HUNTER AS, OJ KELLEY 1946 The growth and rubber content of guayule as affected by variations in soil moisture stresses. J Am Soc Agron 38: 118-134
205. KELLEY OJ, AS HUNTER, CH HOBBS 1945 The effect of moisture stress on nursery-grown guayule with respect to the amount and type of growth and growth response on transplanting. J Am Soc Agron 37: 194-216
206. LLOYD FE 1911 Guayule (*Parthenium argentatum* Gray), a rubber-plant of the Chihuahuan desert. Carnegie Institution of Washington, Washington DC
207. BONNER J 1944 Effect of varying nutritional treatments on growth and rubber accumulation in guayule. Bot Gaz 105: 352-364
208. MITCHELL JW, AG WHITING, HM BENEDICT 1944 Effect of light intensity and nutrient supply on growth and production of rubber and seeds by guayule. Bot Gaz 106: 83-95
209. BUTTERY BR, SG BOATMAN 1976 Water deficits and flow of latex. *In* TT Kozlowski, ed, Water Deficits and Plant Growth, Vol. IV, Academic Press, New York pp 233-289
210. DIJKMAN MJ 1951 Hevea. University of Miami Press, Coral Gables, Florida
211. HAINES WB, EM CROWTHER 1940 Manuring *Hevea* III. Results on young buddings in British Malaya. Empire J Exp Agric 8: 169-184
212. HAINES WB, E GUEST 1936 Recent experiments on manuring *Hevea* and their bearing on estate practice. Empire J Exp Agric 8: 300-324

213. DEMAIN AL 1968 Regulatory mechanisms and the industrial production of microbial metabolites. Lloydia 31: 395-418
214. DREW SW, AL DEMAIN 1977 Effect of primary metabolites on secondary metabolism. Annu Rev Microbiol 31: 343-356
215. HUTTER R 1982 Design of culture media capable of provoking wide gene expression. In JD Bu'lock, LJ Nisbet, DJ Winstanley, eds, Bioactive Microbial Products: Search and Discovery, Academic Press, London pp 37-50
216. WEINBERG D 1978 Secondary metabolism: regulation by phosphate and trace elements. Folia Microbiol 23: 496-504
217. CHEW FS, JE RODMAN 1979 Plant resources for chemical defense. In GA Rosenthal, DH Janzen, eds, Herbivores, Their Interaction with Secondary Plant Metabolites, Academic Press, New York pp 271-307
218. HICKMAN JC 1975 Environmental unpredictability and plastic energy allocation strategies in the annual Polygonum cascadense (Polygonaceae). J Ecol 63: 689-701
219. PITELKA LF 1977 Energy allocation in annual and perennial lupines (Lupinus: Leguminosae). Ecology 58: 1055-1065
220. McKEY D 1979 The distribution of secondary compounds within plants. In GA Rosenthal, DH Janzen, eds, Herbivores, Their Interaction with Secondary Plant Metabolites, Academic Press, New York pp 55-133
221. MOREHART AL, GL MELCHIOR 1982 Influence of water stress on Verticillium wilt of yellow-poplar. Can J Bot 60: 201-209
222. PURKAYASTHA RP, R MUKHOPADHYAY 1976 Level of amino acids and post-infectional formation of anti-fungal substances in relation to susceptibility of rice plants against Helminthosporium oryzae at different nitrogen supply. Z Pflkrankh 83: 221-228
223. SCHOENEWEISS DF 1978 Water stress as a predisposing factor in plant disease. In TT Kozlowski, ed, Water Deficits and Plant Growth, Vol. V, Academic Press, New York pp 61-99
224. JANZEN DH 1979 New horizons in the biology of plant defense. In GA Rosenthal, DH Janzen, eds, Herbivores, Their Interaction with Secondary Plant Metabolites, Academic Press, New York pp 331-350

225. FEENY P 1976 Plant apparency and chemical defense. Recent Adv Phytochem 10: 1-40
226. RHOADES DF, RG CATES 1976 A general theory of plant antiherbivore chemistry. Recent Adv Phytochem 10: 168-213
227. KHANNA P, TJ NG 1972 Effect of phenylalanine and tyrosine on growth and alkaloid production in *Datura metel* L. tissue cultures. Ind J Pharm 34: 42-44
228. TABATA M, N HIRAOKA 1976 Variation of alkaloid production in *Nicotiana rustica* callus cultures. Physiol Plant 38: 19-23
229. LINDSEY K, MM YEOMAN 1983 The relationship between growth rate, differentiation and alkaloid accumulation in cell cultures. J Exp Bot 34: 1055-1065
230. NOZZOLILLO C 1979 On the role of anthocyanins in vegetative tissues: a relationship with stem length in pea seedlings. Can J Bot 57: 2554-2558
231. MOONEY HA, C CHU 1974 Seasonal carbon allocation in *Heteromeles arbutifolia*, a California evergreen shrub. Oecologia 14: 295-306
232. HANOVER JW 1966 Environmental variation in the monoterpenes of *Pinus monticola* Dougl. Phytochemistry 5: 713-717
233. DEMAIN AL 1972 Cellular and environmental factors affecting the synthesis and excretion of metabolites. J Appl Chem Biotechnol 22: 345-362
234. DEMAIN AL, YM KENNEL, Y AHARONOWITZ 1979 Carbon catabolite regulation of secondary metabolism. *In* AT Bull, DC Ellwood, C Ratledge, eds, Microbial Technology: Current State, Future Prospects, Cambridge University Press, Cambridge pp 163-185

SPECIES INDEX

SUBJECT INDEX